O ASI Series

d Science Institutes Series

presenting the results of activities sponsored by the NATO Science
e, which aims at the dissemination of advanced scientific and technological
e, with a view to strengthening links between scientific communities.

s is published by an international board of publishers in conjunction with the
entific Affairs Division

Sciences   Plenum Publishing Corporation
sics   London and New York

thematical and   D. Reidel Publishing Company
sical Sciences   Dordrecht and Boston

avioural and   Martinus Nijhoff Publishers
ial Sciences   Dordrecht/Boston/Lancaster
lied Sciences

nputer and   Springer-Verlag
tems Sciences   Berlin/Heidelberg/New York
logical Sciences

Applied Sciences — No. 108

5 ms/m
delay
in typical
dielectric-fill
cable

Fast Electrical and Optical Measurem

Volume I

NA

Adva

A Ser
Comm
know

The S
NATC

A
B

C

D

E

F

G

Fast Electrical and Optical Measurements

Volume I – Current and Voltage Measurements

edited by

James E. Thompson, PhD.

Chairman, Electrical Engineering Department
University of Texas at Arlington
Arlington, Texas 76019, USA

Lawrence H. Luessen

Head, Directed Energy Branch
Naval Surface Weapons Center
Dahlgren, Virginia 22448, USA

Editorial Committee:

Anthony K. Hyder, Jr., PhD.
Associate Vice President for Research
Auburn University, Auburn, Alabama, USA

Millard F. Rose, PhD.
Director, Space Power Institute
Auburn University, Auburn, Alabama, USA

Magna Kristiansen, PhD.
P.W. Horn Professor of Electrical Engineering
Texas Tech University, Lubbock, Texas, USA

Susie M. Anderson
EG&G Washington Analytical Services
Dahlgren, Virginia, USA

1986 **Martinus Nijhoff Publishers**
Dordrecht / Boston / Lancaster
Published in cooperation with NATO Scientific Affairs Division

Proceedings of the NATO Advanced Study Institute on Fast Electrical and Optical
Diagnostic Principles and Techniques, Il Ciocco, Castelvecchio Pascoli, Italy, July
10-24, 1983

Library of Congress Cataloging in Publication Data

NATO Advanced Study Institute on Fast Electrical
   and Optical Diagnostic Principles and Techniques
   (1983 : Castelvecchio Pascoli, Italy)
   Fast electrical and optical measurements.

   (NATO ASI series. Series E, Applied sciences ;
108-109)
   "Proceedings of the NATO Advanced Study Institute
on Fast Electrical and Optical Diagnostic Principles
and Techniques, Il Ciocco, Castelvecchio Pascoli,
Italy, July 10-24, 1983"--T.p. verso.
   "Published in cooperation with NATO Scientific
Affairs Division."
   Includes index.
   1. Electric measurements--Congresses. 2. Optical
measurements--Congresses. I. Thompson, James E.
II. Luessen, Lawrence H. III. North Atlantic Treaty
Organization. Scientific Affairs Division. IV. Title.
V. Series.
TK277.N37 1983   621.37   86-700
ISBN 90-247-3296-4 (set)
ISBN 90-247-3294-8 (v. 1)
ISBN 90-247-3295-6 (v. 2)

ISBN 90-247-3294-8 (this volume)
ISBN 90-247-2689-1 (series)
ISBN 90-247-3296-4 (set)

Distributors for the United States and Canada: Kluwer Academic Publishers,
190 Old Derby Street, Hingham, MA 02043, USA

Distributors for the UK and Ireland: Kluwer Academic Publishers, MTP Press Ltd,
Falcon House, Queen Square, Lancaster LA1 1RN, UK

Distributors for all other countries: Kluwer Academic Publishers Group, Distribution
Center, P.O. Box 322, 3300 AH Dordrecht, The Netherlands

Printed in The Netherlands

PREFACE

An Advanced Study Institute on Fast Electrical and Optical Diagnostic Principles and Techniques was held at Il Ciocco, Castelvecchio Pascoli, Italy, 10-24 July 1983. This publication is the Proceedings from that Institute.

The Institute was attended by ninety-seven participants representing the United States, West Germany, the United Kingdom, Switzerland, Norway, the Netherlands, Italy, and France.

The objective of the Institute was to provide a broad but comprehensive presentation of the various measurement and analysis techniques that can be employed to investigate fast physical events, nominally in the sub-microsecond regime. This requires both an understanding of the basic principles underlying the diagnostic employed and its limitations, and a knowledge of the practical techniques available to obtain reliable and repeatable data. This Institute was thus structured to begin tutorially, followed by more practical techniques, demonstrations, and discussions.

The Institute was divided into the following major sections: (1) Overview of Applications and Needs; (2) Voltage and Current Measurements; (3) Data Acquisition; (4) Grounding and Shielding; (5) Fast Photography; (6) Refractive Index Measurements; (7) X-ray Diagnostics; (8) Spectroscopy; and (9) Active Optical Techniques. This Proceeding has been divided into two separate volumes. Volume 1, **Current and Voltage Measurements**, includes Sections (1) through (4) above; Volume 2, **Optical Measurements**, includes Sections (5) through (9).

In addition, three sessions were made up of presentations contributed by participants which summarized various research efforts employing high-speed diagnostics; these presentations have not been included in this Proceedings. Several companies also displayed and demonstrated high-speed diagnostic equipment.

We are grateful to a number of organizations for providing the financial assistance that made this Institute possible. Foremost is the NATO Scientific Affairs Division, which provided the single most major contribution for the Institute. In addition, the following US sources made contributions: the Naval

Surface Weapons Center, Air Force Office of Scientific Research, Office of Naval Research, Air Force Wright Aeronautical Laboratories, Army Electronics Research and Development Command, and Naval Research Laboratory.

We would also like to thank the staff of Il Ciocco, particularly Gian Piero Giannotti, its manager, and Bruno Giannasi, the Hotel's Conference Coordinator and the main reason for the on-site success of the Institute. The phrase "Ask Bruno" is now firmly imbedded in the memory of all in attendance for those two wonderful weeks in Tuscany. Our thanks to Dr. Mario di Lullo and Dr. Craig Sinclair, the present Directors of the NATO ASI Program, and Dr. Tilo Kester and his wife Barbara, of the Publications Coordination Office, for their suggestions, patience, and overall help.

Particular thanks to our Associate Editors who willingly volunteered to assist us in reviewing and editing the manuscripts: Anthony Hyder of Auburn University, Kris Kristiansen of Texas Tech University, and Frank Rose, formerly of the Naval Surface Weapons Center, and now keeping Dr. Hyder company at Auburn University. To our Organizing Committee, lecturers, and participants - we couldn't have done it without you. And a special thanks to our wives Lynn L. and Elizabeth T., for their assistance during the Institute; and to Linda Maynard of the Universtiy of South Carolina's Office of Continuing Engineering Education, for her efforts before, during, and following the Institute.

And finally, our appreciation to those most responsible for the actual production of this Proceedings. First, to the EG&G Washington Analytical Services Center Office at Dahlgren, Virginia, which had the task of centrally retyping every lecturer's manuscript and producing a camera-ready document for delivery to the publisher. To Susie M. Anderson, the Supervisor for Word Processing, and Christie K. Wood - thank you for a job well done. Second, to Martinus Nijhoff Publishers, especially Henny Hoogervorst, who had to contend with numerous delays in the delivery of the final manuscript, but never lost her sense of purpose or humor during the past two years.

Lawrence H. Luessen
Naval Surface Weapons Center
Dahlgren, Virginia

James E. Thompson
University of Texas
Arlington, Texas

September 1985

CONTENTS

VOLUME 1:   CURRENT AND VOLTAGE MEASUREMENTS

OVERVIEW OF APPLICATIONS AND NEEDS

VOLTAGE AND CURRENT MEASUREMENTS

DATA ACQUISITION

GROUNDING AND SHIELDING

VOLUME 2: OPTICAL MEASUREMENTS

FAST PHOTOGRAPHY

REFRACTIVE INDEX MEASUREMENTS

X-RAY DIAGNOSTICS

SPECTROSCOPY

ACTIVE OPTICAL TECHNIQUES

OVERVIEW OF APPLICATIONS AND NEEDS

OVERVIEW OF APPLICATIONS AND NEEDS

M. F. Rose
A. K. Hyder

M. Kristiansen

Auburn University
Auburn, AL
USA

Texas Tech. University
Lubbock, TX
USA

1. INTRODUCTION

There is a rapidly growing recognition of the importance of early-time processes in physical phenomena, since long-time behavior of many processes of practical interest are governed by fundamental phenomena which occur in the sub-microsecond time scale. Typical of the areas in which this early time history appear important are the breakdown and conduction processes in vacuum, gases, liquids, and solids; electromagnetic radiation production and propagation; inertial and magnetic confinement fusion devices; throughout pulsed power technology; and in the performance of systems such as electric-discharge lasers and particle-beam accelerators.

Until relatively recently, many of these early-time processes could not be observed with the speed and temporal resolution needed. Often, the first few picoseconds/nanoseconds of a process present to the potential observer a dynamic, rapidly evolving situation which demands measurements with time resolutions unheard of even five years ago. But measurements of a broad class of properties with resolutions in the picosecond-to-nanosecond time frame is a _sine_ _qua_ _non_ to understanding these physical processes.

All high-speed instrumentation techniques have taken advantage of breakthrough developments in electronics, electronic materials, and data processing. Devices such as image converters and intensifiers, photomultiplier tubes, analog/digital converters, and optical data links have revolutionized instrumentation technology, and not only improved accuracy but reduced

laborious data reduction to a few strokes on the keyboard of a
modern mini-computer. Now, even the first few picoseconds of
phenomena are becoming accessible. This early time regime presents to an observer a wealth of information which is no longer
hidden beyond the limitations of the diagnostic device.

2. PRINCIPLES AND TECHNIQUES

The optical and electrical diagnostic principles and techniques presented at this ASI are applicable to a broad range of
physical phenomena - from conventional electrical parameter measurements, to the more sophisticated electro-optical techniques
for determining fields, particle densities, and energy distribution functions. They each share one parameter in common: the
fast time frame and time resolution (typically ps to ns) required
for the understanding of how early-time processes determine and
influence later time behavior.

An excellent example to illustrate the need for the diagnostic techniques described in this Proceedings is the spark
gap switch. In its simplest form, it consists of a dielectric
housing, metallic electrodes, and a suitable insulating gas filling the interelectrode space. Either by field emission or by a
trigger process, electrons appear in the interelectrode gap.
Under the action of the applied electric field, the initiating
electron produces others by impact/ionization phenomena until an
avalanche is formed. This avalanche quickly bridges the gap,
usually on the ps-ns time scale, collapsing the switch impedance
to some value characteristic of the 'on' state. During this
short time, a multispecies plasma is formed, consisting of material from the electrode, and ions formed as a result of the
breakdown of the insulating gas. If high energy is involved,
shock waves are formed which can produce mechanical failure in
addition to the wear and corrosive action of the ions in the
plasma.

In order to understand the switch behavior, it is necessary
to measure many statistical phenomena on the nanosecond scale,
such as formative time, electron temperature, plasma species, and
their evolution as a function of time. From measuring and understanding these phenomena, we establish engineering guidelines
which allow the specification of trigger parameters, lifetimes,
erosion rates, prefire probability, jitter, etc., all of which
are important in an engineering sense for large machines designed
to reapeatedly (and repeatably) operate in high power experiments.

The sophistication of the techniques, as well as the complexity of the general classes of phenomena discussed herein,

require both an understanding of the basic principles underlying the diagnostic and its limitation, and a knowledge of the practical technique which must be employed to obtain reliable and repeatable data.

3.   CONSIDERATIONS

While many industrial processes and applications require high-speed diagnostics, practical considerations by the military have been a major driving force in their development. Detonation physics, weapons' effects (both conventional and nuclear), and inertial confinement fusion have accelerated the development of fast optical techniques capable of nanosecond time resolution. For instance, X-ray diagnostics are necessary to allow imaging through obscuring media and to investigate the collapse of fuel pellets for fusion events.

Weapons-effect simulation, inertial confinement fusion, and directed energy technology have stimulated the development of large electrical machines which produce terawatts of power on the microsecond time scale. The power flow must be controlled with great precision, and often the machines have size and weight constraints which necessitate their operation near the limits imposed by catastrophic failure. Often the power train is controlled by many switches which must function with nanosecond jitter. Since terawatts of power implies megavolts and mega-ampere currents, there is a continuing need to develop new methods to measure these quantities, particularly new techniques which are nonintrusive.

The factors limiting the performance of these machines are also of interest to the industrial community. The most prominent limits are electrical breakdown and subsequent recovery, plasma chemistry, and electrical conductivity and its evolution as a function of time. These phenomena are of considerable interset in themselves and are the subject of major research efforts in both industry and university laboratories. By studying these phenomena, on the subnanosecond time scale, there is evolving an understanding of the intricate and complex processes which determine the late-time behavior of many devices. Picosecond time resolution allows essentially instantaneous measurement of plasma parameters such as ion/neutral densities, ion/neutral temperatures, electron-distribution functions, excited states, electron density, and electron temperature. A strong basic understanding of such phenomena is critical to the development of devices which depend on these parameters to either produce an effect or have a catastrophic limit based upon their value.

In closing this chapter, your attention is directed to
Appendix A,* which is the result of the Institute's participants
collectively ranking the various diagnostic techniques using
factors such as time resolution, range of parameter to be measured, ease of use, cost, and accuracy. Because of the number of
factors involved in a given experiment, Appendix A should only
be used as a guide, but it is useful in describing the relative
merits of one technique over another. The reader is directed to
the technique's respective chapter for a more precise description
and examples of applications.

---

*In Volume II of this work

VOLTAGE AND CURRENT MEASUREMENTS

ELECTRO-OPTICAL MEASUREMENT TECHNIQUES

Robert E. Hebner

Electrosystems Division
National Bureau of Standards
Washington, DC
USA

1. INTRODUCTION

A common motivation for using optical techniques to measure voltages and currents is the need for electrical isolation of the measurement system from the system which is being measured. Ground current effects can be minimized because optical coupling eliminates the wires connecting the two systems. In addition, the effects of radiated electromagnetic interference on the measurement system can be mitigated by physically separating the measuring system from the higher power system which is the source of the interference. A second important motivation for optical measurement techniques is that they can provide increased precision and accuracy over conventional electrical measurements.

An optically coupled system to measure voltage or current is conventionally configured in one of two ways. In the first, the output of a transducer* can be converted to an optical signal, transmitted, received and decoded, and the electrical parameter of interest reconstructed. This paper deals with a second approach which reduces the size, the expense, and the complexity of the transducer. This approach is based on one of three electro-optical effects: Pockels, Kerr, or Faraday. Using these effects in appropriate materials and configurations, the parameter to be

---

*As used in this paper, transducer refers to a primary sensing element for a given electrical parameter which scales the parameter so that its magnitude is compatible with a data acquisition system.

measured changes the optical transmission characteristics of the
sensor. Thus, the electrical signal provides direct modulation
of the light beam without the use of a divider or a shunt.

A generic electro-optical system to measure electrical quantities is shown in Fig. 1. The light source and the detector are
generally operated in a remote location to minimize interference.
The voltage or the current to be measured produces an electric or
a magnetic field. This field interacts with the optically active
material to change the light beam.

The Kerr, Faraday, and Pockels effects relate to specific
interactions between the field to be sensed and the optically
active material. To differentiate among these effects, it is
convenient to consider the general expression,

$$\Delta n = \alpha + \beta F + \gamma F^2 + \ldots, \tag{1}$$

where Δn is a change in the index of refraction of the material,
F is an electric or a magnetic field and α, β, and γ are constants. If F denotes an electric field and α and β are the only
non-zero coefficients, then Eq. (1) describes the Pockels effect.
If F denotes an electric field and α and γ are the only non-zero
coefficients, then Eq. (1) describes the Kerr effect. Finally,
if F denotes a magnetic field and α and β are the only non-zero
coefficients, then Eq. (1) describes the Faraday effect.

The next section describes some of the features of each of
these three effects. The order in which the effects are treated-
Kerr, Faraday, and Pockels--has been chosen to provide a logical
progression in the presentation. Section 3 describes the optical
systems used for electrical measurements. The last section
outlines some applications of these effects in the measurement of
voltage and current.

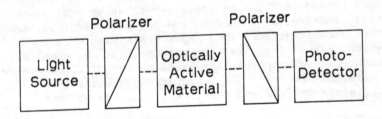

Fig. 1.   Generalized block diagram of an electro-optical system
          to measure voltage or current.

2.   DESCRIPTION OF THE OPTICAL EFFECTS

2.1 Kerr Effect

The basic equation describing the electro-optic Kerr effect is

$$n_{||} - n_{\perp} = \lambda BE^2,\tag{2}$$

where $n_{||}$ is the component of the index of refraction parallel to the applied field and $n_{\perp}$ is the component perpendicular to the field, λ is the wavelength, B is the Kerr coefficient, and E is the electric field. As noted in the discussion of Eq. (1), the material exhibiting the Kerr effect can, in principle, be optically anisotropic, i.e., there can be a difference between the refractive indices even in the absense of a field. Most materials, however, are optically isotropic, so $\alpha = 0$. To simplify the presentation, only the isotropic case is considered in the remainder of this discussion.

Conceptually, the Kerr effect at low frequency arises because the molecules in the optically active material tend to align with the applied electric field. This alignment produces an anisotropy in the index of refraction. Thus, a light beam polarized in the direction of the applied field and a beam polarized perpendicular to that direction will propagate through the material with different velocities. Depending upon the circumstances, the Kerr coefficient can be calculated from more fundamental models of atomic, molecular, and fluid structure with varying degrees of success [1,2,3].

If the Kerr effect is to be used to measure electrical quantities, it is important to identify those parameters which may affect the performance of the measurement system. One of these is the dependence of the Kerr coefficient on the frequency of the applied electric field. The frequency dependence of the Kerr coefficient generally is the same as that of the relative permittivity, as they both result from molecular alignment. The relative permittivity of nitrobenzene [4], a commonly used fluid in Kerr effect systems, is shown as a function of frequency in Fig. 2. This plot is typical of the frequency dependence of the relative permittivity for a polar molecule. Measurements taken for applied fields produced by direct voltage, by alternating voltages in the frequency range from 60 Hz to 10 kHz, and by microsecond pulses have indicated that the Kerr coefficient of nitrobenzene is constant to within 1% over this frequency range [5]. In addition, pulse measurements in the nanosecond range have not indicated any frequency dependence [6]. Thus, the "dc

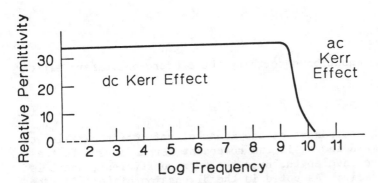

Fig. 2.   Relative permittivity of nitrobenzene as a function of
           frequency on a logarithmic scale. The term "dc Kerr
           effect" is used to describe those cases for which the
           applied field produces the low frequency relative per-
           mittivity, and the term "ac Kerr effect" is used to
           describe measurements at higher frequencies.

Kerr effect" appears to be relatively insensitive to the frequency of the applied field over a fairly wide range of frequencies. At low frequencies, the permittivity results from the
orientation of the molecule. At high frequencies, the molecule
can no longer reorient before field reversal and the remaining
permittivity is due to the redistribution of the electrons within
a molecule. The Kerr effect can be and has been used in each of
these domains. Jargon has developed which calls the Kerr effect
due to reorientation of the molecule the "dc Kerr effect" and
that due to the electronic redistribution, the "ac Kerr effect."
This designation can be confusing in that the "dc Kerr effect"
dominates up through gigahertz frequencies.

     Another distinction which must be drawn is between the
applied electric field and the sensing electric field. The
sensing field is the lower intensity electric field associated
with the light beam which is used to detect the difference between the indices of refraction. The applied fields are of
higher intensity and may be an external electrode system for the
lower frequencies or by a high power laser for the higher frequencies.

     There is some dependence on the frequency (or wavelength) of
the sensing light beam. Away from an absorption line, the product of the Kerr coefficient and the wavelength is approximately
constant. In the vicinity of an absorption line, however, the
product is no longer constant. There have been no reports,
however, of any system in which dispersion over a laser line was
a confounding factor in an electro-optical voltage measurement.

As the Kerr coefficient relates the degree of molecular alignment to the applied field, it is, as would be expected, a function of temperature. In general, the temperature dependence is given by an expression of the form:

$$B = \alpha_0 + \alpha_1 T^{-1} + \alpha_2 T^{-2} + ..., \qquad (3)$$

where T is the temperature and the α's are constants. For nitrobenzene, a change of temperature of 1°C results in an apparent change of 0.25% in the applied voltage [7].

2.2 Faraday Effect

The governing equation for the Faraday effect can be written

$$n_r - n_\ell = \lambda V H / \pi, \qquad (4)$$

[8] where n_r is the index of refraction for right circularly polarized light, n_ℓ is the index for left circularly polarized light, λ is the wavelength of the sensing light beam, V is the Verdet constant,* and H is the magnetic flux density. It should be noted that the direction of light propagation is the same as the applied magnetic field direction, whereas for the Kerr effect, the direction of light beam propagation was orthogonal to the applied electric field.

It is convenient to divide this discussion into two parts. The first deals with materials which have no net magnetic moment, i.e., diamagnetic materials, while the second deals with materials possessing a magnetic moment, i.e., paramagnetic materials.

In diamagnetic materials, the orbital and spin moments of the electrons cancel. So, in the absence of an external magnetic field, the net magnetic moment is zero. If an external field is applied, however, the electronic motion is modified to minimize the total field. The refractive index is sensitive to the electronic redistribution, and the Faraday effect results. Because the Faraday effect in diamagnetic materials does not depend on molecular orientation, it is relatively independent of temperature.

The Faraday effect in paramagnetic materials arises from the interaction between the permanent dipole moment of the molecules and the external field. In the absence of the field, the moments are unaligned. If the external field is applied, the moments

---

*The so-called Verdet constant is not a constant but a quantity analogous to the Kerr coefficient.

tend to align with the field and the indices of refraction are
sensitive to this alignment. Because the Faraday effect in this
case is due to molecular alignment, it has a larger temperature
dependence than it does in diamagnetic materials. In addition,
the Verdet constant is generally larger in paramagnetic than it
is in diamagnetic materials.

In analogy to the Kerr effect, in the Faraday effect, the
Verdet constant depends on the frequency of the applied field
when the reciprocal of the frequency is near the molecular reorientation time, depends upon temperature when molecular reorientation dominates, and exhibits a significant dependence on the
wavelength of the sensing light only in the vicinity of an absorption line.

In practice, Faraday-effect current sensors are typically
diamagnetic glasses doped with a paramagnetic material if a large
Verdet constant is required. But it should be emphasized that
the Faraday effect also occurs in plasmas, in semiconductors,
ferrites, and antiferromagnets [9]. In some applications these
may be the preferred sensors.

2.3 Pockels Effect

The fundamental equation which describes the Pockels effect
is

$$\Delta n = n_o^3 \, r_{ij} \, E, \tag{5}$$

where r_{ij} is the appropriate tensor element, n_o is the index of
refraction in the absence of an applied field, and E is the
applied field. The description of a device employing the Pockels
device usually involves tensor notation because of the natural
symmetries of the system. Specifically, the crystaline material
has the natural symmetry of its lattice structure. It is necessary to relate the crystal symmetry to the directionality of both
the orienting field and the sensing field. Tensor algebra is a
convenient method to express these relationships.

The Pockels effect itself is a direct electro-optic effect
which exists in addition to indirect electro-optical effects,
e.g., piezoelectric and elasto-optic effects [10]. A generalized
response of a Pockels effect device is shown in Fig. 3. The
mechanical effects can be classified into elasto-optic effects,
which include the photoelastic effects or the piezo-optic effect,
Brillouin scattering, and acousto-optic diffraction, and roto-
optic effects.

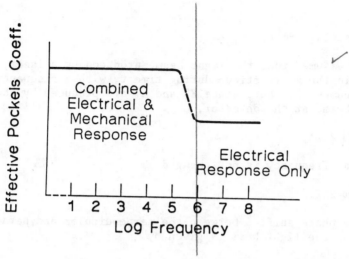

Fig. 3. The effective Pockels coefficient as a function of the frequency of the applied field. At low frequencies both electrical and mechanical effects produce an optical phase shift, while at higher frequencies the phase shift is a result of a purely electrical response.

Electrical measurements using the Pockels effect are generally performed at frequencies either well above or well below the transition between the combined effect and a purely electrical effect. In each of these regions, the response may be temperature dependent, so some technique to mitigate the effect of temperature variation must be employed [11].

3. PROPERTIES OF THE OPTICAL SYSTEM

Recall for the Kerr effect, Eq. (2),

$$n_{\parallel} - n_{\perp} = \lambda B E^2 . \tag{6}$$

The difference between the indices of refraction, $n_{\parallel} - n_{\perp}$, is not measured directly. Rather, the phase shift is measured between the electric field component of the light beam polarized in the direction of the orienting field, $E_{\parallel}$, and the component polarized in an orthogonal direction, $E_{\perp}$.

In general, the spatial and temporal variation of the light beam can be expressed as

$$E_{\parallel} = E_{0\parallel} \sin(k_{\parallel} x_{\parallel} - w_{\parallel} t), \tag{7}$$

and

$$E_\perp = E_{0\perp} \sin(k_\perp x_\perp - w_\perp t).$$ (8)

where it is assumed that the light beam, produced by a laser, propagates in the x-direction, during time t, with a frequency w and a wave number k. But, since $E_\parallel$ and $E_\perp$ are produced by the same laser beam, at the detector,

$$|k_\parallel| = |k_\perp| = |k| = 2\pi/\lambda,$$ (9)

and since the frequency is not changed

$$w_\parallel = w_\perp = w = 2\pi f.$$ (10)

So, the phase shift ϕ between the perpendicular and parallel components of the light beam is given by

$$\phi = (kx_\parallel w t)(kx_\perp w t),$$ (11)

or

$$\phi = k(x_\parallel - x_\perp).$$ (12)

The optical path lengths may be written to incorporate the changes in the indices of refraction

$$x_\parallel = L_0 + \int_0^L n_\parallel d\ell$$ (13)

and

$$x_\perp = L_0 + \int_0^L n_\perp d\ell$$ (14)

where L_0 is the geometric length of the optical path in any region in which $n_\perp = n_\parallel$, i.e., outside of the electro-optic medium and the applied field; L is the length of the path elsewhere; and $d\ell$ is a differential element along the path. Then,

$$\phi = (2\pi/\lambda) \int_0^L (n_\parallel - n_\perp) d\ell,$$ (15)

or, using Eq. (2),

$$\phi = 2\pi \int_{0}^{L} BE^2 d\ell \ . \tag{16}$$

It is conventional to express Eq. (16) as

$$\phi = 2\pi BE^2 \ell' , \tag{17}$$

when the applied field E is produced by a parallel-plate electrode system. The parameter ℓ' is then the effective path length that includes the effect of fringing fields from the electrode edges. So the phase shift for the Kerr effect ϕ_k, can be written,

$$\phi_k = 2\pi BE^2 \ell' \tag{18}$$

and, from a similar analysis, the phase shift for the Faraday effect, ϕ_f, is

$$\phi_f = 2VH\ell' , \tag{19}$$

and the phase shift for the Pockels effect, ϕ_p, is

$$\phi_p = 2\pi n_o^3 r_{ij} E\ell'/\lambda. \tag{20}$$

Generally, the measuring device uses an optically active medium to produce a phase shift but the quantity which is detected is the intensity of the transmitted light beam. To relate the phase shift to the intensity, either one of the two standard mathematical methods for describing the effect of an optical element on a polarized light beam can be used. The two methods are called the Jones calculus and the Mueller calculus [12-15]. Both of these calculii represent the optical element as a matrix and the light beam as a vector. As an introduction to these two approaches, recall that the electric field vector of E of a polarized light beam can be expressed.

$$\vec{E} = E_y \vec{e}_y + E_z \vec{e}_z , \tag{21}$$

where it is assumed that the light beam is propagating in the x direction and $\vec{e}_y$ and $\vec{e}_z$ are unit vectors. The two components can be written

$$E_y = a \cos(kx - \omega t), \tag{22}$$

and

$$E_z = b \cos(kx - \omega t + \delta), \tag{23}$$

where δ is the phase angle between the two components. These two equations can be written in terms of a new variable $\theta = kx - \omega t$.

$$E_y = a \cos\theta, \tag{24}$$

and

$$E_z = b \cos(\theta + \delta). \tag{25}$$

In the Jones calculus, the vector describing the light beam Λ is

$$\Lambda = \begin{bmatrix} E_x \\ E_y \end{bmatrix} = \text{Re} \begin{bmatrix} ae^{i\theta} \\ be^{i(\theta+\delta)} \end{bmatrix} . \tag{26}$$

Recognizing that the factor $e^{i\theta}$ can be ignored in the description of the effect of optical elements and making it implicit that the physical displacement of the electric field vector is given by the real parts of the matrix elements, we obtain the conventional form of the Jones vector

$$\Lambda_j = \begin{bmatrix} a \\ be^{i\delta} \end{bmatrix} . \tag{27}$$

To use the Jones calculus, each optical element is represented by a 2 x 2 matrix and the effect of light propagation through the optical system is described by matrix multiplication.

The second method, which will be used in the remainder of this discussion describes the light beam using a four-element vector, the elements of which are the Stokes parameters,

$$\Lambda_s = \begin{bmatrix} I \\ M \\ C \\ S \end{bmatrix} = \begin{bmatrix} a^2 + b^2 \\ a^2 - b^2 \\ 2ab \cos\delta \\ 2ab \sin\delta \end{bmatrix} . \tag{28}$$

In Eq. (28), the Stokes parameters are I, M, C, and S and are defined in terms of Eqs. (24) and (25). Eq. (28) demonstrates one of the reasons that this calculus is convenient for the present discussion, i.e., the intensity if the quantity of interest and it is an explicit matrix element. The Mueller matrices for various optical components are tabulated and/or derived in the references above. For this discussion, we will need the following matrices for Kerr, Pockels, and Faraday effect devices, for linear polarizers, and for quarter wave plates. Ideally, Kerr and Pockels effect devices are linear retarders, and the appropriate matrix is (where the fast axis is oriented at 90° with respect to the axis)

$$[M_{p,k}] = \begin{bmatrix} 1 & 0 & 0 & 0 \\ 0 & 1 & 0 & 0 \\ 0 & 0 & \cos\phi_{k,p} & -\sin\phi_{k,p} \\ 0 & 0 & \sin\phi_{k,p} & \cos\phi_{k,p} \end{bmatrix} , \quad (29)$$

where ϕ_k is defined by Eq. (18) and ϕ_p by Eq. (20). Faraday effect devices are circular retarders, so

$$[M_f] = \begin{bmatrix} 1 & 0 & 0 & 0 \\ 0 & \cos\phi_f & \sin\phi_f & 0 \\ 0 & -\sin\phi_f & \cos\phi_f & 0 \\ 0 & 0 & 0 & 0 \end{bmatrix} , \quad (30)$$

where ϕ_f is defined by Eq. (19). The matrices for a quarter-wave plate $[M_{\lambda/4}]$ and for a polarizer $[M_\rho]$ are

$$[M_{\lambda/4}] = \begin{bmatrix} 1 & 0 & 0 & 0 \\ 0 & \cos^2 2\theta & \cos 2\theta\sin 2\theta & -\sin 2\theta \\ 0 & \cos 2\theta\sin 2\theta & \sin^2 2\theta & \cos 2\theta \\ 0 & \sin 2\theta & -\cos 2\theta & 0 \end{bmatrix} , \quad (31)$$

$$[M_\rho] = 1/2 \begin{bmatrix} 1 & \cos 2\theta & \sin 2\theta & 0 \\ \cos 2\theta & \cos^2 2\theta & \cos 2\theta\sin 2\theta & 0 \\ \sin 2\theta & \cos 2\theta\sin 2\theta & \sin^2 2\theta & 0 \\ 0 & 0 & 0 & 0 \end{bmatrix} , \quad (32)$$

where θ is the angle that the transmission axis of the polarizer or the fast axis of the quarter-wave plate makes with the z-axis.

For a simple Kerr system, the output light vector $\{\Lambda_o\}$ is given by

$$\{\Lambda_o\} = [M_\rho(135°)] \, [M_k] \, [M_\rho(45°)] \, \{\Lambda_i\}, \quad (33)$$

where $\{\Lambda_i\}$ is the input light vector. For this discussion it is assumed that the input beam is unpolarized and has unit intensity, i.e.,

$$\{\Lambda_i\} = \{1,0,0,0\} . \quad (34)$$

Performing the calculation indicated in Eq. (33) yields

$$\{\Lambda_o\} = 1/2\{\sin^2(\phi_k/2),0,\sin^2(\phi_k/2),0\} . \quad (35)$$

So the normalized intensity is

$$I/I_m = \sin^2(\phi_k/2), \quad (36)$$

where I_m is the maximum transmitted intensity. Using Eq. (18),
Eq. (36) becomes

$$I/I_m = \sin^2((2\pi BE^2\ell')/2),\qquad(37)$$

or

$$I/I_m = \sin^2(\pi(E/E_m)^2/2),\qquad(38)$$

where

$$E_m = (2B\ell')^{1/2}.\qquad(39)$$

We note that if we assume that the orienting field is provided by
a parallel-plate electrode system, we can assume

$$E = V/d,\qquad(40)$$

and

$$E_m = V_m/d,\qquad(41)$$

as we anticipated in the discussion preceding Eq. (18). Equations (38) and (39) can then be written

$$I/I_m = \sin^2(\pi(V/V_m)^2/2)\qquad(42)$$

and

$$V_m = d/(2B\ell')^{1/2},\qquad(43)$$

where V_m is called the cell constant because its value is determined by the Kerr coefficient of the material filling the cell
and by the cell geometry. A significance of the cell constant
can be seen from Fig. 4 which plots the transfer function of a
Kerr system. When $V = V_m$, Eq. (42) says that the argument of the
sin function is $\pi/2$. From Fig. 4, it can be seen that $V = V_m$
is the lowest voltage which produces maximum transmission through
the optical system.

It should be emphasized that the transfer function shown in
Fig. 4 is highly nonlinear. The nonlinearity, however, is such
that the system becomes more sensitive as the voltage is increased. This property is important in accurate pulse measurements and has been exploited in the design of a peak reading
voltmeter and in the software to reconstruct the voltage waveform
from the optical measurement [16,17].

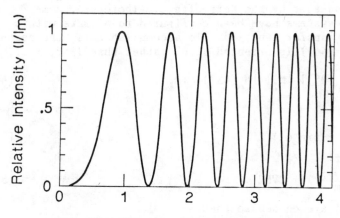

Fig. 4. The normalized transmitted intensity as a function of
         the normalized electric field applied to a Kerr cell.

A Faraday effect system can be configured in the same manner as
the simple Kerr system, (Eq. (33)), i.e.,

$$\{\Lambda_o\} = [M_\rho(135°)] \ [M_f] \ [M_\rho(45°)] \ \{\Lambda_i\}. \tag{44}$$

Performing the matrix multiplication yields

$$I/I_m = \sin^2(\phi_f/2), \tag{45}$$

and from Eq. (18),

$$I/I_m = \sin^2 VH\ell'. \tag{46}$$

In an analogy to the Kerr effect discussion, Eq. (45) can be
written

$$I/I_m = \sin^2(H/H_m), \tag{47}$$

where

$$H_m = 1/V\ell' \tag{48}$$

Equation (47) is plotted in Fig. 5. An important distinction between the Faraday and the Kerr effects is that the Faraday effect is a linear function of the field while the Kerr effect is quadratic. Thus, the Faraday effect does not have the high field sensitivity possessed by the Kerr effect. Therefore, some Faraday effect systems have been configured to operate with a nearly linear transfer function. To accomplish this, the second polarizer (Eq. (44)) is oriented at 0° rather than 135°, i.e.,

$$\{\Lambda_o\} = [M_\rho(0°)] \ [M_f] \ [M_\rho(45°)] \ \{\Lambda_i\}. \tag{49}$$

Then,

$$I/I_m = (1 + \sin \phi_f)/2 \tag{50}$$

or for small ϕ_f,

$$2I/I_m \sim 1 + 2VH\ell'. \tag{51}$$

So, if V and ℓ' are chosen so that

$$VH\ell' < 0.25 \text{ rad}, \tag{52}$$

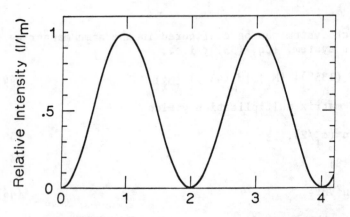

Fig. 5. The normalized transmitted intensity as a function of the normalized magnetic field applied to a Faraday effect sensor.

the error in assuming a linear relationship between the optical intensity and the magnetic field intensity will not exceed 1%. When a system is operated in this mode, present jargon occasionally characterizes the system as being optically biased to operate in the linear mode. The motivation for this jargon can be seen in Fig. 5. In that figure $I/I_m = 0$ when $H = 0$. From Eq. (51), however, with the polarizer at $0°$, $I/I_m = 1/2$, when $H = 0$. Thus, adjusting the polarizer from $0°$ to $135°$ biases the system response to an approximately linear segment of the sine wave.

Because the Mueller matrix for the Pockels effect is the same as that for the Kerr effect, the normalized intensities have the same functional form, i.e., for the optical system

$$\{\Lambda_o\} = [M_\rho(135°)] \ [M_p] \ [M_\rho] \ [M_\rho(45°)] \ \{\Lambda_i\}, \tag{53}$$

$$I/I_m = \sin^2\phi_p/2, \tag{54}$$

or, from Eq. (20),

$$I/I_m = \sin^2(\pi n^3_o r_{ij} \ E\ell'/\lambda). \tag{55}$$

Like the Faraday effect (Eq. (47)), this expression contains the field to the first power. So this system, too, is frequently biased to operate with a linear relationship between the applied field and the transmitted intensity. For example, if the optical system can be described by the equation:

$$\{\Lambda_o\} = [M \ (135°)] \ [M_{\lambda/4}(90°)] \ [M_p] \ [M_{\lambda/4}(0°)] \ x$$

$$[M \ (45°)] \ \{\Lambda_i\} \ , \tag{56}$$

then

$$2I/I_m \cong 1 + \phi_p = 1 + \pi n^3_o r_{ij} E\ell'/\lambda. \tag{57}$$

4.   APPLICATIONS

Detailed discussions of systems using these effects are found in the literature and in the remainder of this paper. The purpose of this discussion is to present a few examples to demonstrate the applications and operation of electro-optic and magneto-optic devices.

Figure 6 shows the transmittance as a function of position of a parallel-plate Kerr cell, subjected to a high voltage pulse. The numbers shown are the appropriate values of E/E_m, Eq. (38),

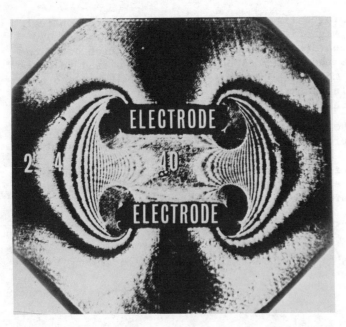

Fig. 6.   Transmittance of a parallel plate Kerr cell subjected
           to a high voltage pulse.

and the nonuniformity between the plates and along a line connecting them is due to fringing fields at the ends of the
electrodes [18]. If a photodetector were set to detect the
transmittance at the center of the interelectrode region, a response like that shown in Fig. 7 could be used to determine the
peak voltage or the entire waveform.

        The calibration is typically performed at a low frequency or
dc, a situation in which space-charge distortion of the field is
possible [19]. This effect can be mitigated because the electric
field can be measured as a function of position. Under this
condition,

$$V = \int_{o}^{d} \vec{E} \cdot d\vec{\ell}, \tag{58}$$

where V is the potential difference between the plates and d is
the plate separation. Thus, one can calibrate the system by
equating the integral of the electric field to the applied voltage. It should also be noted that, from Gauss's Law,

$$\nabla \cdot \vec{E} = \rho/\varepsilon, \tag{59}$$

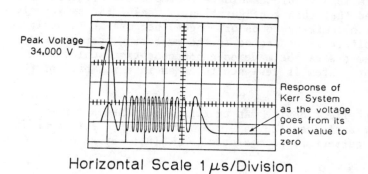

Peak Voltage
34,000 V

Response of
Kerr System
as the voltage
goes from its
peak value to
zero

Horizontal Scale 1 μs/Division

Fig. 7. Response of a Kerr system as a high voltage pulse goes
from its peak value to zero.

the space charge density ρ can be determined. This measurement
approach has been used to improve the understanding of space
charge in such fluids as water and transformer oil [20,21]. A
portable Kerr system [22] for the measurement of high voltage
pulses is shown in Fig. 8. The complete system is less than
50 cm long and can be used up to 100 kV. Operated in this mode,

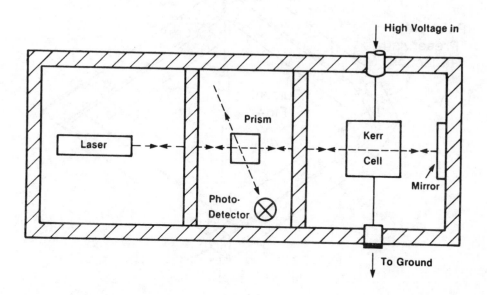

Fig. 8. Block diagram of a portable Kerr system used to measure
high voltage pulses.

the Kerr system is a viable candidate to replace a divider. It should be noted that this configuration is useful for relatively slow pulses (i.e., microseconds or longer) because it is designed so that the voltage is connected to the measuring system via a lead. Measurements can be made on nanosecond time scales when the measurement system is designed to be an integral part of the pulse generator.

A Faraday effect device can be configured like a conventional pulse transformer [23,24], i.e., so that the active medium encircles the current to be measured, Fig. (9). By Ampere's circuital law,

$$\phi = \int \vec{B} \cdot d\vec{\ell} = \mu_o \int_S \vec{J} \cdot d\vec{a} \quad , \tag{60}$$

the magnetic field, as measured by the Faraday effect, can be related to the current flowing through the sensor. It should be noted that instead of the glass block shown in Fig. (9), the sensor could be a fiber optic sensor. This configuration is attractively simple in that a fiber optic cable can serve as both the

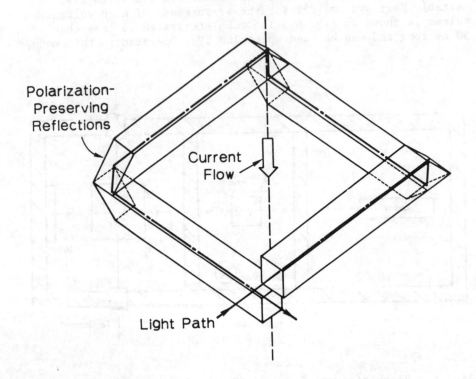

Fig. 9. A Faraday effect system to measure pulsed current.

signal transmission system and as the transducer. At present,
the residual birefringence of the cable limits the accuracy and
resolution of this approach.

A Pockels effect device, as shown in Fig. (10), has been
developed to measure radiated electric fields [11] in the range
of 1.0 V/m at 10^8 - 10^9 Hz. One lithium tantalate crystal has a
small antenna to improve the coupling to the field. The second
crystal is rotated with respect to the first and it compensates
for thermal effects in the sensing crystal. This example emphasizes that electro-optic systems can be used to measure relatively low signal levels as well as measuring the kilovolt or
megampere signals described above.

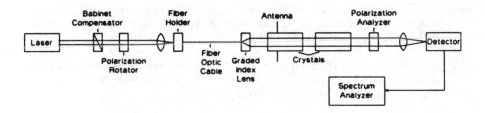

Fig. 10. A Pockels effect system to measure rf fields.

5.   SUMMARY

Electro-optical and magneto-optical devices have been and
can be developed to measure electric and magnetic fields, voltage, current, and space charge. The systems are not generally
available commercially so the present applications are those in
which the potential benefits outweigh the cost of system development. It is anticipated that the present trend toward increased
optical coupling in complex data acquisition systems will stimu-
late the development and application of electro-optic and magneto-optic sensors.

REFERENCES

1. W. S. C. Chang, Principles of Quantum Electronics, Reading,
    MA: Addison-Wesley, 1969, pp. 232-233.

2. E. Fredericq and C. Houssier, Electric Dichroism and Elec-
    tric Birefringence, Oxford: Clarendon Press, 1973,
    pp. 1-27.

3. A. D. Buckingham, "Electric Birefringence in Gases and Liq-
    uids" in Molecular Electro-Optics, C. T. O'Konski, Ed.,
    New York: Marcel Dekker, 1976, pp. 27-62.

4. G. L. Clark, "Dielectric Properties of Nitrobenzene in the
    Region of Anomalous Dispersion," J. Chem. Phys., vol. 25,
    pp. 125-129, 1956.

5. R. E. Hebner, "Calibration of Kerr Systems Used to Measure
    High Voltage Pulses," National Bureau of Standards Report,
    NBSIR 75-774, 1975.

6. E. E. Bergmann and G. P. Kolleogy, "Measurements of Nanosec-
    ond HV transients with the Kerr Effect," Rev. Sci. Instrum.,
    vol. 48, pp. 1641-1644, 1977.

7. R. E. Hebner and M. Misakian, "Temperature Dependence of the
    Electro-Optic Kerr Coefficient of Nitrobenzene," J. Appl.
    Phys., vol. 50, pp. 6016-6017, 1979.

8. R. E. Hebner, R. A. Malewski, and E. C. Cassidy, "Optical
    Methods of Electrical Measurement at High Voltage Levels,"
    Proc. IEEE, vol. 65, pp. 1524-1548, 1977.

9. L. M. Roth, "Faraday Effect", Concise Encyclopedia of Solid
    State Physics," R. G. Lerner and G. L. Trigg, Eds. (Addison-
    Wesley: Reading, MA) p. 88, 1983.

10. D. F. Nelson, Electric, Optic, and Acoustic Interactions in
    Dielectrics (Wiley: New York), pp. 283-353, 1979.

11. J. C. Wyss, M. Kanda, D. Melquist, and A. Ondrejka, "Optical
    Modulator and Link for Broadband Antennas" Conf. Precision
    Electromagnetic Meas. (IEEE Cat. No. 82 CH 1737-6), pp. P-16
    and P-17, 1982.

12. W. A. Shurcliff, Polarized Light, Production and Use
    (Harvard University Press: Cambridge, MA), pp. 15-31, 1962.

13. D. Clarke and J. F. Grainger, Polarized Light and Optical
    Measurement, (Pergamon Press: New York), pp. 17-41, 1971.

14. A. Gerrard and J. M. Burch, Introduction to Matrix Methods in Optics, (Wiley: New York), pp. 179-239, 1975.

15. B. A. Robson, The Theory of Polarization Phenomena, (Clarendon Press: Oxford), pp. 7-24, 1974.

16. D. C. Wunsch and A. Erteza, "Kerr Cell Measuring System for High Voltage Pulses," Rev. Sci. Instrum., vol. 35, pp. 816-20, 1964.

17. R. E. Hebner, E. C. Cassidy, and J. E. Jones, "Improved Techniques for the Measurement of High Voltage Impulses Using the Electro-optic Kerr Effect," IEEE Trans. Instrum. Meas., vol. IM-24, pp. 361-366, 1975.

18. P. D. Thacher, "Optical Effects of Fringing Fields in Kerr Cells," IEEE Trans. Elec. Insul, vol. EI-11, pp. 40-50, 1976.

19. E. C. Cassidy, R. E. Hebner, M. Zahn, and R. J. Sojka, "Kerr Effect Studies of an Insulating Liquid under Varied High-Voltage Conditions," vol. EI-9, pp. 43-56, 1974.

20. M. Zahn and T. Takada, "High Voltage Electric Field and Space-charge Distributions in Highly Purified Water," J. Appl. Phys., vol. 54 pp. 4762-75, 1983.

21. E. F. Kelley and R. E. Hebner, "Electro-Optic Measurement of the Electric Field Distribution in Transformer Oil" IEEE Trans. Power Appar. Sys., vol. PAS-102, pp. 2092-7, 1983.

22. R. E. Hebner and S. R. Booker, "A Portable Kerr System for Measurement of High Voltage Pulses," in Proc. IEEE SOUTH-EASTCON, vol. 1, pp. 3A-1-1 to 3A-1-5, 1975.

23. J. Katzenstein, W. Caton, and G. M. Wilkinson, "The Measurement of Pulsed Electric Currents by the Faraday Effect," in Measurement of Electrical Quantities in Pulse Power Systems, R. H. McKnight and R. E. Hebner, Eds., U. S. Nat. Bur. Stands. Spec. Publ. 628, pp. 277-288, 1982.

24. G. A. Massey, J. C. Johnson, and D. C. Erickson, "Laser Sensing of Electric and Magnetic Fields for Power Transmission Applications," SPIE, vol. 88, pp. 91-96, 1976.

ELECTRO-OPTICAL AND MAGNETO-OPTICAL STUDIES OF COLD CATHODE ELECTRON BEAM GUN DISCHARGES

M. Hugenschmidt

Deutsch-Französisches Forchungsinstitut
Saint-Louis
France

1. INTRODUCTION

In high power pulse technology, various types of controlled electrical gas discharges, such as those used in high power lasers, are in a continuing state of evaluation. Controlled electrical discharges, however, may also be of use in other systems as opening switches in electromagnetic mass drivers. Electron beam guns have, therefore, been conceived and developed for many of these purposes since about 1970. Both thermionic and cold cathode systems have been constructed and investigated [1,2, 3,4]. These cover a large range of moderate to high electron beam current densities. Thermionic guns are mainly applied at low current densities below 1 A/cm^2. Considerably higher densities, up to several tens to hundreds of A/cm^2, are attained with cold cathode guns. Voltages usually range from several hundreds to several thousands of kV.

The present work, at ISL Saint-Louis, is aimed at developing a cold cathode electron beam gun for preionizing or stabilizing high power laser discharges. As the gun shall be used in a multigas laser system, the performance characteristics have to be adapted to the special requirements of the various laser gases. In contrast, electron beam sources that are commercially available are always designed exclusively for one special type of laser gas. The multigas-laser capability thus requires a detailed knowledge of the current densities, the acceleration voltages and the pulse durations involved. These are strongly related to the electrical parameters, the currents and voltages applied to the gun.

These currents and voltages can be measured by applying conventional techniques, including resistive or capacitive voltage
dividers and magnetic pick-up loops. These techniques, however,
are, in most cases, strongly sensitive to disturbances due to
high-frequency oscillations, noise or effects of stray capacitances.

As already pointed out by several authors, these difficulties may largely be overcome by using magneto-optical and electro-optical effects. Electrical measurements based on the
Faraday and the Kerr effects have, therefore, been applied to
optimize the above mentioned, low inductance, electron beam device [5,6]. Some results will be given to reveal the usefulness
of these methods that can be adapted to any high voltage circuit
with minor changes.

2.   EXPERIMENTAL SETUP

The electron beam gun under investigation consists of two
chambers with two large area windows of 0.9 m x 0.1 m each. Various window materials have been tested such as thin foils of
titanium or aluminum (several tens of μm thick), as well as hy-
brid windows containing a thin plastic foil (20 μm), covered with
a thin sheet of aluminum (5 μm). The electron guns are usually
operated at pressures of 10^{-4} mbar. The cold cathodes consist of
parallel arrays of molybdenum blades held in an aluminum structure. The cathode-anode separation could be varied. Most of the
experiments, however, have been performed with relatively short
gaps of about 3.5 to 4 cm between the blades and the window
support grid which acts as anode. This is expecially of interest
in the short pulse mode, if electron pulses of only a few hundreds of ns are required. The two guns are driven by a four
stage Marx generator with voltages up to 400 kV and stored energies up to 3.2 kJ. The Marx bank, in the series of experiments
to be reported in the present paper, was electrically connected
to the gun chambers by a short, parallel plate, transmission
line. In this line, the Kerr cell and the Faraday glass rod
could easily be mounted close to the high voltage feed-throughs
of the cathodes. A schematic drawing is given in Fig. 1.

For the electro-optical and magneto-optical measurements of
the electron beam gun performance, the two laser chambers,
mounted on specially designed rails, were removed. These chambers are made of stainless steel. They are to be used for a
transverse electrical discharge configuration with an overall
discharge volume of slightly more than 20 l and a length of the
active volume of 2 m. The main discharges are fed by suitable
sustainer capacitor banks. Both the capacitances and the

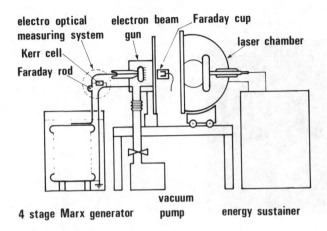

Fig. 1.   Experimental setup.

charging voltages of the sustainer can be varied in such a way
that the specific electrical energies and the E/p-values (E =
electrical field strength, p = pressure of the laser gas mixtures) can be adapted to the different types of laser gases.
Similar considerations hold for the optical components of the
laser cavities that have to be suitably chosen, according to the
wavelengths of the laser emission. As gas mixtures containing
fluorine are included in the program, all mechanical parts of the
system including the bellows, the mirror mounts and the gas
handling system are made of stainless steel. The electron beam
currents that are extracted from the gun through the foil windows
are measured by means of a Faraday cup probe. This probe was
placed near the window. It was also operated at low pressures of
about 10^{-4} mbar. The entrance window consists of a 15 μm thick
foil. This probe can be laterally displaced, so that both the
reproducibility and the current distribution across the window
surface can be measured. A typical curve of the electron beam
current density, as measured with the Faraday cup probe, is given
in Figure 2. These currents are determined from the voltages
across a low inductive 4.3Ω resistor, mounted inside the evacu-
ated part of the shielded housing. Pulse halfwidths under the
present experimental conditions are of the order of 300 ns to
400 ns. Peak current densities of 4 to 5 A/cm^2 were thus attained. These are average values across the 7 cm^2 Faraday probe
entrance window. As already mentioned, the relatively short

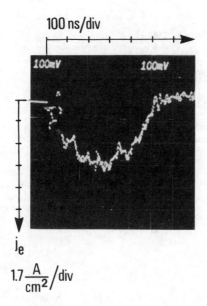

100 ns/div

100mV   100mV

j_e

$1.7 \frac{A}{cm^2}/div$

Fig. 2. Electron beam current densities.

pulse halfwidths are determined by the short closure time the
electrons need to cross the cathode anode gap of the vacuum discharge.

A first estimate on the spatial distribution of the electrons extracted through the two gun windows is obtained from
simple time integrated photographs of the electron gun output.
From these recordings it can qualitatively be concluded that the
guns are capable of providing homogeneously distributed electron
current densities. It can thus be expected that they are well
suited for initiating, dissociating, preionizing or sustaining
large volume high pressure discharges for various lasing gases.

3.   ELECTRO-OPTICAL AND MAGNETO-OPTICAL MEASUREMENTS

Linear and quadratic electro-optical effects, such as the
Pockels or the Kerr effects, as well as magneto-optical effects
such as the Faraday effect are well known and have been used for
various purposes for many years. One major application has been
their use in high-speed shutters both in fast photography and in
laser physics and technology. Their applicability in high voltage and high current measurements have been pointed out by many
investigators.

3.1 Basic Relations

A great advantage is that optical measuring systems are not physically coupled to the high voltage circuits neither by resistances nor by capacitances or inductances. Due to this fact, a high precision can be achieved. As the relaxation times of most of the optical effects are rather short (these times are, for example in the case of the Kerr effect, in the pico-second range) short rise times are achievable in these measurements. The bandwidths are usually only limited by the detectors used or by other electronic components of the measuring system such as amplifiers, oscilloscopes or digitizers.

Some of the most important equations needed in electro-optical Kerr measurements are given in Fig. 3. As known, the differences of the electric field induced refractive indices ($n_{||}, n_{\perp}$) of the ordinary and extraordinary beams, respectively, are found to be proportional to the square of the E-field amplitude. In the crossed polarizer configuration, the equation relating the light intensities I_1 and I_2 to the high voltages applied is given by the sin-square law as indicated in Fig. 3. The half-wave voltages $U_{\lambda/2}$ are determined by the geometrical dimensions (lengths l of the electrodes and distance a of the electrode gap)

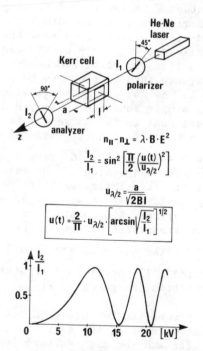

$$n_{||} - n_{\perp} = \lambda \cdot B \cdot E^2$$

$$\frac{I_2}{I_1} = \sin^2 \left[\frac{\pi}{2} \left(\frac{u(t)}{u_{\lambda/2}} \right)^2 \right]$$

$$u_{\lambda/2} = \frac{a}{\sqrt{2Bl}}$$

$$u(t) = \frac{2}{\pi} \cdot u_{\lambda/2} \cdot \left[\arcsin \sqrt{\frac{I_2}{I_1}} \right]^{1/2}$$

Fig. 3. Basic relations concerning the Kerr effect.

of the Kerr cell and by the Kerr constant B. In our experiments,
nitrobenzene has been used as a Kerr liquid. The curve of the
ratio I_2/I_1 of the transmitted light of a HeNe-laser, as a function of the high voltage applied to the Kerr cell, is given in
the lower part of Fig. 3. As much higher voltages than a few
tens of kV are encountered in the electron beam gun discharges,
the Kerr cell had to be connected to the transmission line by
means of a low inductance $CuSO_4$ voltage divider, having at least
a voltage division ratio of 10 : 1. As can be seen from Fig. 3,
the plane of polarization of the laser beam entering the Kerr
cell has to be rotated by 45 degrees with respect to the horizontal axis of the electric field. The light intensities I_2 are
recorded with a fast avalanche photodiode (BPW 28) having a rise
time of < 1 ns and a fast oscilloscope (TK 7844) or a transient
digitizer (R 7912), respectively. Both the oscilloscope and the
transient digitizer have a bandwidth of 500 MHz. The mode beat-
ing signal at 250 MHz, due to the partial selflocking of the
longitudinal modes in the HeNe-laser cavity, had to be suppressed
electrically by using a narrow band pass filter centered around
250 MHz.

Figure 4 summarizes some basic relations required if the
Faraday effect is used for measuring electric currents. A glass
rod having a high Verdet constant V is inserted in a single loop
coil in the high voltage transmission line of width 1. A magnetic field H is thus induced if a current flows from the Marx
bank to the electron beam gun load. The angle α, through which
the plane of polarization of a linearly polarized beam entering
the glass rod will be rotated, can then be shown to be proportional to the magnetic field H. As for the geometrical configuration chosen, the following simple equation holds $\int H \, dl = i$.
The angle α is directly proportional to the current i to be
determined. The Verdet constant of the Schott glass (SF 57)
used in our investigations is $V = 1.43 \cdot 10^{-3}$ degrees/A. This
value holds for $\lambda = 632.8$ nm. The dispersion of the materials
have to be taken into account if other lasers emitting other
wavelengths are used. If the analyzer is rotated by an angle of
$\beta = 45°$ with respect to the polarizer, the equation relating the
intensity ratio $I_{2\alpha}/I_{20}$ of the transmitted light to the angle α
reduces to $I_{2\alpha}/I_{20} = 1 - \sin 2\alpha$. The currents are then deduced
mathematically from the experimentally determined light intensity
ratios by using the arcsin-relation, as given in Fig. 4.

Due to the geometrical size of the two electron beam gun
modules, including the driving Marx bank, the spark switches, the
parallel plate transmission line and the high-voltage isolator
feed-throughs, the overall inductance of the circuit is rather

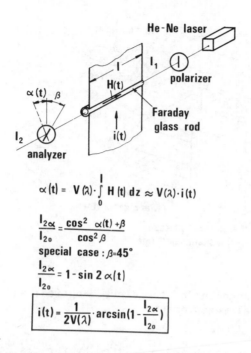

$$\alpha(t) = V(\lambda) \cdot \int_0^l H(t)\, dz \approx V(\lambda) \cdot i(t)$$

$$\frac{I_{2\alpha}}{I_{2o}} = \frac{\cos^2 \alpha(t) + \beta}{\cos^2 \beta}$$

special case : $\beta = 45°$

$$\frac{I_{2\alpha}}{I_{2o}} = 1 - \sin 2\alpha(t)$$

$$\boxed{i(t) = \frac{1}{2V(\lambda)} \cdot \arcsin\left(1 - \frac{I_{2\alpha}}{I_{2o}}\right)}$$

Fig. 4. Basic relations concerning the Faraday effect.

high. It is of the order of several μH. The current rise times can thus be expected to be much slower than the corresponding voltage rise times. Photo-multipliers with response times of several ns (for example RCA 931 A) proved, therefore, to be fast enough for measuring Faraday rotation current signals.

3.2 Experimental Investigations

The simultaneously applied Kerr and Faraday measurements, the basic principles of which have been discussed in the previous section, allow the electrical parameters to be optimized. They also allow a remote control of the gun performance as a part of a complete laser system. Such remote measurements proved to be most important for safety reasons, as the electron beam dis-charges under study are emitting a considerable amount of x-rays in each shot. Figure 5 shows schematically the main parts of the Marx generator, as well as the optical measuring system. The Kerr cell and the Faraday rod are supplied by a single 15 mW HeNe-laser using a 50% reflecting and 50% transmitting beam splitting plate. By the use of some additional mirrors that are

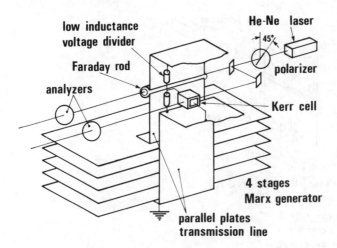

Fig. 5.   Schematic arrangement of the electro-optical and
          magneto-optical measurements.

not shown in Fig. 5, the two beams modulated by the discharges
are fed to the two detector systems, placed apart in a screen
room. A more recent setup will be used in the future in which
the optical signals are transmitted over even larger distances by
using fiber optic couplings between the analyzers and the detectors. These signals will then be digitized and stored in the
memory of a transient recorder allowing the data to be transferred to a PDP11 computer for processing and calculating the
electric currents and voltages involved.

     Figure 6 shows the Faraday and Kerr cell signals of a typi-
cal discharge, as recorded with the transient digitizer (R 7912).
For quantitative evaluation, both traces are simultaneously recorded with the dual beam oscilloscope (TK 7844). To improve the
temporal resolution, especially during the initial phase of the
discharges, even higher speeds of the time base of 50 ns/div.,
down to 10 ns/div., have been used. The current signals were
directly calibrated to yield the angle of the rotation of the
plane of polarization α. The calibration could be accomplished
by manually rotating the polarizing prism and relating the angle
to the measured intensities. The temporal shape of the currents
is thus directly proportional to the Faraday signal from which
the current is deduced by the simple relation $i(t) = \alpha(t)/V(\lambda)$.

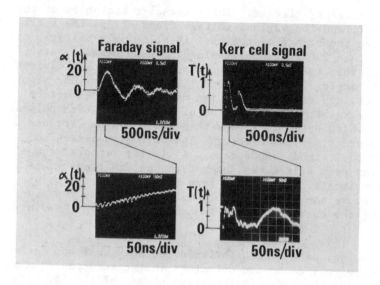

Fig. 6. Faraday and Kerr cell signals.

The calculation of the voltages from the measured Kerr cell signals is slightly more complicated, as the $\sin^2$- law relating the transmitted light intensities and the electric field strengths, has to be taken into account. As can be seen from Fig. 6, the Kerr signal rise time is extremely short; the first half-wave maximum is attained within 10 ns. The signal furthermore undergoes several oscillations that have to be carefully analyzed by considering the $(\frac{I_2}{I_1}$-versus-U) curve shown in the lower part of Fig. 3.

3.3 Quantitative Evaluation and Discussion

As an example, the quantitative evaluation of the recorded signals, discussed in the previous section, is shown in Fig. 7. As already pointed out, the electric current follows the Faraday signal. It attains its first maximum of 14 kA after a time delay of 500 ns and decays in a strongly damped oscillating discharge. From these measurements, the inductance of the overall electrical circuit can be determined to be 1.2 μH.

The voltage applied to the cold cathodes of the guns can be seen to rise much faster. The maximum is attained after 60 ns and decreases to zero within a few hundreds of ns. According to the charging voltage applied to the Marx generator in this series of measurements, peak voltages up to 360 kV should be expected in

an idealized, lossless circuit. The internal resistances of the
source, however, and the various voltage drops across the spark
switches or the inductance of the circuit allow only voltages of
slightly more than 220 kV to be attained. These measurements
clearly reveal that the losses in such pulsed high power devices
are not at all neglegible and that considerable errors, or wrong
interpretations, may result if precise measurements can not be
performed. The rather short duration of the high voltage pulses,
the values of which decay to zero within about 600 ns, are due to
the short closure times the electrons inside the guns need to
short-circuit the cathode anode gaps. A similar temporal shape
of the electron beam pulses can, therefore, be expected. This is
confirmed experimentally as can be seen from the lower part of
Fig. 7 which shows the simultaneously measured Faraday cup signal, revealing the current density transmitted through the windows.

As the optical path lengths between the Kerr cell and Faraday rod on one side and the two detectors in the screened room on
the other side are the same, the relative phases of the currents
and voltages are supplied automatically. The results, as shown
in Fig. 7, thus allow determination of the temporal variation of
the plasma impedance. As can be seen in Fig. 8, the impedance
drops from $Z = \infty$ to a few Ohms within some hundreds of nanoseconds. These values decrease to zero after about 800 ns. In the
following, negative impedances, which are typically found in low
pressure discharges, determine the discharge characteristics.
They are rapidly decreasing, approaching $Z = -\infty$ at $t = 1\,200$ ns,
the instant the current changes from the positive to the negative
half-wave. In its later development, the impedance again is
found to be positive as both the currents and the voltages have
then changed their sign.

Another most important parameter that can be deduced from
these current and voltage measurements is the electrical power
that is dissipated by the strongly non-linear load. The temporal
variation of the power can also be seen from Fig. 8. These
values are steeply rising within 250 ns. Peak values of more
than 1.5 GW are thereby attained. As the voltages in our present
experiment decrease more rapidly than the currents, the power
dissipated decreases to zero in the same time as the voltages.
From 800 ns to 1,200 ns, negative power values can be observed,
revealing that during this time interval energy flows back from
the electron beam gun load to the driving Marx generator.

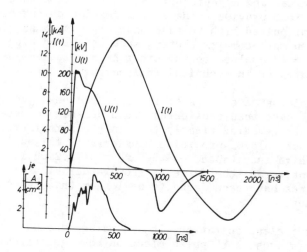

Fig. 7.   Temporal shape of the electrical current and voltage
          as well as the extracted electron beam pulses.

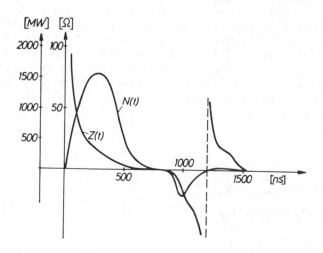

Fig. 8.   Time variation of the gun plasma impedance and the
          dissipated power.

4.   CONCLUDING REMARKS

Electro-optical and magneto-optical measurements used in
these investigations provide valuable information on the electrical parameters of pulsed high voltage discharges such as electron
beam guns or even more complex systems such as pulsed high energy
lasers. They can be applied to any other high voltage measuring
problem in research or in technical engineering.

As an example, some studies of the electron beam gun system
developed at ISL have been considered and discussed in detail.
These measurements revealed clearly that internal resistances or
inductances are capable of providing considerable losses. Due to
the relatively high inductance, of the order of 1.2 µH in the
special case considered, the currents grow rather slowly. The
current rise times have been found to be longer than the electron
gun closure times.

For more efficient operation, it had to be concluded that
the inductance in such short pulse electron beam gun circuits has
to be considerably reduced. This can be achieved by replacing
the parallel plate transmission line, connecting the Marx generator with the gun chambers, by four specially designed coaxial
cables. These have been built and successfully tested in the
meantime. For measuring purposes, however, it proved to be
disadvantageous since a Kerr cell could not be introduced into
the coaxial configuration. To overcome these difficulties, a
large bandwidth, Ohmic, voltage divider, yielding a ratio of
10,000 : 1, with a bandwidth of more than 500 MHz, has been developed. The electrical low voltage signals are then used to
modulate a laser diode which is part of a fiber optic coupled
transmitter receiver system. The data can then also be transmitted over great distances without being affected by noise or
high frequency pick-up. For further processing and numerical
evaluations, similar considerations hold as in the case of the
Kerr cell measurements.

REFERENCES

1. C. Fenstermacher, Entropie 89-90, p. 53 (1979).

2. G. K. Loda, "Advances in Laser Technology," SPIE vol. 138,
    p. 69 (1978).

3. R. A. Gerber, E. L. Patterson, J. of Applied Physics 47,
    p. 3524 (1976).

4. N. G. Basov, F. M. Belenov, V. A. Danilychev, A. F. Suchkov
    Sov. Phys. Usp. 17, p. 705 (1975).

5. J. E. Thompson, M. Kristiansen, M. O. Hagler, IEEE Trans-
    actions on Instrumentation and Measurement 25, p. 1 (1976).

6. J. E. Thompson, Proc. of the Workshop on Measurements of
    Electrical Quantities in Pulse Power Systems, Boulder (1981)
    edited by R. H. McKnight, R. E. Hebner Jr., issued (1982).

FIBER OPTIC MAGNETIC FIELD AND CURRENT SENSORS

George Chandler

Los Alamos National Laboratory
Los Alamos, NM
USA

1. INTRODUCTION

The development of fiber optics technology for communications included the investigation of environmental effects on the properties of the fibers. One result of those investigations was the exploitation of the effects in sensors [1,2]. Many of the applications involve the strain-optic characteristics. This discussion will center, however, on the use of the Faraday effect.

In any optical material, the angle of rotation F of the plane of polarized light in the presence of a magnetic field $\underset{\sim}{H}$ is governed by the Verdet constant V of the material and described by

$$F = V\int \underset{\sim}{H} \cdot d\underset{\sim}{l}. \tag{1}$$

If the fiber makes a closed loop around a current I then

$$F = VI. \tag{2}$$

The Verdet constant V has the value 4.68×10^{-6} rad A^{-1} for silica, at 633 nm [2].

The first use of optical fiber-Faraday rotation current sensing was aimed at the electric power industry [3,4]. Recently, workers at Los Alamos National Lab [5,6] have accomplished the detection of large currents in fast fusion and other machines. The technique shows promise but is is not without difficulties.

2. PROPERTIES OF OPTICAL FIBERS

Most optical fiber is made for communications applications. There are three general types, illustrated in Fig. 1. Most sensor work uses single-mode fiber which is primarily intended for high-bandwidth, long-distance communications, although some fibers are being produced specifically for sensor applications or with properties which enhance their suitability for both types of use [8-12].

The elementary explanation of the propagation of light in an optical fiber is in terms of total internal reflection at the interface between the core and the cladding. The index of the core glass is higher than that of the cladding by a small amount Δn (.001 to .05). Figure 2 illustrates the geometry of the explanation. As the angle of incidence θ, on the core-clad interface is increased, it reaches a critical angle θ_{crit} above which all the light is reflected and stays in the core. The angle ϕ_{crit} at which that light ray is incident on the face of the fiber is a significant parameter. The numerical aperature (NA) is the sine of this angle, which is related to the indices of refraction by

$$\sin \phi_{crit} \equiv NA = (n_1^2 - n_2^2)^{1/2}. \tag{3}$$

The figure shows an unguided ray which partially escapes into the cladding. This ray propagates in the clad for a short distance but is quickly attenuated in well-designed fibers. The part remaining in the core is also attenuated after several more reflections.

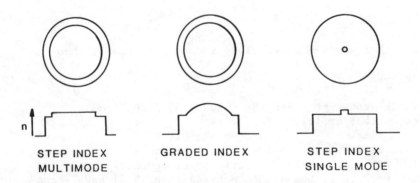

STEP INDEX   GRADED INDEX   STEP INDEX
MULTIMODE   SINGLE MODE

Fig. 1.   Illustrating the three basic types of optical fiber in terms of the relative dimensions of core and clad, and the profile of the index of refraction.

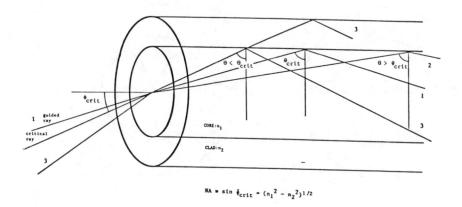

NA = sin ϕ_{crit} = $(n_1^2 - n_2^2)^{1/2}$

Fig. 2. Illustrating the total internal reflection model.

A more sophisticated and useful model results from solving Maxwell's equations with appropriate boundary conditions. The radial dependence of the normal guided modes of the fiber is described by Bessel functions: Bessel functions of the first kind within the core, and modified Bessel functions of the second kind in the cladding (see Fig. 3). The azimuthal dependence is described by sine or cosine functions, and the Z or longitudinal fields are very small and ignored in mode visualization. Interesting features of fibers for sensor purposes are generally results of the radial dependence. The radial distribution of the fields is of primary importance in explaining and predicting the performance of optical fibers. Most sensor work (but not all) is done with single mode fibers, the core of which is so small that it propagates only the lowest order mode. This paper will be concerned only with single mode fibers.

2.1 Strain Effects

The effective length Z of the optical path in a fiber can be varied by subjecting the fiber to stress. Three important effects of strain can be discussed with the aid of the following general equation:

$$\Delta Z = \frac{\partial Z}{\partial L}\Delta L + \frac{\partial Z}{\partial n}\Delta n + \frac{\partial Z}{\partial \beta}\Delta \beta, \qquad (4)$$

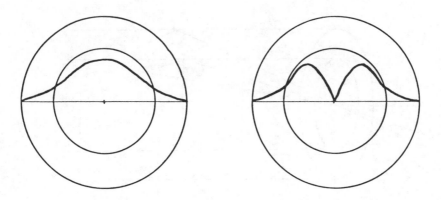

Fig. 3.   Sketch of the radial dependence of the fields in the
          lowest order modes of an optical fiber.

where ΔL is the change in physical length, determined by the
Young's modulus and the Poisson's ratio of the material; Δn is
the change in the index of refraction, involving the photoelastic
constants (or Pockel's coefficients); and $\Delta\beta$ is the change in the
propagation constant of the fiber, determined by the dependence
of that constant on the dimensions of the guiding structure.

     Strain effects are monitored by putting the fiber in one arm
of a Mach-Zehnder interferometer. A transducer is fabricated by
bonding the fiber to a device which produces strain in response
to some desired field; for example a piece of magnetostrictive or
piezoelectric material, or an acoustically optimized mandrel.
References 1 and 2 describe this well-developed technology in
detail.

2.2 Faraday Rotation

     If the fiber were a perfectly homogeneous, measurement of
the angle of rotation would be straightforward. Unfortunately,
the ellipticity of the fiber core, stresses frozen into the fiber
during drawing, and stresses imposed from the environment all
combine to produce linear birefringence, signified by δ. Analysis of the output state of polarization in the presence of both
Faraday rotation (which is also described as induced circular
birefringence) and distributed linear birefringence is fairly
complicated. Smith [3] does the analysis for the somewhat idealized but useful case illustrated in Figure 4.

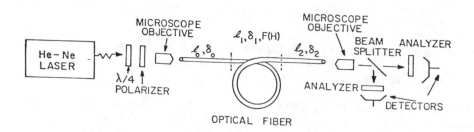

Figure 4.   Sketch of current measurement apparatus showing the
            regions of the fiber for Smith's analysis.

There are three regions of the fiber: region 0 is a "lead-
in" zone, with linear birefringence δ_0 and not exposed to any
magnetic field; region 1 is the field-sensing zone with linear
birefringence δ_1 and induced circular birefringence F; and region
2 is a "lead-out" zone with no circular birefringence and linear
birefringence δ_2. If linearly polarized light is injected with
the plane of polarization parallel to one of the fiber birefringent axes, and if the output light is analyzed with polarizers
axes oriented at $\frac{\pi}{4}$ radians to the output polarization axes, then
the quantity

$$T = \frac{I_1 - I_2}{I_1 + I_2} \, , \tag{5}$$

where $I_{1,2}$ are the intensities at the two detectors, is given by

$$T = -2\cos(\delta_2 + \eta) \, \sin\chi \, \sin\frac{\phi}{2} \, (\cos^2\frac{\phi}{2} + \cos^2\chi \, \sin^2\frac{\phi}{2})^{1/2} \, , \tag{6}$$

where

$$(\frac{\phi}{2})^2 \equiv (\frac{\delta_1}{2})^2 + F^2 , \tag{7}$$

$$\tan\chi \equiv \frac{2F}{\delta_1} \, , \tag{8}$$

and

$$\eta \equiv \tan^{-1}(\cos\chi \; \tan\frac{\phi}{2}) \; . \tag{9}$$

Equation (6) simplifies considerably when $F \gg \delta_1$ and δ_2, to

$$T \sim -\sin 2F. \tag{10}$$

This condition can be achieved in two ways: by having very large fields which produce very large F, or by "biasing" F with induced circular birefringence [13,14]. The latter is accomplished by twisting the fiber [13,14].

The effect of linear birefringence is illustrated in Fig. 5. It can be seen that for values of F of a few hundred degrees or less the response of the fiber system is dramatically dependent on the amount of linear birefringence present in the sensor zone and the lead-out zone (note that δ_0 does not appear in Eq. (6)).

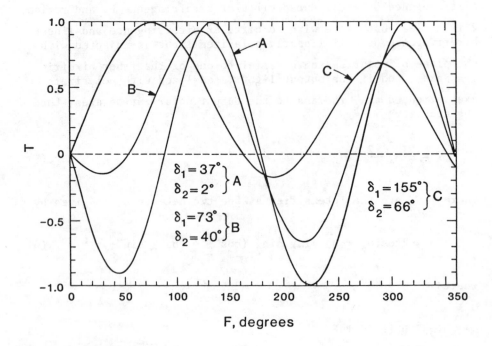

Fig. 5. Plots of T vs F for different values of $\delta_1 + \delta_2$ having the same value of $\cos(\delta_0 + \delta_1 + \delta_2)$.

3.   EXPERIMENTAL RESULTS

Figure 6 is a sketch of the apparatus built to measure the current in the High Density Z-Pinch (HDZP) of Hammel, et al., at Los Alamos [15].   In this experiment, two electrodes are in a pressure vessel containing two atmospheres of hydrogen gas.   A coaxial line applies a high voltage pulse across the electrodes. A narrow (~100 microns diameter) current channel is created by a laser beam which ionizes the gas.   The question arose as to whether all the current was contained within the visible channel. The apparatus shown in Fig. 6 was devised to determine this.

The fiber encircles the current channel as shown, with the two straight sections being about 15 mm apart.   The idea was to move the fibers around between shots and determine whether the amount of current enclosed changed over distances on the order of millimeters, indicating that current was being carried in a larger channel than predicted.   This part of the experiment has not yet been completed but some interesting results have been obtained.

The input optics in Fig. 6 produce linearly polarized light, the plane of polarization of which can be rotated about the fiber

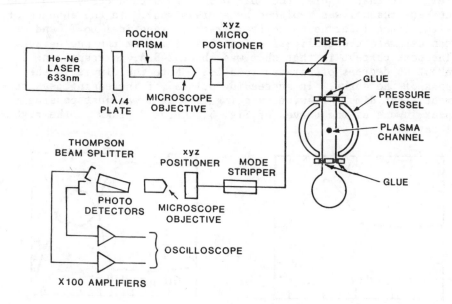

Fig. 6.   Sketch of Faraday rotation current measurement diagnostic on HDZP.

axis. The fiber used in this experiment is made with very low
intrinsic linear birefringence (1.5°/meter) and for single-mode
operation at 633 nm (He-Ne) and above [8]. The end of the fiber
is cleaved and held in a motor-driven micropositioner, at the
focal point of a microscope objective. About 25% of the light
incident on the microscope objective is injected into the fiber.
A mode stripper at the output end removes light injected into the
cladding by the laser and by the plasma light from the HDZP. The
output beam is imaged on two photodetectors located behind a
Thomson beam-splitting prism. The photocurrents are directly
proportional to I_1 and I_2 in Eq. (5), and are amplified and recorded on a dual-beam oscilloscope. By rotating the plane of
polarization of the input light and recording the polarization of
the output, it is possible to locate the fiber axes at both input
and output ends [3], and to measure the cosine of the total linear birefringence. No way is known to measure the birefringence
directly. Figure 5 is a plot of Eq. (6) for the experiment shown
in Fig. 6, for three possible values of δ_1 and δ_2 corresponding
to the measured value of $\cos(\delta_1 + \delta_1 + \delta_2)$.

Figure 7 shows the traces of I_1 and I_2 on the right and
the current. as detected by an inductive current probe on the
left. In that figure, the HDZP was fired with a dummy load (the
current channel was replaced by a brass rod). In the absence of
significant laser noise, either trace is identical to T, and we
can estimate the qualitative features of T from the photograph.
The peak current in that shot was about 248kA, corresponding to
about 66 degrees of Faraday rotation. Note the dimple in the
peak of the curve. In succeeding pulses, of increasing peak current, that dimple grew, convincing us that T was just passing a
peak in one of the curves of Fig. 5. Curve C peaks at the right
location.

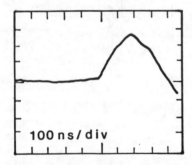 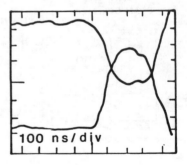

100 ns/div   100 ns/div

Fig. 7.   Inductive current probe (left) and traces of I_1 and
          I_2 on HDZP with dummy load.

Unexpected benefits accrued from the use of this novel technique. Figure 8 is a trace taken during a malfunction of the system. The current switched from the fiber optic traces, but only a small glitch marked the event on the conventional diagnostic.

Finally, in Fig. 9, we see the result of a pressure wave striking the fibers. This was confirmed by a Schlieren photograph taken with a nitrogen laser at about 600 nsec. The effect is apparently to increase the light lost from the fiber, possibly by modulating the index of refraction at the core-clad interface, and not to change the birefringence which would cause the traces to move in opposite directions,

In recent weeks we wrapped a fiber [10] 7 turns around the ZT-40, reversed-field, pinch at Los Alamos, in a location to measure the toroidal current. Using the same apparatus shown in Fig. 6, except that the outputs were recorded on digitizers, instead of an oscilloscope, we obtained the traces in Fig. 10. The top traces are I_1 and I_2, the center trace is T calculated from I_1 and I_2, and the bottom is the toroidal current from a conventional diagnostic. No analysis has been performed on this data yet. Figure 11 compares T from three shots with different currents: 80 kA, 120 kA, and 180 kA.

An example [6] of the use of this technique with very large currents and hence very large F is shown in Figs. 12 and 13. Figure 12 is the raw scope trace, and Fig. 13 compares the current from another probe, with the current determined by counting fringes in Fig. 12, and applying a calibration factor of 250°/mA.

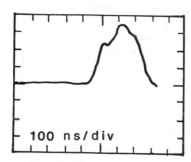
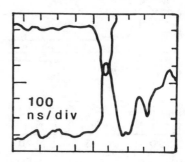

Fig. 8.   Traces recorded during shot when current switched out of the prescribed channel.

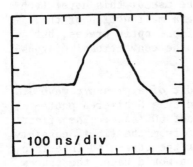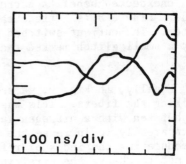

Fig. 9. Showing the effect of a pressure wave striking the
         fiber.

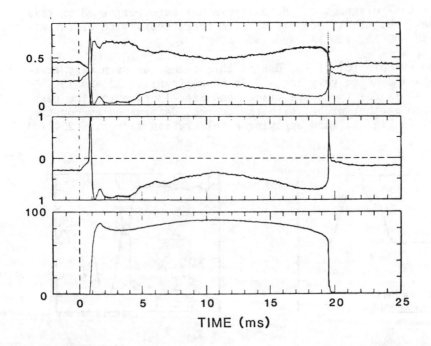

Fig. 10. Top I_1 and I_2; center, T; bottom, toroidal current
          from ZT-40 shot.

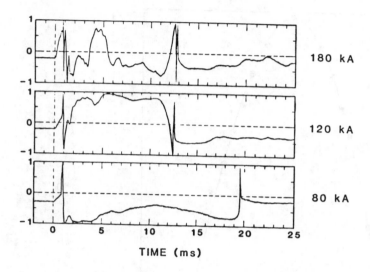

Fig. 11.   T from three ZT-40 short at different currents.

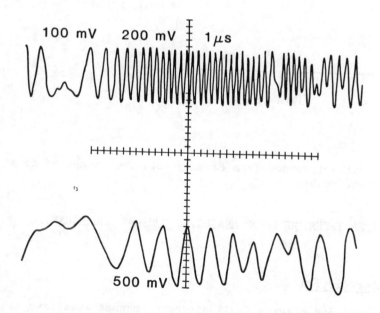

Fig. 12.   Raw data from a 5 megampere source.

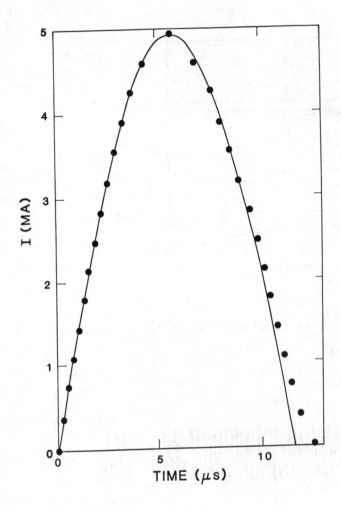

Fig. 13. Current determined from Faraday rotation, compared to a
          current probe.

4. OTHER FIBER TECHNIQUES FOR MEASURING CURRENT AND MAGNETIC FIELDS

4.1.4 Magnetostriction

One may envision placing small pieces of magnetostrictive
material in the region of a magnetic field to map it as a function of time, using optical fiber to measure the strain as was

described briefly above. Magnetometers have been constructed using metal-coated fiber, but eddy currents limit their frequency response. We propose to use non-conducting ferrites which will not suffer from eddy currents. The time response is limited, however, by the speed of sound in the ferrite and the fiber. Using millimeter-size pieces of ferrite and grinding away part of the fiber cladding, we calculate that such a sensor should have a response time on the order of hundreds of nanoseconds.

4.2 SAGNAC Interferometer

Figure 14 is a sketch of a Sagnac interferometer for measuring current [17]. Circularly polarized light is injected by means of a beamsplitter into both ends of a fiber looped around a current. The emerging beams interfere on a detector. The magnetic field will be parallel to the light in one direction and antiparallel in the other. In this situation, the Faraday effect is to increase the index of refraction for one and decrease it for the other, causing a phase shift between the two interfering beams.

4.3 Polarization Optical Time Domain Interferometry [18] (POTDR)

This is an adaptation of the popular technique for determining the condition of a fiber by monitoring the reflections, from imperfections and discontinuities, of a pulse introduced at one end. Figure 15 shows the essential features. The reflected beam is passed through a polarization analyzer. The output is a measure of the polarization-inducing environment of the fiber as a function of distance along the fiber. This becomes, then, a technique for looking at the distributions of such fields. Fields which might be mapped this way include magnetic (Faraday effect), electric (Kerr effect), strain, and temperature.

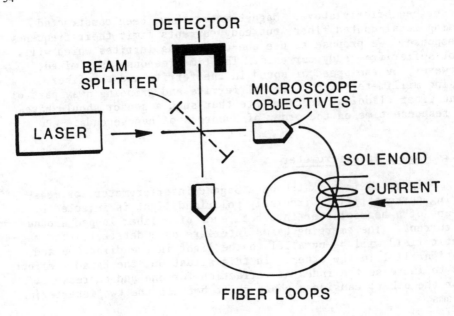

Fig. 14.   Sketch of Sagnac interferometer for measuring currents.

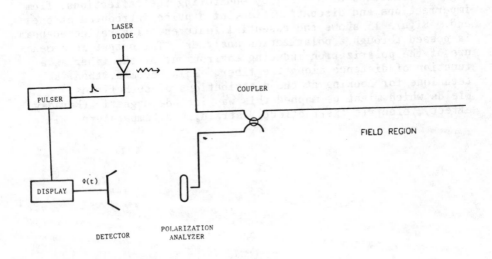

Fig. 15.   Sketch of POTDR scheme.

REFERENCES

1. T. G. Giallorenzi, et al, "Fiber Optic Sensor Technology
    Overview," IEEE J. Quantum Electronics QE-18,626, 1982.

2. C. M. Davis, et al, Fiber Optic Sensor Technology Handbook,
    1982, Dynamic Systems Inc., 8200 Greensboro Drive,
    Suite 500, McLean VA 22102, USA.

3. A. M. Smith, "Polarization and Magnetooptic Properties of
    Single-Mode Optical Fiber," Appl. Opt. 17, 1, 1978.

4. A. Papp and H. Harms, "Magnetooptical Current Transformer,"
    Appl. Opt. 19, 22, p3279, 1980.

5. G. I. Chandler and F. C. Jahoda, "Current Measurements by
    Faraday Rotation in Single-Mode Fibers," Proc. Los Alamos
    Conf. on Optics '83, to be published by SPIE.

6. L. Veeser et al, "Measurements of Megampere Currents with
    Optical Fibers," Proc. Los Alamos Conf. on Optics '83, to be
    published by SPIE.

7. J. E. Midwinter, Optical Fibers for Communications, John
    Wiley, 1979.

8. York Technology Ltd., Chaucer Industrial Park, Eaton Lane,
    Winchester Hants SO23 7RU, England 0962 64295, or 357 Nassau
    St., Princeton NJ 08540 USA. (609) 924-7676.

9. Andrew Corp., 10500 W. 153rd St., Orland Park IL 60462,
    USA. (312) 349-3300.

10. Lightwave Technologies, Inc., 6737 Valjean Ave., Van Nuys CA
    91406 USA. (213) 786-7873.

11. EOTEC Corp., 200 Frontage Rd., West Haven CT 06516 USA.
    (203) 934-7961.

12. ITT Electrooptics Div., 7635 Plantation Rd., Roanoke VA
    24019 USA. (703) 563-0371.

13. S. C. Rashleigh and R. Ulrich, "Magneto-optic Current
    Sensing with Birefringent Fiber," Appl. Phys. Lett. 34,
    768-770, 1979.

14. R. Ulrich and A. Simon, "Polarization Optics of Twisted
    Single-Mode Fiber," Appl. Opt. 18, 2241-2251, 1979.

15. J. E. Hammel, J. S. Shlachter, and D. W. Scudder, "Recent
     Results on Dense Z-Pinches," to be published in Nuclear In-
     strumentation Methods.

16. A. Yariv and H. V. Winsor, "Proposal for Detection of Mag-
     netic Fields Through Magnetostrictive Perturbation of Opti-
     cal Fibers," Opt. Lett. $\underline{5}$, 3, 87-89. 1980.

17. H. J. Arditty, M. Papuchon et al, "Current Sensor Using
     State-of-the-Art Interferometric Techniques," Tech. Dig. of
     Third Int'l Conf. on Integrated Optics and Optical Communi-
     cations, San Francisco 1981. Paper WL3, published by Opti-
     cal Society of America. IEEE catalog no. 81 CH 1649-3.

18. A. J. Rogers, "Polarization Optical Time-Domain Reflectrome-
     try: a Technique for the Measurement of Field Distribu-
     tions," Appl. Opt. $\underline{20}$, 6, 1060-1074, 1981.

AN ELECTRO-OPTICAL TECHNIQUE FOR MEASURING HIGH FREQUENCY FREE SPACE ELECTRIC FIELDS

J. Chang and C. N. Vittitoe

Sandia National Laboratories
Alburquerque, NM
USA

1. INTRODUCTION

Electro-optical voltage measurement techniques, based on the Kerr and Pockels effects, offer a number of advantages over conventional resistive or capacitive voltage dividers. As with all electron analog sensors, their performance is affected by EMI, ground loops, and noise pickup; and their frequency response is limited by stray capacitance and inductance associated with the divider circuits. Optical analog sensors are not susceptible to these perturbations and are often considered superior [1]. One of the most outstanding characteristics of optical analog sensors is their inherent band width; for example, the intrinsic response time of Kerr fluids is measured in picoseconds [1]. In practice, however, the cells are placed between electrodes and the associated capacitance usually limits the sensor band width to less than 1 GHz. In addition, electrodes may introduce electrical breakdown along the surface of the cell. These concerns are contributing factors to the under-utilization of the electro-optic techniques in pulse power research and systems.

We have explored the possibility of using bulk electro-optical components, without electrodes, in a dielectric medium to measure electric fields. This approach will recover the lost band width and, at the same time, remove the electrode-introduced breakdown limitation. Of course, the field strength measured is still limited by the breakdown strength of the dielectric medium. An optical analog sensor based on this approach could be placed in any of the dielectric media, in a pulsed power system, to make electric field measurements. For example, it could be placed in

the oil, oil and water interface, water, and the water vacuum
interface. It could also be used in air or in vacuum to make rf
and microwave measurements.

In the following, we will describe the techniques and a
proof-of-principle experiment in which a free standing, KDP,
crystal was used in free space to measure the output of a 3 GHz
pulsed magnetron.

2.   PRINCIPLE OF OPERATION

The technique used is similar to conventional laser modulation techniques. A linearly-polarized HeNe or Argon ion laser
beam is passed through a 1 cm cube, KDP crystal with the direction of polarization at 45° with respect to the electrically-in-
duced birefringent axes. A second polarizer, orientated at 90°
with respect to the incident beam polarization, is placed on the
output side of the KDP crystal. The net result of an electric
field acting on the crystal, along the beam direction, is to ro-
tate the polarization of the beam proportional to the amplitude
of the electric field. Due to this rotation, the transmitted
light is no longer polarized at 90° to the second polarizer and
a component of the transmitted light is allowed to pass through.
A schematic representation of such a setup is shown in Fig. 1.
To record the amplitude modulated light output from the second
polarizer, a high-speed Hamamatsu [2] electro-optical streak
camera, with a 35 ps time resolution, is used. For a modulation
voltage $V_m \sin\omega_m t$ acting on the crystal, the output light intensity I_o is related to the input light intensity I_i by [3]

$$\frac{I_o}{I_i} = \sin^2\left(\frac{\Gamma_m}{2} \sin\omega_m t\right) ,$$

where Γ_m is $\pi(v_m/V_\pi)$ and V_π is the voltage required to cause a
90° rotation of the beam polarization. V_π is measured to be 8.36
kV for KDP at HeNe wavelength 632.8 nm [4]. Using this technique, a measurement of the output to input intensity ratio gives
a direct determination of V_m which is related to the E component
of the rf field by the length of the crystal.

To make a frequency measurement, the arrangement described
has to be modified to include a 1/4-wave plate between the crystal and the second polarizer. This serves the purpose of placing
the operating point in quadrature at 50% modulation. This
changes the intensity ratio to

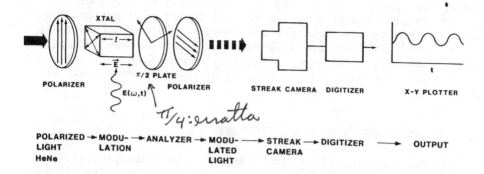

Fig. 1. Schematic of the electro-optic technique.

$$\frac{I_o}{I_i} = \sin^2 \left(\frac{\pi}{4} + \frac{\Gamma_m}{2} \sin \omega_m t \right)$$

$$= \frac{1}{2} \left[1 + \sin \left[\left(\Gamma_m \sin \omega_m t \right) \right] \right].$$

If Γ_m can be made small, i.e., $\ll 1$, the relationship becomes, by expansion,

$$\frac{I_o}{I_i} \sim \frac{1}{2} \left[1 + \Gamma_m \sin \omega_m t \right].$$

The modulation frequency on the intensity ratio is, therefore, a direct copy of the rf frequency.

Since V_π is effectively a measure of the crystal constant and is well known, the use of this technique requires no detailed calibration and only relative measurements need to be made. The only requirement being that the measurements be made in the linear response portion of the detector. The detector in this case, is the streak camera. An additional constraint on this technique is a frequency limitation resulting from the consideration of the time required for a photon to transit the crystal, compared to the rf wave period. Obviously, for the situation where the photon transit time is much longer than the rf wave period, modulation intensity and frequency are no longer linearly related to the rf field amplitude and frequency. For a 1 cm cube KDP, for example, the upper frequency limit is ~5 GHz. However, this also implies a 0.25 cm thick KDP will have a frequency limit at 20 GHz.

3.   THE PROOF-OF-PRINCIPLE EXPERIMENT

In the experiment, we sought to show, in principle, that using the electro-optical technique we could measure directly the rf frequency and power. The KDP crystal was held by epoxy to a 0.25 inch diameter dielectric rod which, in turn, was mounted on a tripod. In this arrangement, the KDP crystal was placed at 11° to the axis of the magnetron horn and 1.5 m away. The angular position placed the crystal at the peak of the TM_{01} radiation pattern. The tripod, as well as the laser and other optical components, were shielded from rf radiation by Echosorb walls with only the crystal and part of the supporting dielectric rod exposed. With this setup, first the frequency was measured and then, after removing the 1/4-wave plate, the modulation amplitude was measured.

The transmitted light was conducted by a 1 mm diameter optical fiber to the streak camera in a screen room, 10 m away. The whole system was first checked for linearity by measuring the streak camera output against different input light intensities. Then a series of null tests were made to check for possible effects due to EMP and x-ray background. We found that, with the magnetic field on the magnetron source turned off, i.e., no microwave output, no modulation nor light transmission were detected. In addition, we also found no modulation when we moved the crystal to a position where the E field component was perpendicular to the laser beam direction and the crystal modulation axis.

For the frequency measurement, the streak camera was set to 10 ns full sweep to achieve maximum time resolution (~25 - 30 ps). A typical modulation pattern, as recorded by the streak camera, and output to an x-y plotter are shown in Fig. 2. The average period measured over 11 cycles is 321 ps which gives a frequency measurement of 3.11 GHz.

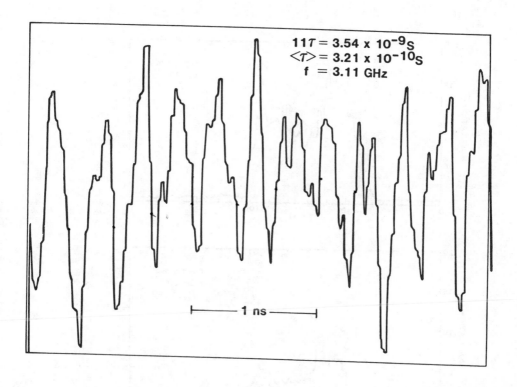

$11\tau = 3.54 \times 10^{-9}$ s
$\langle \tau \rangle = 3.21 \times 10^{-10}$ s
f $= 3.11$ GHz

⊢————— 1 ns —————⊣

Fig. 2.   Recorded modulation about the 50 percent point.   The
average period over 11 cycles is 321 ps, giving a source
frequency of 3.11 GHz.

For the modulation amplitude measurements, the streak camera
was set to 500 ns sweep so that the whole microwave pulse could
be recorded.   Typical modulated light pulses are shown in Fig. 3
for two distances.   As can be seen, the output pulse has a
two-humped shape and a full width, half maximum (FWHM) of approximately 55 ns.   The square of the measured fields, which is proportional to power, follows roughly the r^{-2} dependence (Fig. 4).
The determination of r^{-2} dependence is crucial in making certain
that the polarization rotation in the crystal has not gone beyond
$\pi/2$ radians and the modulation amplitude has not gone beyond V_{π}.
To precisely determine the power, the field pattern in the crystal, and the coupling efficiency between the free space rf field
and the crystal have to be known.

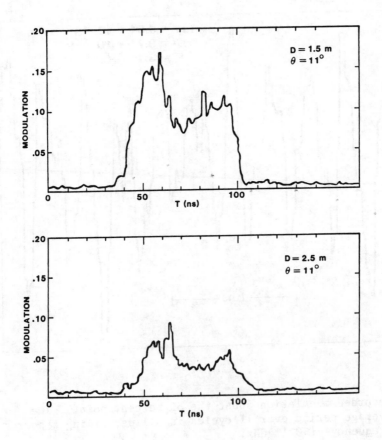

Fig. 3.   Modulation envelope at 1.5 m and 2.5 m.

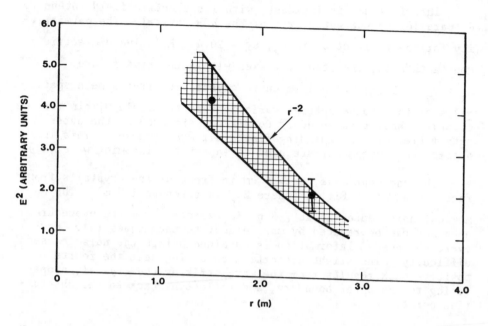

Fig. 4. Peak power in arbitrary units versus distance and compared to r^{-2} fall off.

3.1 Data Analysis

Wanted: $\vec{E}$-fld in free space at the position of KDP·Xstal rather than that in KDP Xstal

Relating the incident free space field to the modulation
voltage in the KDP crystal is a problem of a plane wave interacting with a cubic dielectric. In solving this problem, we have
used a three dimensional computer code.

The interaction between the incoming signal and the KDP
crystal is modeled using a finite-difference approximation to
Maxwell's equations. The incoming wave is assumed to start
at t = 0 and to be of the form $E_i = E_{in} \sin\omega t$, where $\omega = 2\pi f$;
f = 3.1 GHz. The application of time-domain, finite-difference
techniques to scattering problems has been described elsewhere [5]. The motivation for applying this finite-difference
approach to the dielectric-interaction problem is that it is well
suited for obtaining detailed information about electric field
variation within the crystal and for modeling the physics of the
problem.

The rf signal is incident, with its electric field intensity
E_z parallel to the optic axis of the KDP crystal. The permittiv-
ity for the z direction is ε_z, $\varepsilon_z = 20.4$. Relative dielectric
constants along the other two axes within the crystal are
$\varepsilon_x = \varepsilon_y = 45$ [3]. The 1 cm cubic crystal requires a mesh spacing
~ 1/4 cm to resolve spatial variations of the field within the
crystal. Radiation boundary conditions are used at the outer
extremities of the expanding finite-difference mesh. (This is
outside the Huygen surface used to launch the incoming wave [5]).

The incident wave, at 4.6 cm in front of the crystal's front
face, is given in Fig. 5 (where E_{in} is taken as 1 V/m). The numerical noise entering at 7-8 ns is expected from the procedures
used and can be reduced by more elaborate techniques [6]. However, the needed information is obtained before the noise causes
difficulty. The slight distortion occurring near the fourth
positive peak results from the wave striking the crystal, propagating to the outer boundary, and reflecting back to the observation point.

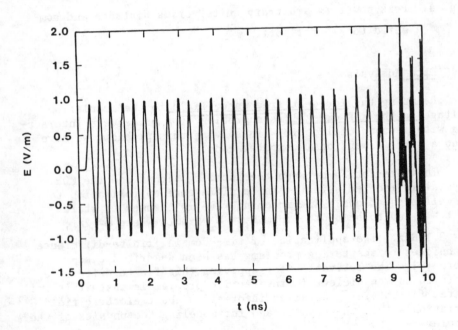

Fig. 5.   Electric field intensity for incoming wave.
          $E_{in} = 1$ V/m.

Observation points at 4 positions along the laser path, through the crystal center, have recorded the variation of E_z inside the crystal. At positions ±1/8 cm from the crystal center, the temporal variation of E_z is given in Fig. 6. Along the laser path inside the crystal, the E_z amplitude is 1/4 to 1/5 its exterior value. The corresponding static ratio for a dielectric

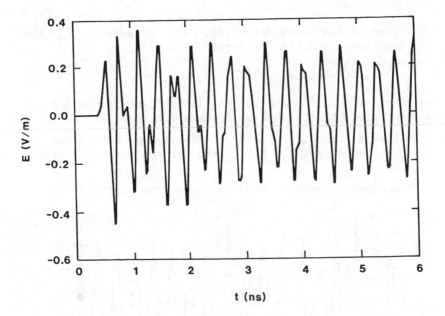

Fig. 6. E_z at ± 1/8 cm from KDP crystal center.

sphere is $3/(\varepsilon + 2)$ or $1/7.5$ for the KDP crystal [7]. With the E_z data, numerical integration of $E_z dz$ along the laser path generates the voltage $V(t)$ across the crystal. Earlier, this voltage was represented as $V_m \sin\omega_m t$. The finite-difference approach provides a more detailed representation. Because of the finite crystal size, the E_z encounters a time-shift as the laserbeam traverses the crystal. Between observation points separated by $3/4$ cm, the E_z encounters a time shift of 0.75×1.5 cm/(3×10^{10} cm/s) = 3.75×10^{-11} s (since the KDP index of refraction at optical wavelengths is 1.5) [3]. This shift is $> 10\%$ of the oscillation period and is significant. The resulting $V(t)$ variation is seen in Fig. 7.

Modulation of the laser intensity, after passage through the two plane polarizers and the KDP crystal, is given by:

$$I_0/I_i = \sin^2\phi, \text{ where } \phi = \pi V(t)/(2V_\pi).$$

The plane of polarization for the laser light is rotated through the angle ϕ by the applied voltage $V(t)$. I_i is the intensity of the incident laser beam. Inserting our calculated $V(t)$ and $V_\pi = 7.05$ kV for the KDP crystal at the argon-ion laser

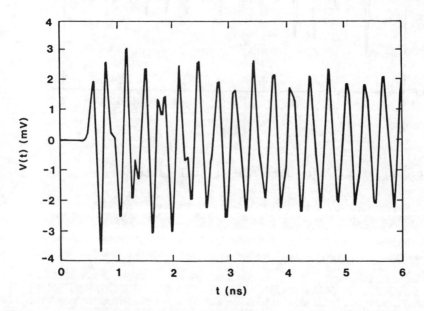

Fig. 7.   Voltage across 1 cm KDP crystal, as measured across laser paths. $E_{in} = 1$ V/m.

wavelength (514.3 nm), yields Fig. 8. For this illustration the incident amplitude is taken as 15 kV/cm.

Comparison of calculations with experimental data requires the detached analysis of how the modulated laser beam is recorded by the streak camera. The streak camera divides a streak record into 256 time incremental records with each being an integral of the input intensity over a fixed small time interval. Therefore, the recorded quantity is proportional to the total number of photons incident on the camera during that time interval. This quantity is given by:

$$Q_m = \int_t^{t+\tau} I_i \sin^2\left(\frac{\pi}{2} \frac{V(t)}{V_\pi}\right) dt = \int_t^{t+\tau} I_i \sin^2\phi\, dt$$

where τ is the integration time of each time increment.

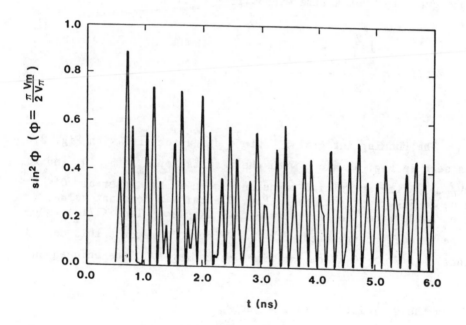

Fig. 8. Modulation of the laser intensity I_o/I_i; $E_{in} = 15$ kV/cm.

Prior to the actual measurement, the streak camera is calibrated by setting the second polarizer at a number of angular settings different from the null point of 90°. The recorded light quantity is given by:

$$Q_c = \int_t^{t+\tau} I_i \sin^2\theta\, dt = (I_i \sin^2\theta)\tau$$

where θ is the angular displacement of the second polarizer from 90°. I_i then can be defined as

$$I_i = \frac{Q_c}{(\sin^2\theta)\tau}$$

to compare with these experimental quantities, the crystal modulation needs to be determined by using the calculated $V(t)$ as shown in Fig. 8. For this purpose, we determine the time averaged quantity over a time window from t_1 to t_2.

$$\langle I/I_i \rangle = \frac{\int_{t_1}^{t_2} \sin^2\phi\, dt}{\int_{t_1}^{t_2} dt} \, .$$

The running integral $\int_0^t \sin^2\phi\, dt$, associated with Fig. 8, is seen in Fig. 9, along with the results for E_m = 5, 10, and 20 kV/cm. After an initial transient, seen in the interval $0 < t < 2$ ns in Fig. 9, the slope in Fig. 9 attains a constant value, i.e. aside from the small-time-scale ripples. The $\langle I/I_i \rangle$ is then uniquely correlated with E_m. Figure 10 records this dependence. At $E_o > 20$ kV/cm, at times exceeding $\pi/2$, the uniqueness is lost.

Using the Q_m relation, $\langle I/I_i \rangle$ becomes

$$\langle I/I_i \rangle = \frac{Q_m}{I_i \tau} \, .$$

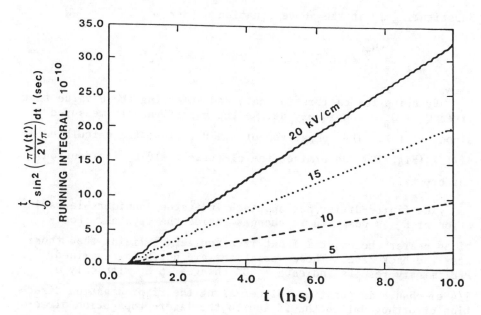

Fig. 9.   Primary integral of $\sin^2 \tau$.

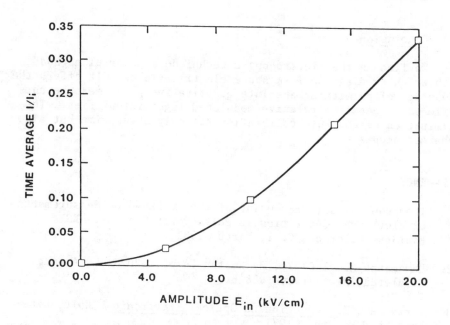

Fig. 10.   Modulation variation with incident rf amplitude.

Substituting I_i in the above equation we derive,

$$\langle I/I_i \rangle = \frac{Q_m}{Q_c} (\sin^2\theta) \ .$$

By using the calibration data and selecting the θ value that gives $Q_c = Q_m$, we uniquely derive the experimentally measured slope, $\langle I/I_i \rangle$. The comparison of $\sin^2\theta$ against the calculated $\langle I/I_1 \rangle$ (Fig. 10), determines the electric field E_m incident on the crystal.

The finite-difference approach indicates considerable variation of E_z as position is changed inside the crystal. Positions nearer the crystal front face see larger fields than those near the back face. These variations are not being examined. Preliminary results indicate > 30% changes in E_z with only a 1/4 cm change in position, either along the rf propagation direction or orthogonal to the rf and to the laser-propagation directions.

Thus laser-path variation can be used to expand the range of applicability for the measurement technique.

3.2 Conclusion

We find in the electro-optic technique a powerful method with which to diagnose free space electric fields. It offers the advantage of permitting absolute quantitative power measurements while only measuring relative modulated light intensities. The elimination of absolute calibration not only saves time but also enhances accuracy.

REFERENCES

1. Thompson, J. E., Proceeding of the Workshop on Measurements of Electrical Quantities in Pulse Power Systems, NBS, Boulder, Colorado, P. 1, March, 1981.

2. Hamamatsu Model C1370, Hamamatsu Corp., 420 South Avenue, Middlesex, New Jersey 08846.

3. Yariv, A., Introduction to Optical Electronics, Holt, Rinehart, and Winston, 1971.

4. Adhave, R. S., J. Opt. Soc. Am. <u>59</u>, 414, (1969).

5. D. E. Merewether and R. Fisher, Finite Difference Solution of Maxwell's Equations for EMP Applications, Defense Nuclear Agency Report DNA 5301F, April 22, 1980.

6. L. T. Simpson, R. Holland, and S. Arman, Treatment of Late-Time Instabilities in Finite-Difference EMP Scattering Codes, IEEE Trans Nuc Sci. NS-29, pp. 1943-1948, December 1982.

7. J. D. Jackson, <u>Classical Electrodynamics</u>, John Wiley and Sons, Inc., 1975, p. 151.

ELECTROMAGNETIC SENSORS AND MEASUREMENT TECHNIQUES

Carl E. Baum

U.S. Air Force Weapons Laboratory
Kirtland Air Force Base
Albuquerque, New Mexico
USA

1. INTRODUCTION

This paper reviews a subject which has developed over the past two decades: the optimal design of sensors for transient (or broadband) electromagnetic measurements and proper installation of the sensors involving concepts from topology and symmetry. This has been reviewed in the following three important papers listed below:

1. C. E. Baum, E. L. Breen, J. C. Giles, J. P. O'Neill, and G. D. Sower, "Sensors for Electromagnetic Pulse Measurements Both Inside and Away From Nuclear Source Regions", Sensor and Simulation Note 239, Jan. 1978, IEEE Trans. Antennas and Propagation, Jan. 1978, pp. 22-35, and IEEE Trans. EMC, Feb. 1978, pp. 22-35.

2. C. E. Baum, "Sensors for Measurement of Intense Electromagnetic Pulses", Sensor and Simulator Note 271, June 1981, and Proc. 3rd IEEE International Pulsed Power Conference, Albuquerque, NM, June 1981, pp. 179-185.

3. C. E. Baum, E. L. Breen, F. L. Pitts, G. D. Sower, and M. E. Thomas, "The Measurement of Lightning Environmental Parameters Related to Interaction with Electronic Systems", Sensor and Simulation Note 274, May 1982, and IEEE Transactions on Electromagnetic Compatibility, May 1982, pp. 123-137.

In this section, the three papers have been combined into one, so that the reader may have it all in this reference volume.

There is yet another smaller review paper.

C. E. Baum, E. L. Breen, C. B. Moore, J. P. O'Neill, and G. D. Sower, "Electromagnetic Sensors for General Lightning Application", Proc. Lightning Technology, NASA Conf. Pub. 2128 and FAA-RD-80-30, April 1980, pp. 85-118.

However, this can be considered as updated version of 3 above and is thus also contained in the present paper.

The measurement of transient and broad-band electromagnetic fields and related parameters is important for various kinds of electromagnetic environments. Although there are some specific differences in the non-EM-field part of such environments, the basic EM-field part is common to them all. Often one is aided by the presence of uniform isotropic media in which to measure such fields, and perhaps the presence of highly conducting boundaries near which to measure the fields. However, there can be more difficult media to consider in the cases of the various environments of interest. In the case of the nuclear electromagnetic pulse (EMP), one has the difficult medium of the nuclear source region with nuclear radiation, source current density, and non-linear and time-varying air conductivity. In the case of natural lightning in the stroke itself (as in the direct strike to an aircraft) there is a nonlinear and time-varying air conductivity (or corona). In pulse-power machinery, where media are often stressed near breakdown, similar nonlinear and time-varying conductivity phenomena occur. So, there are some similarities in the electromagnetic sensor design requirements and some differences as well.

While the basic concepts of electrically-small antennas for measuring electromagnetic fields are quite old, optimization of these in a transient or broad-band sense is relatively recent. The historical motivation for these developments resides in the nuclear electromagnetic pulse (EMP) program. More recently, such sensors have been used for measuring lightning environments, including (in some cases) appropriate modifications. EMP simulators [93] have used electrical pulsers [94] as the sources for the simulators proper (or antennas). In this context, the electromagnetic sensors of our concern have been often used and even included within the actual pulse-power machinery.

The measurement of electromagnetic-field parameters is at once both simple and complex. From one point of view, one need only make an elementary application of the Maxwell equations. The resulting quasi-static concepts give some basic design approaches for electromagnetic-field sensors. However, this is

only the beginning; things are not as simple as they may first
seem. A more sophisticated point of view recognizes that numerous complications cloud such a picture. One exception relates
to the limitations of quasi-static theory; one often wants the
maximum sensitivity (and corresponding size). Another exception
relates to the complexity of environmental conditions in which
the electromagnetic fields are accompanied by other electromagnetic parameters such as source currents, nonlinear conductiv-
ities, etc. which are not present under the classical assumptions
concerning field measurements. Both of these exceptions are
sometimes encountered in lightning electromagnetic measurements.

In measuring the nuclear electromagnetic pulse, one has to
often deal with distributed electromagnetic quantities such as
electric and magnetic fields, current densities, charge densities, and conductivity, as well as integral quantities such as
voltage and current. As indicated in Fig. 1, there are four
kinds of distributed quantities which are directly related by
Maxwell's equations and constitutive equations [53]. This cyclic
set of physical quantities is related to the nuclear source via
the source current density J_c and the ionization source density S_e (electron-ion pairs per m^3s). It is these quantities,
or combination of them, or simple transformations of them that
one wishes to measure. The sensor problem is then basically how
to measure these. (See the papers in the EMP-Special Issue [109]
by Longmire concerning EMP environments, Higgins, Marin and Lee
concerning SGEMP, and Baum concerning EMP simulators.)

Lightning electromagnetic measurements have some similarity
to electromagnetic measurements under other conditions. Much of
the technology discussed here has evolved initially through the
nuclear electromagnetic pulse program. EMP environmental measurements include cases in which a large bandwidth is desired and
cases in which source region considerations are important (including nuclear radiation, source current density, and nonlinear
and time-varying air conductivity). While there are some differences in the case of lightning, many of the design considerations
are the same or similar.

Another application of electromagnetic sensors is in pulse-
power generators where large impulsive voltages and currents and
associated electromagnetic fields are present. Such pulse-power
equipment finds application in nuclear and lightning environmental simulation (among various applications). The design techniques for electromagnetic sensors have been applied here as
well, both for large bandwidth and for source-region problems (as
in associated charged-particle beams).

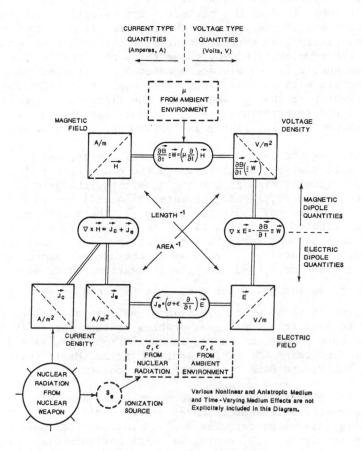

Fig. 1. Diagram of the Basic Electromagnetic Quantities for
Source Region EMP Environment and Interaction.

The first topic presented summarizes the sensor designs.
There are, in addition, some special factors to be considered
when using such sensors. The topology of the conducting cables
must be integrated with the topology of the other conductors in
the measurement situation. In some cases one can utilize the
symmetry inherent in the sensors, combined with symmetry in the
fields and/or measurement cables, to minimize the effects (noise)
of the scattering of certain field components by the sensor and
cables.

2.   SOME BASICS

Summarizing some of the basic aspects of electromagnetic sensor design, first we have our definition of a sensor as a special kind of antenna with the following properties.

1.   It is an analog device which converts the electromagnetic quantity of interest to a voltage or current (in the circuit sense) at some terminal pair for driving a load impedance, usually a constant resistance appropriate to a transmission line (cable), terminated in its characteristic impedance.

2.   It is passive.

3.   It is a primary standard in the sense that, for converting fields to volts and current, its sensitivity is well known in terms of its geometry; i.e., it is "calibratable by a ruler". The impedances of loading elements may be measured and trimmed. Viewed another way, it is, in principle, as accurate as the standard field (voltage, etc.) in a calibration facility. (A few percent accuracy is usually easily attainable in this sense.)

4.   It is designed to have a specific convenient sensitivity (e.g., $1.00 \times 10^{-3} m^2$) for its transfer function.

5.   Its transfer function is designed to be simple across a wide frequency band. This may mean "flat", in the sense of volts per unit field or time derivative of field, or it may mean some other simple mathematical form that can be specified with a few constants (in which case more than one specific convenient sensitivity number is chosen).

A first important category of such sensors is the electric-field sensors. Figure 2 shows the basic topology of such a sensor (two separate conductors connected to a terminal pair) and its equivalent-circuit representation (valid for electrically small sensors). The three basic sensor parameters are related as

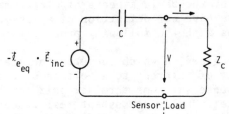

a. Thevenin Equivalent Circuit

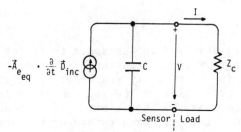

b. Norton Equivalent Circuit

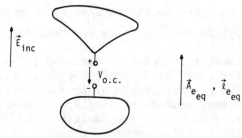

c. Electric Dipole Sensor

Fig. 2. Electrically Small Electric Dipole Sensor in Free Space

$$\vec{A}_{e_{eq}} = \frac{C}{\varepsilon} \vec{\ell}_{e_{eq}}$$

$\vec{A}_{e_{eq}} \equiv$ equivalent area

$$\tag{1}$$

$\vec{\ell}_{e_{eq}} \equiv$ equivalent length (or height)

$C \equiv$ capacitance

so that only two of the basic parameters are independent. Note that if, in addition to the medium permittivity ε, there is a conductivity σ, then a conductance G appears in parallel with the capacitance C in the equivalent circuit.

The basic parameters of the magnetic-field sensors are indicated in Fig. 3. The basic topology of such a sensor is a loop broken to connect to a terminal pair. The basic sensor parameters are related as

$$\vec{A}_{e_{eq}} = \frac{L}{\mu}\vec{\ell}_{h_{eq}}$$

$$\vec{A}_{h_{eq}} \equiv \text{equivalent area}$$

$$\vec{\ell}_{h_{eq}} \equiv \text{equivalent length} \tag{2}$$

$$L \equiv \text{inductance}$$

Again, only two of the basic parameters are independent. The medium permeability μ is often that of free space, μ_o.

An important question, relating to these kinds of sensors, is which type is best for a certain kind of application. Such questions are usually cast into an efficiency format in the sense of most output per unit input. Here one must recognize the broadband character of the measurement problem so that output should also include an appropriate bandwidth in its definition.

One concept of historical and technical interest [22] is that of equivalent volume that has the formulas for electric

$$V_{e_{eq}} = \frac{\varepsilon_o}{C}\vec{A}_{e_{eq}} \cdot \vec{A}_{e_{eq}} = \frac{C}{\varepsilon_o}\vec{\ell}_{e_{eq}} \cdot \vec{\ell}_{e_{eq}} = \vec{A}_{e_{eq}} \cdot \vec{\ell}_{e_{eq}}$$

$$V_{h_{eq}} = \frac{\mu_o}{L}\vec{A}_{h_{eq}} \cdot \vec{A}_{h_{eq}} = \frac{L}{\mu_o}\vec{\ell}_{h_{eq}} \cdot \vec{\ell}_{h_{eq}} = \vec{A}_{h_{eq}} \cdot \vec{\ell}_{h_{eq}} \tag{3}$$

and magnetic dipole sensors, respectively. The equivalent volume is based on the energy extracted from the incident field and delivered to the load. This equivalent volume can be divided by geometrical volume to give a dimensionless efficiency. This geometrical volume might be a specified volume into which the sensor is to fit; the better sensor design has the better efficiency.

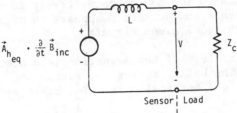

a. Thevenin Equivalent Circuit

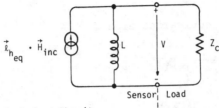

b. Norton Equivalent Circuit

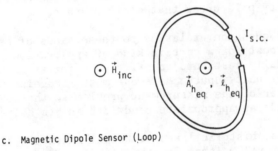

c. Magnetic Dipole Sensor (Loop)

Fig. 3. **Electrically small magnetic dipole sensor in free space.**

This type of definition is appropriate for cases in which the critical frequency $1/(Z_c C)$ and Z_c/L (for constant resistance Z_c) is within the electrically small regime, and the basic limitation on the sensor design is size.

Several of the sensor types discussed herein are not constrained directly by physical dimensions but by upper frequency response (ℓ_c or f_c which might be interpreted as a characteristic time t_c) for which the approximation of the response being proportional to the time derivative type of field quantity dotted into an equivalent area breaks down. The sensor size can be made as large as possible to obtain sensitivity for a given bandwidth. As the sensor size is increased, the approximation of an electrically small sensor breaks down at the highest frequencies of interest. One defines then the characteristic frequency or time according to when the ideal dot product and derivative response is in error by some specified amount. The resulting figure of merit is found to be [52]

$$\Lambda_e = \left(\frac{Z_c}{Z_o}\right)^{1/2} \left|\vec{A}_{eq}\right| \ell_c^{-2} \qquad \Lambda_h = \left(\frac{Z_o}{Z_c}\right)^{1/2} \left|\vec{A}_{h_{eq}}\right| \ell_c^{-2} \qquad (4)$$

for electric and magnetic dipole sensors, respectively, where the wave impedance of free space is

$$Z_o = \left(\frac{\mu_o}{\varepsilon_o}\right)^{1/2} \qquad (5)$$

and Z_c is the assumed frequency independent load resistance, typically the characteristics impedance of a transmission line. For this purpose, we have introduced a characteristic length (noting that the high-frequency limitation tends to be related to transit times on the structure) as

$$\ell_c = ct_c = \frac{c}{\omega_c} , \qquad (6)$$

where c is the speed of light, thus putting the bandwidth in length units. The figure of merit is of the form sensitivity times (bandwidth)^2, a quantity which is not a function of sensor size but only a function of the design, shape, and impedance loading distribution. The definition of this figure of merit is based on power delivered to the load Z_c which places electric and magnetic sensors on a common basis for comparison. Note that in this form the figure of merit applies to sensors in free space (or uniform isotropic media).

The various sensors in their free space designs can usually be mounted on ground planes by cutting them in half along an appropriate symmetry plane. The figures of merit for a given type of design are different in these two situations. In this paper we refer the figures of merit for each design type to their free space (full sensor) versions. Note that a particular sensor design, in a ground plane version, may have a different equivalent area and drive a different load impedance, although both are simply related to the free space versions.

A recent paper [87] goes into the question of the definition and measurement of this figure of merit. By a spherical harmonic expansion of the sensor response, the effects of angular and frequency response can be separated to give two ways to define the upper frequency response, one based on the frequency response of the appropriate dipole term and a second based on the frequency where the contribution of the higher order multipole terms become significant. The ideal spherical resistive sheets used in this paper can be chosen so as to attain a figure of merit (using $1/\sqrt{2}$ of the ideal frequency response to define the upper frequency response) of almost 2.0. However, other types of designs (ideal and/or practical) may achieve even higher figures of merit.

Measurements have been performed on the response of various D-dot sensors and B-dot sensors (the designs being described later). The results of these measurements [86] indicate that $\Lambda_{.707}$ (meaning frequency response defined by $1/\sqrt{2}$ or .707 of the ideal low-frequency behavior) may be as good as about 3 in these practical sensors.

Current sensors are a special category. An important class relies on an integral form of one of Maxwell's equations as

$$\oint_C \vec{H} \cdot d\vec{\ell} = \int_S \vec{J}_t \cdot dS \equiv I_t , \tag{7}$$

where $\vec{J}_t$ is the total current density passing through the surface S bounded by the contour C. As indicated in Fig. 4 the basic sensor concept is to measure the magnetic field (or usually its time derivative) at many places around an area through which the current of interest flows. Appropriately summing (or averaging) these measurements experimentally gives the total current through the area. Note that the total current density is just

$$\vec{J}_t = \nabla \times \vec{H} = \vec{J}_c + \vec{J}_\sigma + \vec{J}_\varepsilon \tag{8}$$

including source current density J_c (e.g., the Compton current density in an EMP source region), the conduction (J_σ), and displacement (J_ε) current densities which may be even nonlinear in some circumstances. In linear, time-invariant, isotropic media we have

$$\vec{J}_t = \vec{J}_c + \sigma\vec{E} + \frac{\partial}{\partial t} \vec{D} \tag{9}$$

$$\vec{D} = \varepsilon\vec{E} .$$

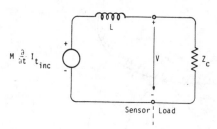

$$I_{t_{inc}} = \vec{A}_{t_{eq}} \cdot \vec{J}_{t_{inc}} \quad \text{for Measurement of Distributed Current Density}$$

a. Norton Equivalent Circuit

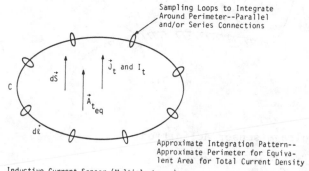

b.   Inductive Current Sensor (Multiple Loops)

Fig. 4. **Electrically small inductive current sensor in free
         space.**

This type of sensor also has a self-inductance L, which is
in general not the same as the mutual inductance M. One could
modify the Thevenin equivalent circuit of Figure 4 into a Norton
form. However, this would bring the self-inductance L (which is
in general not as accurately known as M) into the sensor sensitivity, when operated in the short-circuit mode.

Voltage sensors are closely associated with electric-field
sensors. Electric-field sensors typically measure the potential
(voltage) between two conductors (highly conducting compared to
the total medium conductivity) and relate this potential to the
electric field through an equivalent length as in Eq. (1). Here
we need only the potential difference itself. However, like an
electric-field sensor, voltage sensors have bandwidth restrictions related to the definition of potential. As in Fig. 5,
one has the voltage as a path integral of the electric field as

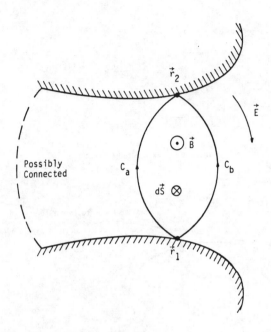

Fig. 5.   Voltage sensor definition.

$$V = - \int_C \vec{E} \cdot d\vec{\ell} , \tag{10}$$

where C connects points $\vec{r}_1$ and $\vec{r}_2$ on two separate conductors. However,

$$\vec{E} = - \nabla\Phi - \frac{\partial}{\partial t} \vec{A}$$

$\Phi \equiv$ scalar potential

$\vec{A} \equiv$ vector potential

$$\tag{11}$$

giving (for stationary C)

$$V = \int_C (\nabla\Phi) \cdot d\vec{\ell} + \frac{\partial}{\partial t} \int_C \vec{A} \cdot d\vec{\ell}$$

$$= [\Phi(\vec{r}_2) - \Phi(\vec{r}_1)] + \frac{\partial}{\partial t} \int_C \vec{A} \cdot d\vec{\ell} . \tag{12}$$

Now, if one has more than one possible contour, say C_a and C_b, then the corresponding voltages, V_a and V_b, are in general different as

$$V_a - V_b = - \int_{C_a} \vec{E} \cdot d\vec{\ell} + \int_{C_b} \vec{E} \cdot d\vec{\ell} = - \int_{C_a - C_b} \vec{E} \cdot d\vec{\ell}$$

$$= \frac{\partial}{\partial t} \int_{C_a - C_b} \vec{A} \cdot d\vec{\ell}$$

$$= \frac{\partial}{\partial t} \int_{S_{a,b}} \vec{B} \cdot d\vec{S} , \tag{13}$$

$S_{a,b}$ = surface bounded by contour $C_a - C_b$,

which is derived from either of

$$\nabla \times \vec{E} = - \frac{\partial}{\partial t} \vec{B} \qquad (\vec{B} = \mu \vec{H})$$

or,

$$\vec{B} = \nabla \times \vec{A} . \tag{14}$$

Typically the contour of concern is near some electrical connection. However, one must be careful (at high frequencies especially) of regarding conductors as equipotentials for voltage measurements. Note that this definition even allows for the measurement of the voltage between two points on the same conductor (e.g., a loop) when $\Phi_a = \Phi_b$.

3. SENSORS FOR USE OUTSIDE OF SOURCE REGIONS

In the simpler situation that there are no source currents in and around the sensor and the local medium is well behaved (typically free space), the problem is considerably simplified. This simpler situation allows one to pay more attention to accuracy details and bandwidth optimization. Several important kinds of sensor designs for electric and magnetic fields and current have been developed and used and are discussed here.

3.1 D-dot and Electric Field Sensors

3.1.1 D-dot sensors. The D-dot sensor is used to measure the
time-rate-of-change of electric flux density. The sensor's response is described by the Norton equivalent circuit of Fig. 2.
The frequency domain response of the sensor is given by

$$\tilde{V}(s) = \frac{\varepsilon s \tilde{\vec{E}}_{inc}(s) \cdot \vec{A}_{e_{eq}} Z_c}{1 + s Z_c C} \tag{15}$$

and for frequencing where $\omega \ll 1/(Z_c C)$ the response can be simply
expressed as

$$\tilde{V}(s) \underset{\sim}{\sim} \varepsilon s \tilde{\vec{E}}_{inc}(s) \cdot \vec{A}_{e_{eq}} Z_c . \tag{16}$$

It is of primary importance that an accurate determination of
sensor area can be made. For that reason, only sensor geometries
with accurately calculable areas are used. Sensor capacitance,
as a design parameter, need not be known so accurately, but it
should be a low value as it shunts the load resistance and determines the high frequency response.

The hollow spherical dipole (HSD) sensor design [38], [43],
[51], [59], [84] uses the geometry of a sphere with a narrow slot
around the equator. The slot is resistively loaded by the signal cables. The sensors shown in Fig. 6 are the HSD-2B(R) and
HSD-4A(R). Each consists of two hemispherical shells mounted on
a ground plate. Signal current from each hemisphere flows to the
ground plate through four equally spaced 200 Ω strip lines. The
four strip lines from each hemisphere join at the center of the
base of each hemisphere and then continue along a 50 Ω coaxial
cable. The two 50 Ω coaxial cables are contained inside the output stem which extends radially out in the plane of the center
plate to a twinaxial connector. The signals from the two hemispheres produce a differential signal which is then carried by
standard 100 Ω twinaxial cable [10], [85]. The sensors in Fig. 6
are for use in making free space measurements. A single-ended
version of the sensor, for use on a conducting ground plane, consists of just one-half of the above sensor and the signal is carried by a single 50 Ω coaxial cable. The sensitivity of the HSD
sensor is expressed as an equivalent area. The area is shown to
be $\vec{A}_{e_{eq}} = 3\pi a^2$ where a is the sensor sphere radius [38], [43].

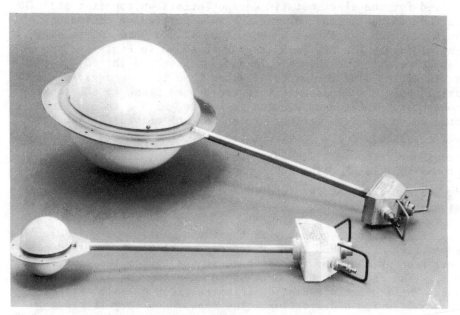

Fig. 6. HSD-2B(R) and HSD-4A(R) D-dot sensor.

The Λ_{10-90} figure of merit (using the characteristic time as the 10-90 rise time) for the HSD-2B(R) is .078. HSD sensors have been fabricated with equivalent areas of .1 and .01 m^2 in both differential (as in Fig. 6) and single-ended versions.

An improved sensor geometry, from the standpoint of figure of merit, is the asymptotic conical dipole (ACD). The ACD sensor geometry is determined by a method described in [33], [84]. The particular shape used to date is derived from a line charge λ (z) on the z axis given by

$$\lambda(z) = \begin{cases} \lambda_o \text{ for } 0 < z < z_o \\ -\lambda_o \text{ for } 0 > z > -z_o \\ 0 \text{ for } z = 0 \\ 0 \text{ for } z > z_o \end{cases}$$

$$\lambda_o > 0 \quad , \quad z_o > 0 \ . \tag{17}$$

The potential distribution for the above charge distribution is solved for the electrostatic equipotentials surrounding it. The surface of the sensor corresponds to a particular equipotential surface which approaches a 100 Ω cone at its base in its differential form. The ACD-1A(R) sensor is shown in Fig. 7. The design details for this single-ended sensor are in [60]. It consists of the sensor element attached to a 50 Ω semi-rigid, coaxial cable which passes within the ground plane to the coaxial connector. The sensor element is covered with a thin dielectric dome which provides weather protection and mechanical support.

The sensor has an equivalent area of 1×10^{-4} m^2 and an upper frequency response > 7.6 GHz. Sensor element capacitance to ground is 1.16 pF. The ACD sensors have been fabricated with equivalent areas of 10^{-3} and 10^{-4} m^2 in both differential and single-ended versions.

An additional geometry is the flush plate dipole (FPD). The geometries of the HSD and ACD sensors cause electric field enhancement which is most pronounced at the top of the sensing element. The enhancement is 3 times for the HSD and larger for the ACD. The flush plate dipole minimizes field enhancement and chances for field distortion. The sensor geometry is shown in Fig. 8. It is basically a conducting disk, centered in a circular aperture, in a conducting ground plane [20], [39], [41],

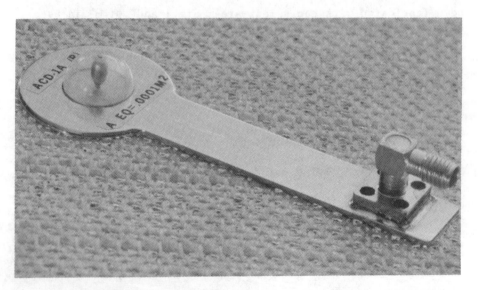

Fig. 7.   ACD-1A(R) D-dot sensor.

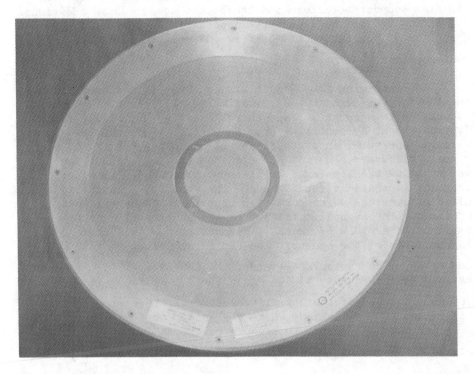

Fig. 8.   FPD-1A D-Dot sensor.

[69], [84]. The signal is taken from the sensor element at four
equally spaced points around its circumference by 200 Ω strip
lines. The strip lines feed two 100 Ω coaxial cables which are
paralleled into a 50 Ω connector. The flat surface of the sensor
is covered by a thin piece of mylar which acts as a weather
cover. The bottom side of the sensor is covered by a conducting
pan to provide a consistent electrical environment as well as to
provide protection. Resistive loops are positioned inside the
cover to absorb energy below to the sensor element. The equivalent area of the FPD is derived in [41] in which the area is

given as a normalized area $A = |\vec{A}_e| / (\pi ab)$ where a and b are the
$\quad\quad\quad\quad\quad\quad\quad\quad\quad\quad\quad eq$

radii of the sensor element and the circular aperture, respectively. For the FPD-1A a = .0508 m, b = .0635 m, A = .988, and
$A_{eq} = .01 \text{ m}^2$. The normalized capacitance is calculated [41] for
various disk and aperture radii and, for the dimensions of the
FPD-1A, that value is 6.8 pF. This value of capacitance, along
with the 50 Ω cable impedance, would give a frequency response of

468 MHz. The presence of the mylar sheet covering the sensor, the disk support structure and the bottom cover add an additional 1.2 pF and reduce the frequency response to approximately 390 MHz. The Λ_{10-90} figure of merit is .08 related to a differential configuration.

The conforming flush plate dipole (CFD) is designed for use on non-flat surfaces such as missile and aircraft skin where it is desired to measure surface charge density [20], [72], [84]. It consists of a conducting disk sensing element, centered in a circular aperture, in a conducting ground plane, as is the previously discussed FPD. It differs from the FPD in that the volume below the sensing element is much reduced with a corresponding increase in capacitance. The sensor is fabricated from pliable materials which permit it to be mounted on cylindrical surfaces. The signal is taken from the sensing element to a small daimeter 50 Ω cable that is sandwiched within the conducting surfaces of the sensor. The cable terminates on a 50 Ω connector. The equivalent area of the CFD-1A is .001 m^2 and it has a corner frequency of 106 MHz. The Λ_{10-90} figure of merit is .00053. Because of the low figure of merit, this sensor's use is limited to applications where the sensor must be mounted on non-flat surfaces. The CFD sensors have been fabricated with equivalent areas of .001 and .01 m^2.

3.1.2 Electric field sensors. The common sensor for measuring electric field intensity is the parallel plate dipole (PPD). The PPD-1A(R) is shown in Fig. 9. This sensor [8], [36], [37], [45], [58], [84] is built in the form of a parallel plate capacitor. The conducting sensor plate is supported above a conducting baseplate by nylon spacers. The output signal is obtained from an attenuating resistor attached to the center of the top plate, in series with a 50 Ω output cable in the sensor base, which terminates in a coaxial connector. Sensor sensitivity is determined by the sensor plate equivalent height and the divider formed by the attenuating resistor and the output resistance. The sensor time constant is determined by the plate capacitance and the total resistance to ground. In order to attain a choice in sensitivity and time constant, the attenuating resistor is placed in an interchangeable resistor mount. The resistor mount has provisions for high frequency compensation of the resistor's stray capacitance. Assuming proper compensation of the attenuating resistor, the frequency domain response of the sensor is given by

Fig. 9. PPD-1A(R) E sensor (exploded view).

$$\tilde{V}_{o.c.}(s) = -\tilde{\vec{E}}_{inc}(s) \cdot \vec{\ell}_{e_{eq}} \frac{Z_c}{R + Z_c} \frac{sC(Z_c + R)}{sC(Z_c + R) + 1} , \qquad (18)$$

where $\vec{\ell}_{e_{eg}}$ is the equivalent height of the sensor, R is the value

of the attenuating resistor, and Z_c is the output cable imped-

ance. The thickness of the base plate used with electric field

sensors causes a field enhancement that must be considered in the

height calculations [37]. The field enhancement above a base

plate is approximated by the factor $(\pi/2)\xi_o$, where ξ_o is the

ratio of the base plate thickness to radius for single-ended

sensors and the ratio of plate thickness to diameter for differen-

tial sensors. Sensor capacitance [36] considers the effects of

top plate thickness and fringing fields as well as the added

capacitance from the nylon spacers. It is a desirable feature of

the parallel plate dipole that a wave polarized normal to the

plate and propagating across the plate is not perturbed by the

sensor. The frequency response for the sensor is determined only
by the output circuitry. The risetime for the PPD sensor to this
type of electric field is less than one nanosecond. The single-
ended PPD sensor is the most commonly used, but a differential
version has been built. It consists of a pair of circular parallel plates mounted on opposite sides of a ground plate. The
signals are taken from the plates in the same manner as for the
PPD-1A(R). The 50 Ω cables from either side are joined in a
twinaxial connector. The electric field sensors have been built
with capacitances of one nF and 200 pF. The attenuating resistors are chosen to give time constants of 1, 10, and 100 microseconds. The sensor top plate is spaced 1 cm above the base
plate. The accuracies of sensor height, capacitance, and resistor values are ±1% for each. The figure of merit rating does not
apply for this sensor. The solid aluminum top plate can support
undesirable resonances under certain excitations. This could be
remedied by the use of a resistive top plate.

3.1.3 Current sensors. The circular, parallel mutual-induc-
tance (CPM), sensor is used to measure the time derivative of the
total current through the aperture of the sensor. The CPM [31],
[61], [65], [67], [84] is an inductive sensor of torodial shape
as illustrated in Fig. 10. The loop turns are oriented to be
sensitive to the component of the magnetic field, H, with respect
to the measurement axis. This sensor has a cross section of
width w, inner radius r_1, and outer radius r_2. The mutual inductance is

$$M = \frac{N \, \mu_r \mu_o \, w \ell n \, \frac{r_2}{r_1}}{2\pi} , \tag{19}$$

where N is the effective number of turns, μ_r is the relative
permeability of the sensor volume, and μ_o is the permeability of
free space. For all CPMs developed to date, M is 10^{-8} H and $\mu_r =$
1. The CPM sensors have been built with aperture diameters of
.1, .2, .5, 1, and 2 meters. The 10-90 response time of the sensors, for on-axis current, varies from 1 ns for the .1 m sensor
to 3 ns for the 2 m sensor. The response time is further degraded by transit time effects for off-axis currents. The accuracy
of the sensor is limited by manufacturing dimensional tolerances
to ±1 percent.

The I-dot, one-turn, insertion unit (I1I) is an inductive
current sensor designed to respond to the time derivative of the
current in the center conductor of a coaxial cable [31], [84].

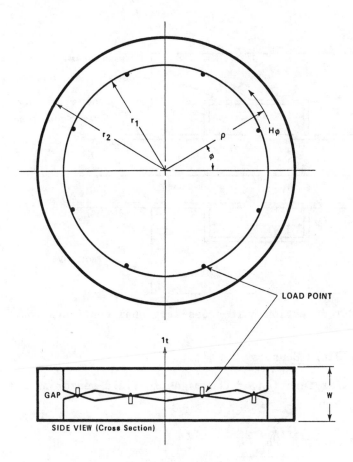

Fig. 10.   Typical CPM sensor geometry.

The sensor is used by inserting it in series with a coaxial cable
for the current measurement.   The I-dot insertion unit operates
on the same principles as the CPM current sensors.   The basic
form of the sensor is shown in Fig. 11.   The sensor is constructed around the coaxial cable with the wall and sensing gap
of the sensor forming an integral part of the outer conductor of
the coaxial cable.   The output signal is picked up by two, 100-Ω
cables whose center conductors cross the sensing gap at diametri-
cally opposite points.   The signals from the two, 100-Ω cables
are joined in parallel to a 50-Ω connector on the sensor wall.
The I1I-1A current insertion unit has a mutual inductance of
5×10^{-9} H and a 10-90 risetime of less than .3 ns.   It is designed for use with 1-5/8 inch, styroflex, 100 Ω cable.   Its
accuracy is limited by manufacturing tolerances to ±3 percent.

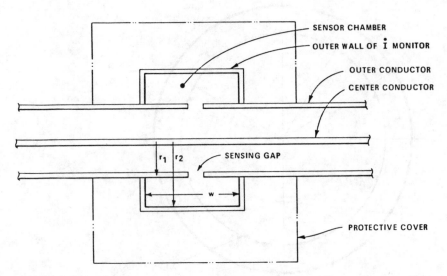

Fig. 11.   I-dot insertion unit cross-sectional configuration.

3.2 Magnetic Field Sensors

The transfer function of the magnetic field sensor of Fig. 3 is given by,

$$\tilde{V}(s) = \frac{s\tilde{\vec{B}}_{inc}(s) \cdot \vec{A}_{h_{eq}} Z_c}{sL + Z_c} .$$

(20)

For frequencies where $\omega \ll Z_c/L$,

$$\tilde{V}(s) = s\tilde{\vec{B}}_{inc}(s) \cdot \vec{A}_{h_{eq}} .$$

(21)

For frequencies where $\omega \gg Z_c/L$,

$$\tilde{V}(s) = \frac{\tilde{\vec{B}}_{inc}(s) \cdot \vec{A}_{h_{eq}} Z_c}{L} .$$

(22)

Using the concept of equivalent length Eq. (2), the above Eq.(22) can be expressed as

$$\tilde{V}(s) = \vec{\tilde{H}}_{inc}(s) \cdot \vec{\ell}_{h_{eq}} Z_c . \tag{23}$$

3.2.1 Multi-gap loop (MGL). The MGL series of magnetic field sensors [25], [26], [57], [70], [73], [76], [77], [84] is used for high-frequency B-dot measurements. Signal distribution for these sensors is shown in Fig. 12. The sensor is built in the form of a right circular cylinder. The cylinder is formed from 1/16 inch printed circuit board material which is etched to provide the gaps and the 200 Ω strip lines shown in Fig. 12. The sensor is divided into four quadrants by axial shorting plates that connect to the cylinder midway between the gaps. The signals from quadrants one and three are combined to form one side of the differential output signal and the signals from quadrants two and four combine to form the other. Combining the signals in this manner minimizes the E-field response. The gaps are formed with the proper angle to form a 200 Ω impedance which improves the sensor risetime. The cylindrical geometry of the MGL sensor

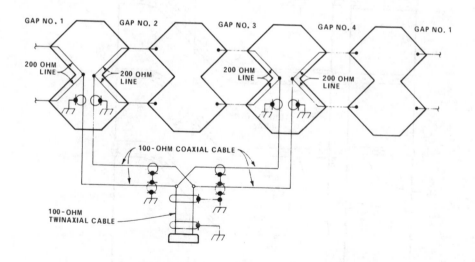

Fig. 12. MGL B-dot sensors (signal distribution).

permits an approximate determination [25] of the effects of the
number of gaps, the cable impedance, the sensor length and orientation of the gaps with respect to the magnetic field. The
single-ended sensors are essentially one-half of the sensor described above except that they consist of two adjacent quadrants
with signals connected in parallel. MGL sensors have been built
with equivalent areas of 10^{-1}, 10^{-2}, 10^{3}, 10^{4}, and 10^{-5} m^2.
Equivalent area is maintained to an accuracy of ±1 percent. The
area accuracy considerations are discussed in [57].

3.2.2 One-conductor, many-turn loop (OML). A single-gap, half-
cylinder loop with four-turn wiring [14], [84], and an equivalent
area of one square meter is available for measurements requiring
more sensitivity. This sensor, designated the OML-1A(A), operates as a derivative output device at frequencies below 3.5 MHz
and has a risetime of about 100 ns. Fig. 13 shows the wiring
diagram. Special triaxial cable, with 25 Ω outer line and 50 Ω

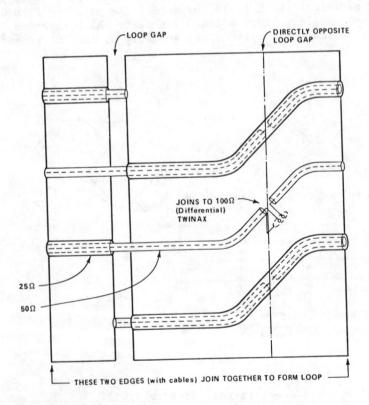

Fig. 13. OML-1A(A) B-dot sensor (expanded diagram).

inner line, is used. The gap voltage is picked off at four
points and carried by the 25 Ω outer lines to two summing gaps in
the 50 Ω internal lines. The two, 50 Ω lines drive the 100 Ω
differential output at the final gap in the cable. The four
voltage pickoff points along the gap were selected experimentally
to optimize frequency response.

3.2.3 Multi-turn loop (MTL). The MTL-1 [24], [27], [46], [47],
[66], [74] is a full-loop (free field), 50 turn sensor with an
equivalent area of 10 m^2 and a B-dot upper frequency response of
approximately 25 kHz. Above 25 kHz the sensor is self-integrating with its useful bandwidth extending to 3 MHz. Above this
frequency, resonances within the complex signal distribution network perturb the output signal. The sensor has an equivalent
length of 0.02 m, and a self-inductance of 6.3 x 10^{-4} H. The
MTL-1 design employs several special features to achieve the 3.5
MHz bandwidth [14]. It has four loop-gap signal pickoffs, and is
wound in two identical 25 turn half-loops, each of which drives
one side of the differential output. Fig. 14 shows the sensor
interior. A shield of resistively loaded loops can be seen on
the outside of the coil. This shield is electrically connected
to eight, axial shorts which help break up resonances and keep
out unwanted electric and magnetic field components. The 50 turn
sensor loop is wound on a fiberglass coil form in a counterwound

Fig. 14. MTL-1A(A) sensor (coil with axial shorts and conducting
          shield in place).

manner with a sequence of over and under crossovers on diametri-
cally opposite sides of the coil. The signal is developed across
the loop gaps and transmitted through 25 Ω outer part of a triaxial cable to the summing junction where they add and are transmitted to the sensor output connectors on a 50 Ω inner part of a
triaxial cable. The two, 50 Ω, half-loop output cables then combine to drive a 100 Ω differential output cable. The 25 Ω and
50 Ω signal cables are part of the sensor loop which is wound
with 50/25 Ω triaxial cable. Midpoint grounds divide each half-
loop into two quarter loops. The midpoint ground structure consists of the sensor stem and connections to the loop structure
at equipotential points and do not affect the desired sensor response. They do, however, decrease the "electrical size" of the
structure for undesired resonances and, hence, improve the high
frequency performance. Additional equipotential points are connected together to further improve the sensor response. Resistors are used in these "interwinding shorts" to dissipate the
resonant energy. Final adjustments of the high frequency response are made by adding a resistive shield around the outside
of the coil to exclude the incident electric field at low frequencies. The quotient of equivalent volume divided by geometric
volume is 1.3.

The MTL-2 has 10 turns, an equivalent area of 10^{-2} m^2, and
equivalent length of 10^{-2}m, a self-inductance of 1.25×10^{-6} H
and an upper frequency response of 12.6 MHz for B-dot operation.
The quotient of equivalent volume divided by geometric volume is
.56.

3.2.4 Octahedral 3-axis loop (O3L) sensor. The O3L-1 sensor
consists of three mutually orthogonal, B-dot loops and was designed for use as a trigger sensor for applications in which the
angle of incidence and polarization of an incident wave are unknown [68]. Each loop has an equivalent area of .20 m^2, a self-
inductance of 1.3×10^{-6} H, a risetime of 7 ns, and an upper
frequency response for B-dot operation of 12 MHz. Each loop,
shown schematically in Fig. 15, consists of four signal gaps,
two moebius summing gaps, and two summing "tee" connections, each
driving one side of a differential output connector [14].

4.   APPLICATION TO FAST TRANSIENT LIGHTNING DATA AWAY FROM
     SOURCE REGION (SOUTH BALDY PEAK)

These techniques have been successfully employed to obtain
high resolution (10 ns) of the electromagnetic fields away from
lightning areas and corona [100, 101]. The measurements are made
at the center of a 30m x 30m, square wire mesh, ground plane

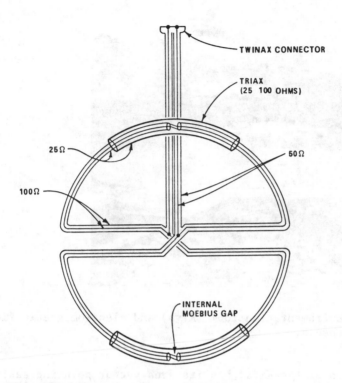

Fig. 15. O3L-1A sensor (loop details for one of the three
          loops).

placed on a relatively flat surface of the earth near the crest
of South Baldy Peak (in New Mexico). Copper ground rods are connected at 3m intervals around the perimeter of the wire mesh and
across its surface. The Kiva (instrumentation enclosure) is
buried at the center of the wire mesh with its roof flush with
the earth and electrically connected to the mesh. Figure 16
shows the site viewed from the approximate south.

        The Kiva is a 4.25m diameter by 2.44m high welded tank, made
from 3mm (1/8") thick sheet steel. The top of the tank is extended about 1m in radius to provide an adequate mounting surface
for the sensors as well as an access area for some of the penetrations into the Kiva. Access to the Kiva is by the stairway
shown in the right of the photo. Shielding integrity of the Kiva
is maintained by a "shielded room" door on the site of the tank.

        .Two, MGL-3 B-dot sensors are shown in Fig. 16. The MGL-3
has an equivalent area of $0.1m^2$. The sensor (B-dot East), on the

Fig. 16.   Instrumentation room (Kiva) and electromagnetic field
           sensors.

upper portion of the Kiva, has its area vector pointing east and
the sensor (B-dot north), to the left of the Kiva, has its area
vector pointing north. B-dot east has the significance that when
B-dot is positive in the east direction, the voltage on the
center conductor of the sensor output connector is positive, and
similarly for B-dot north.

     The ACD-5 D-dot sensor is shown at the front and right of
Fig. 15. The area vector for the ACD-5 is pointing toward
zenith. A D-dot signal that is positive toward zenith will produce a negative voltage on the center conductor of the sensor
output connector. These sensors are illustrated in Figs. 17 and
18, respectively.

     Each sensor is mechanically and electrically connected to
the Kiva surface in several places around the edge of the sensor
baseplate. The output signal from each sensor penetrates the
Kiva top surface by going directly from the sensor output to a
feedthru connector. RG-213 coaxial cable is used inside of the
Kiva. The sensors are mounted on a 1.7 meter radius about the
center of the Kiva. Because of the separation of sensors, arrival times for signals can vary by up to 11 ns, depending on the
azimuth and elevation of the signal source.

Fig. 17.   MGL-3A B-dot sensor.

Fig. 18.   ACD-5A(A) D-dot sensor.

Each of the sensor signals is digitized with a Biomation model 8100 waveform recorder. This waveform recorder has eight bit resolution (256 levels). Each sampled voltage is accurate to 0.4 percent of full scale. The minimum sample interval is 10 ns. The pre-trigger mode of recording is used to permit any desired part of the 2048 recorded samples to be those taken before the recorders are triggered. Typically about 20% of the samples are saved prior to trigger. The recorders are connected with their triggers in parallel. Triggering any recorder by signal causes all of them to trigger.

The recorders are controlled by one HP 9825 mini-computer controller. Upon completion of the recorder digitization, a signal is sent to the HP 9825 that then transfers the shift register data from each recorder, in turn, into the memory of the HP 9825. The HP 9825 then re-arms the recorders in preparation for acquisition of the next signals. If the recorders are triggered within a certain waiting time, the transfer process is repeated and the recorders are re-armed. This process is re-peated and the recorders are re-armed. This process can be repeated several times, depending on the size of the HP 9825 memory. When the waiting time is exceeded without a recorder trigger, the HP 9825 data are transferred to magnetic disc for later analysis. In this mode of operation, an interval of about 25 milliseconds is required to collect a set of data from the four recorders and then to re-arm them for a subsequent acquisi-tion from the same lightning flash. By operating in this mode, it is, in principle, possible to obtain data both on the initi-ating stepped-leader process and on subsequent return strokes all in the same flash.

The field information for lightning is obtained by numeri-cally integrating the digitized time-derivative data. The base-line drift of the recorder and the inability to accurately de-termine the value of the drift results in a ramp error. This problem can be alleviated by integrating the waveforms before sampling.

A timing mark is super-imposed on the field change data to determine where, in the lightning discharge process, the fast waveform recorders are triggered. This is done by triggering a one-shot of about 4 ns duration from the parallel trigger connec-tion of the Biomation 8100.

Figure 19 shows an example of the data obtained. On the 20 µs scale there are many pulses crowded together. Expanding first 7 µs shows some of the wealth of detail in this leader-like data sample. In particular, one begins to get some appreciation of the very fast times involved in such transients and why fast

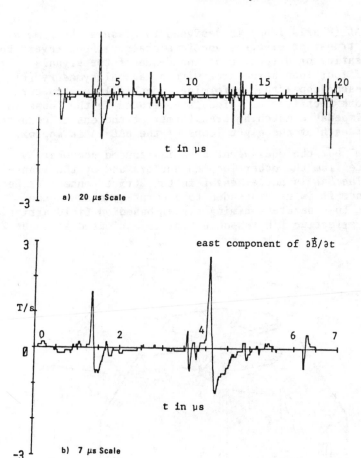

Fig. 19.   Leader example.

sensors and recorders are required to measure it.   This measurement system has found characteristic times for rises even less
than 30 ns [100, 101].

As discussed in [100], the transient waveforms from the
three sensors can be used to determine the direction of incidence
and polarization of the wave, and thereby, locate the source when
used with acoustic data.   This gives some information concerning
the direction and strength of the source currents.   To compare to
this, a time-of-arrival system has been installed using three

8AL-1A B-dot loops at approximately the corners of an equilateral
triangle with about 92m edges. Results are currently being obtained.

The 8AL (8 axis loop) is designed to measure the time of
arrival of transient electromagnetic signals without regard to
the polarization or direction of incidence of the signals. It
consists of four loop structures with octagonal symmetry (Fig.
20), with each loop actually forming a double loop structure. A
diode network rectifies all signals generated in the loops and
drives an impedance matching transformer at the output connector.
The area of each of the eight loops of the 8AL-1A is approximately .2 m^2 but the equivalent area is reduced somewhat by
interference from the other loop conductors and by the transformers. The 8AL is not a sensor in the strict sense (see Section 2) since it is not intended to accurately measure the
fields, but to generate a timing signal based on field arrival
time. The effective L/R response time is estimated at about 7
ns.

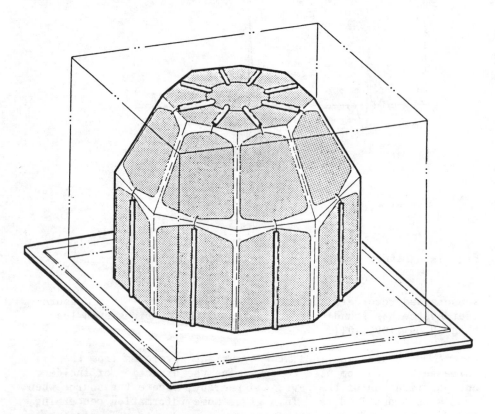

Fig. 20. 8AL-1A time-of-arrival B-dot loop.

5.   SENSORS FOR USE INSIDE A NUCLEAR SOURCE REGION

The environment inside a nuclear source region is rather
inhospitable for electromagnetic measurements [4], [53]. Referring to Fig. 1, we have a distributed source current density
J_c which originates primarily from Compton scatter of γ rays and
photoelectric scatter of X rays (the division not being precise).
These processes occur in air (where present) as well as in the
various materials of the sensor itself [11], [18].

Besides a source current density, we have conduction effects
associated with the surrounding medium and perhaps in the sensor
itself caused by the incident nuclear radiation [11], [17], [18],
[19]. The impact of the air conductivity (in an atmospheric environment) is fundamentally different for electric (capacitive-
conductive) and magnetic (inductive) devices. Since the air conductivity is a nonlinear effect (because of the dependence of the
electron mobility on the electric field), it is imperative that
an electric type sensor not significantly distort the local electric field so as not to change the conductivity [11]. For a magnetic type sensor, the problem is somewhat different. Local
changes in the air conductivity are not as significant. The magnetic field incident on the sensor is more governed by the currents in a volume of space with dimensions of the order of the
radian wavelength (or skin depth) so that the local perturbations
do not matter so much (at least for the lower frequencies) [18].

The problems of concern then include:

a.   source currents in all materials present in the photon
beam creating unwanted current and charge distributions and associated noise sources;

b.   ionization of air, if present, which constrains electric
sensor design to be nondistorting of the local electric field,
and which loads the loop-gaps of magnetic sensors;

c.   ionization of other dielectric materials (including surface tracking) which can load signals in the sensor and associated signal cables.

In reducing the deleterious effects in the nuclear source
region, various general guidelines are useful.

a.   For sensors mounted on test objects, the sensor base
should match the local surface of the test object both in material and shape.

b. Sensor cables should be made of low atomic number materials (both conductors and insulators with nearly matched atomic numbers). Differential signal outputs can also be used to reduce some of the noise signals.

c. Sensor cables should be removed from the radiation environment as soon as possible and shielded with high atomic number material (lead) to reduce the radiation, except very close to the cables where low atomic number materials can reduce the electron emission [54].

d. For magnetic sensors:

In air, they are encapsulated in dielectric with conductivity orders of magnitude lower than that of air under irradiation.

In vacuum, they are made of sparse (grid), low atomic number, materials to reduce electron emission.

e. For electric sensors:

In air or vacuum, they must negligibly distort the electric field in the immediate vicinity to not perturb (significantly) the current density, whether conduction current density as in air, or electron transport in vacuum as in SGEMP [15].

In air or vacuum, they are made of low atomic number materials and in a grid design to reduce electron emission.

5.1 Electric Field Sensors

As discussed in the previous section, the most severe effect of a nuclear source environment on EMP measurements is that on the electric field sensor in air. The nuclear radiation displaces charge into and out of the sensor and the effects associated with air conductivity alter the characteristic response of the sensor. The PMD-1 sensor (Parallel Mesh Dipole) [1], [15], [82], [83], Fig. 21, reduces the effect of the air conductivity by means of having its sensing element constructed of fine wires (.0025 cm diameter in the A and B version, .0013 cm diameter in the C version) as opposed to the solid plate of a parallel plate dipole, so that most of the conduction current is allowed to flow around the dipole conductors instead of through them. This construction also minimizes the electrode mass, while effectively keeping the electrical area nearly constant, thereby minimizing the Compton current from the air onto the electrode and from the electrode into the air. The mesh wire is aluminum or aluminum-magnesium alloy suspended from nylon thread 0.5 cm above the

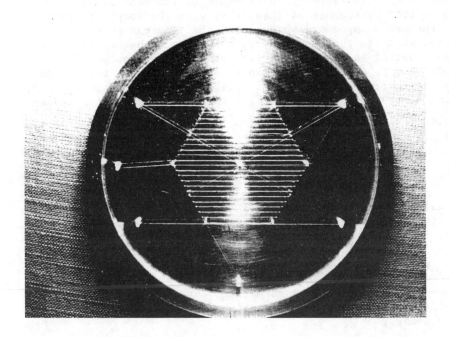

Fig. 21. PMD-1A E sensor.

ground plane. In the quasistatic case, the wire grid lies on an
equipotential plane, so that the equivalent length of the sensor
is also 0.5 cm. The output voltage of the E-field sensor is obtained by measuring, through a sensing resistor, the voltages
across the capacitor/conductor formed by the wire mesh and the
ground plane. The sensing resistor is in series with the terminated signal cable forming a resistive voltage divider [13]. The
voltage across the capacitor will decay with a time constant
determined by the sensor capacity (20 pF) and the sensing resistor except for local conductivity. This time constant must be
long compared to measurement times of interest.

5.2 Magnetic Field Sensors

The maximum air conductivity limits the loop radius to the
order of a skin depth or less at the highest frequency of interest. Below this frequency the air conductivity does not significantly enter into the loop response. Thus, for such a loop, the
nonlinear and time varying character of the air conductivity is

insignificant. However, sensor associated equipment, such as
cables in the air medium, should generally be limited to the same
dimensions to avoid magnetic field distortions which may couple
into the loop. It is possible to minimize the conductivity related effects by the use of insulators with the loop structure.
Also, the cable impedance which loads the loop can be chosen,
together with the loop inductance, to give a frequency response
of the order of the skin depth limitation. The problem of air
conductivity is eliminated in those sensors intended for use in
nonvacuum environments by encapsulating the volume enclosing the
sensing element with an epoxy resin which has a radiation induced
conductivity of less than 10^{-5} (S m^{-1})/(rad s^{-1}). This material
itself is a conductive medium under radiation but several orders
of magnitude lower than sea level air. Those sensors intended
strictly for use in vacuum have as little dielectric in them as
possible.

Each segment of cable acts as a Compton diode in a gamma
radiation field. Negative or positive charge is collected on the
center conductor, depending on the details of the charge transport. In addition, if the cable dielectric is hydrogenous
(e.g., polyethylene), the center conductor will collect neutron
scattered protons, further complicating the picture. The signals
produced by these radiation stimulated currents depend on the
geometry of the sensor and the quality of the differencing techniques used.

The magnetic field sensors used in nuclear environments are
all loop structures [3], [6], [7], [18], [19], [50], [55], [56],
[75], [78], [80] with the signal cables wired in a moebius configuration. The most important kind of these are designated as
CML (Cylindrical Moebius Loop) sensors. These greatly reduce the
common mode radiation noise currents found in the split shield
loop type of sensors and are made to have the same low differential radiation noise level (using symmetrical construction, etc.)
as the split shield loop. A CML sensor can be shown to be a two-
turn loop by tracking current flow from one twinaxial cable lead
to the other (Fig. 22). At frequencies where the magnetic field
does not penetrate the shield of the gap-loading cables, the sensor acts as a single-turn cylindrical loop with a resistive gap,
load given by the total terminating cable impedance. The four
gap, loading coaxial cables in the sensor are properly terminated at the point of coax-to-twinax junction as depicted in
Fig. 21. A voltage, V, at the gap appears as a positive signal
in one pair of 100 Ω, gap-loading cables and as a negative signal
in the other pair. The signal from the gap arrives at the coax-
to-twinax junction at the same time from all gap cables, which
produces a differential mode signal across the balanced twinax.
For a differential signal, the twinax may be considered to be two

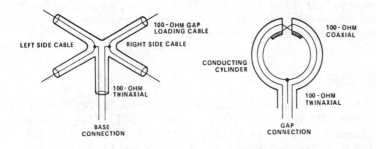

(a) electrical connections

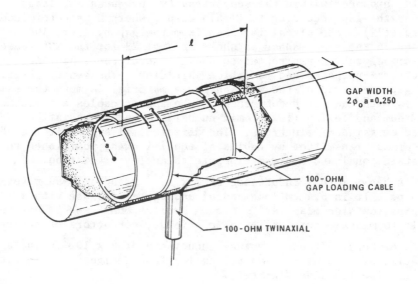

(b) typical loop configuration

Fig. 22. CML sensor

resistors, each of a value of 50 Ω to ground, that properly terminate the 50 Ω parallel combination of the two, 100 Ω coaxial
cables (from each side of the gap). For a given gap voltage, a
signal voltage of twice the amplitude of the gap voltage appears
at the balanced twinax output.

Six models of CML sensors have been designed and fabricated. They vary in equivalent area from 5 x 10^{-3} m^2 to .02 m^2 and have been built with encapsulation for use in air measurements and glyptol coated mesh materials for use in vacuum.

5.3 Current Sensors

Radiation hardened current sensors have been designed which are similar to the CPM series of I-dot probes [30], [31]. These sensors are designed to be part of a specific structure in a way that they will not appreciably affect the current flow on that structure.

5.3.1 Outside moebuis mutual inductance (OMM). Figure 23 shows three, OMM-1A, I-dot sensors assembled into a cylindrical antenna [71], [81]. Surface current flowing along the cylinder axis must pass through the sensor's internal cavity. The changing magnetic field, produced within the sensor cavity, produces a voltage across the gap according to (1/N)M dI/dt, where M is determined by Eq. (19). The signal is taken from the gap by four, 100 Ω cables in the same manner as shown in Fig. 22 for the CML sensor. The signal cables are routed to the inside of the sensor for electrical purposes and radiation shielding. The sensor interior and gap are encapsulated with an epoxy material in much the same way as with the CML B-dot sensor. The sensor cables are made of aluminum and Teflon (low atomic number) to reduce radiation induced emission of electrons. The differential signal from each sensor is transmitted by cables of equal length, positioned to maintain equal and minimum exposure to nuclear radiation. The OMM-1A has a mutual inductance of 2 x 10^{-9} H and a 10-90 risetime < .5 ns. It is 6.4 cm long and of a diameter for use with a 10 cm pipe (outside diameter). A much smaller OMM-2 sensor has been built to measure current in cable shields, conductors or structural members. It has a mutual inductance of 2 x 10^{-9} H and a risetime of < .5 ns. It is 8.9 cm long and designed to use with a 2 cm pipe (outside diameter).

5.3.2 Flush moebius mutual inductance (FMM). The FMM is a radiation hardened sensor for measuring J_n-dot (time derivative of total current density component normal to the surface) [20]. A sketch of the FMM-1A is shown in Fig. 24 [79]. It is wired and encapsulated much like the OMM type of sensor. In this case, however, it has an equivalent area to convert the current density to a current, besides its mutual inductance. It is important to note that the impedance presented by the gap to currents on the

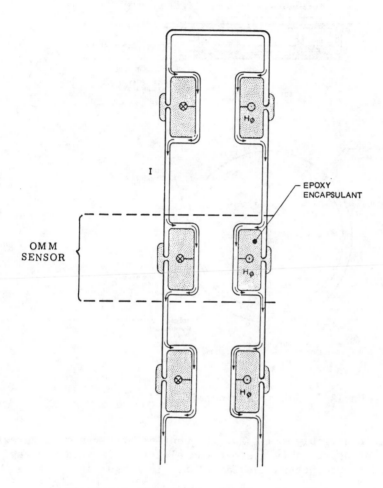

Fig. 23. Cylindrical antenna with OMM sensors.

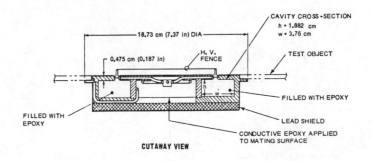

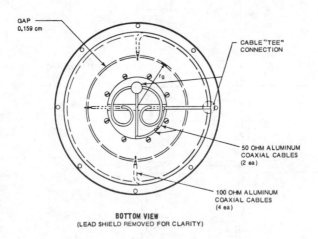

Fig. 24.   FMM-1A J_n-dot sensor details.

surface is negligibly small for frequencies or times of interest
so as to not appreciably distort the electric field in the region
surrounded by the potted annular slot.   In measuring J_{t_n} on a
test surface, the portion leaving the sensor plate must pass
over the surface of the sensor cavity.   This gives rise to a
voltage across the sensor gap expressed in Fig. 4, where M is
determined by Eq. (19) and $\vec{A}_{t_{eq}}$ is determined by [41].   The voltage is taken from the sensor gap by four equally spaced, 100 Ω,
cables in the manner of the CML B-dot sensor of Fig. 8.   Cable
passages and connection points are provided in the sensor structure which is machined from aluminum.   The sensor cavities and
gap are filled with an epoxy as a radiation hardening measure.
The epoxy extends up from the gap to provide increased voltage

standoff and minimize electron transport across the gap. The FMM-1A is approximately 19 cm in diameter. It has a mutual inductance of 5 x 10^{-9} H, an equivalent area of 1 x 10^{-2} m^2, and a 10-90 risetime < 1 ns.

6.   SENSORS FOR USE NEAR LIGHTNING ARCS AND IN CORONA

The environment inside the lightning arc and corona region is rather inhospitable for electromagnetic measurements. It differs from the nuclear source region in that we do not have a distributed source current density originating from Compton scatter of γ rays and photoelectric scatter of X rays occurring in air and in the various materials of the sensor itself. We do have conduction effects associated with the surrounding air medium. The impact of the air conductivity is fundamentally different for electric (capacitive-conductive) and magnetic (inductive) devices. Since the air conductivity is a nonlinear effect (because of the dependence of the electron mobility on the electric field), it is imperative that an electric type sensor not significantly distort the local electric field so as not to change the conductivity. For a magnetic type sensor, the problem is somewhat different. Local changes in the air conductivity are not as significant; the magnetic field incident on the sensor is more governed by the currents in a volume of space with dimensions of the order of the radian wavelength (or skin depth) so that the local perturbations do not matter so much (at least for the lower frequencies).

The problem of concern then is ionization of air, if present, which constrains electric sensor design to be nondistorting of the local electric field and which loads the loop-gaps of magnetic sensors.

In reducing the deleterious effects in the lightning-source-region, various general guidelines are useful.

a.   For magnetic sensor:

The gaps are encapsulated in dielectric to prevent a shunt conductivity across the gap.

b.   For electric sensor or current density normal to a conducting surface:

They must negligibly distort the electric field in the immediate vicinity to not perturb (significantly) the conduction current density.

Various designs of sensors for electric and magnetic fields and currents in lightning-source-regions have been developed and used and are discussed here.

It is generally easier to construct lightning-source-region sensors than it is to construct nuclear source region sensors. The latter is rich in high energy photons (γ and X ray) and also possibly neutrons. This necessitates sensor elements being as sparse as possible and also that all materials used be of as low atomic number as practical so that cross-sections to the irradiation are minimized. This also includes such items as output cables which are specially fabricated and delicate. For lightning sensors we can use other materials such as brass sheet metal and copper-jacketed coaxial cables.

For mounting of lightning-source-region sensors on objects, such as airplanes, a few simple precautions are in order. The sensors should be as flush with the surface of the object as possible so as not to perturb the fields and possibly attract direct attachment to the sensors. Cables between sensors and recording equipment should be shielded as soon as possible, and the recording system enclosed inside a topological electromagnetic shield.

6.1 Electric Field Sensors

As discussed in the previous section, the most severe effect of a lightning source environment on electromagnetic measurements is that on the electric field sensor in air. Effects associated with air conductivity alter the characteristic response of the sensor.

6.1.1 Parallel mesh dipole (PMD). The parallel mesh dipole sensor, Fig. 21, reduces the effect of the air conductivity by means of having its sensing element constructed of fine wires, as opposed to a solid plate of a parallel plate dipole, so that most of the conduction current is allowed to flow around the dipole conductors instead of through them. This type of sensor is discussed in Section 5.1. Note that the PMD-1 is not designed to be mounted on an aircraft where a large air velocity would severely damage it.

6.2 B-dot and H-Field Sensors

The maximum air conductivity limits the loop radius to the order of a skin depth, or less, at the highest frequency of interest. Below this frequency the air conductivity does not significantly enter into the loop response. Thus, for such a

loop, the nonlinear and time varying character of the air conductivity is insignificant. However, sensor-associated equipment
such as cables in the air medium should generally be limited to
the same dimensions to avoid magnetic field distortions which may
couple into the loop. It is possible to minimize the conductiv-
ity-related effects by the use of insulators with the loop structure. Also, the cable impedance which loads the loop can be
chosen, together with the loop inductance, to give a frequency
response of the order of the skin depth limitation. The problem of air conductivity is greatly reduced in these sensors by
encapsulating the volume enclosing the sensing element gap with
an epoxy resin.

6.2.1 Cylindrical moebius loop (CML). A useful magnetic field
sensor for lightning source environments is a loop structure with
the signal cables wired in a Moebius configuration designated as
a Cylindrical Moebius Loop (CML) sensor. These have been built
with encapsulation for use in conductive air measurements. See
Section 5.2 and Fig. 22 for a description of this type of sensor.

6.2.2 Multiturn hardened loop (MHL). Corresponding to the
multiturn loop (MTL) designs discussed in the previous section,
some similar designs have been built with dielectric encapsulation to avoid the conductivity effects. Designated the multiturn
hardened loop (MHL), some designs have already been used for EMP
source region experiments and could be used for lightning experiments.

6.3 Current and Current Density Sensors

     Radiation hardened current sensors have been designed which
are similar to the CPM series of I-dot probes. These sensors are
designed to be part of a specific structure in a way that they
will not appreciably affect the current flow on that structure.

6.3.1 Outside moebius mutual inductance (OMM). This type of
sensor is discussed in Section 5.3 and illustrated in Fig. 23.

6.3.2 Outside core I (OCI). Similar to the OMM, except using a
magnetic core in the toroidal volume for the measured magnetic
flux, is the OCI. This makes the output respond to the current
waveform instead of its time derivative, above some low frequency. Multiturn windings with coaxial cables and Moebius gaps
are also used, thereby employing many of the design features of
the multi-turn hardened loops (MHL). These sensors have been

used in nuclear source regions and are applicable to lightning.
One should be careful of the large low-frequency content of
lightning return stroke pulses and the corresponding possibility
of unwanted saturation of the magnetic cores.

6.3.3 Flush plate dipole (FPD). Special FPD-2 sensors have been
made for aircraft flying in lightning source regions (direct
stroke attachment). Their sensitivity is $A_{eq} = 0.02 \text{ m}^2$. The
performance of these FPD sensors was necessarily degraded from
that of the standard FPD in the following areas: The distance
between the sensing element and the bottom pan was reduced because of size limitation on the sensor location. The volume between the element and the pan was filled with dielectric to both
eliminate the possibility of conductive air within the sensor and
to provide a firm mounting for the element in the aircraft wind-
stream. The gap area was also thoroughly encapsulated with dielectric. All of the above steps serve to increase the sensor
capacitance with the result that the bandwidth is decreased to
70 MHz. Versions of this hardened sensor were also made with
rectangular sensing elements to fit specified areas. Both circular and rectangular versions were made with curved surfaces, including the sensing element, to conform to the aircraft shape so
as to not create electrical and airflow proturbances. The figure
of merit for these special FPD sensors is thereby reduced to
.005.

6.3.4 Flush moebius mutual inductance (FMM). With external
geometry similar to the FPD, the FMM has been used in nuclear
source regions because of its low impedance (a small inductance)
loading the central plate. The signal pickoff is like the OMM,
to which the response is similar. This sensor responds to the
time derivative of the total current, thereby emphasizing the
high frequencies.

6.3.5 Inside moebius mutual inductance (IMM). In this case the
slot is around the inside of the toroid. This last type is a
direct example of the use of such devices in pulse power machinery, since it was developed for measuring electron beam currents
inside circular conducting cylinders [95].

7.   LIGHTNING ENVIRONMENTAL MEASUREMENTS ON AIRCRAFT

In order to define lightning environmental criteria for aircraft, appropriate to the vulnerability of the electronics, one
needs measurements of the lightning environment under inflight
conditions. This raises the question of how to perform such
measurements and which measurements are most appropriate. Two
situations are of interest: a near miss and (more important) a
direct strike.

For the case of a near miss (say hundreds of meters or somewhat larger) one can first use data available from measurements
on the ground. For the important fast transients as in [99],
[100], the fields can be extrapolated back to the sources and
fields closer to the source using the Green's-function techniques
in [100].

On an aircraft, one can, in principle, measure the electromagnetic fields incident on the aircraft. However, such measurements are very difficult due to the aircraft scattering. If one
places sensors on the aircraft skin or near it, one must recognize that, in general, the aircraft is now part of the sensor and
introduces the aircraft resonances into the sensor response.
These can introduce errors which vary the response by an order of
magnitude over the frequencies of concern. Furthermore, this
wild response varies strongly with angle of incidence and polarization of the incident wave. Unfolding the true incident fields
is a difficult inverse-scattering problem. In such a measurement, one has the surface fields (or surface current and charge
densities) for a particular airplane which cannot be directly
applied to another airplane. This situation leads to surface
electromagnetic parameter measurement on many aircraft. A similar, but even more complex, problem occurs if one measures the
internal response at selected points on a particular aircraft.

If one wishes to measure the fields incident on an aircraft,
then one must be very careful how he goes about it. Specifically, one should find spatial locations with respect to the aircraft such that the aircraft scattering can be neglected, at
least for particular field components. Most aircraft have an
electromagnetic symmetry plane (as far as external scattering is
concerned) passing through the fuselage from nose to tail and
approximately perpendicular to the wings. Incident and scattered
fields can be divided into two independent parts designated symmetric and antisymmetric with respect to this symmetry plane
[89]. Placing a magnetic sensor on this plane for measuring a
magnetic field parallel to the plane makes it totally insensitive
to the symmetric scattering (which includes the large fuselage
resonances). Rotating the sensor axis one can try to cross
polarize it to the antisymmetric scattering from say the wings

and horizontal stabilizers (if present); for this purpose, the
wings and horizontal stabilizers should be in the same direction
from the sensor. This places the sensor off the nose or tail of
the aircraft. Selecting, say, the nose to avoid the close proximity of the horizontal stabilizers, one can construct a conducting boom (which can shield the sensor cable(s)) protruding from
the nose a fuselage diameter or two to minimize the local antisymmetric magnetic field scattered from the fuselage. This is
not an easy measurement. Alternatively, one might have a boom or
trailing package extended sufficiently far from the tail. One
might also begin by orienting the sensor to be insensitive to the
antisymmetric scattering, leading to similar conclusions concerning sensor location. This has been covered in detail in a recent
paper [88].

In a direct-strike situation, one cannot separate incident
and scattered fields due to the nonlinear arc and corona processes. One is forced to measure surface electromagnetic parameters, and hence to measure these on various aircraft. One would
like at least to bound these electromagnetic parameters for various aircraft skin locations to be able to apply the results, at
least approximately, to new untested aircraft designs. The sensors discussed in this section are appropriate for these measurements. Since one is usually limited as to number of waveform
measurements on a particular aircraft, then one must be concerned
with where to place the sensors. Total-normal-current-density
sensors (as well as electric-field sensors) would first be placed
near the aircraft extremities (nose, tail, and wing tips) where
such fields are intense. Magnetic sensors could be placed near
the wing/fuselage intersection through which the currents from
the lightning arcs (which tend to attach/detach near the extremities) will pass, noting that the currents into such a junction
approximately sum to zero (Kirchoff approximation). Current sensors can be built into likely extended attachment/detachment
positions such as the pitot tube.

Note that, even in direct-strike conditions, one can at
least approximately separate the exterior aircraft response from
the interior response if the skin has some shielding properties.
The exterior region, as well as some of the penetrations, will be
affected by the nonlinear corona [108]. It is these aspects that
require special measurements. The response of the aircraft interior can be considered somewhat separately using concepts from
electromagnetic topology [90].

7.1 WC-130 Experiments

A program, joint between the Air Force Wright Aeronautical Laboratories and the National Oceanic and Atmospheric Administration, has instrumented a WC-130 aircraft as indicated in Fig. 25. Some results are given in [102]. Most recently [103] some direct strikes have been obtained.

The latest CML design is specifically intended for skin current B-dot measurements on aircraft flying in and near thunderstorms. The ground plate mounts flush with the aircraft skin as it replaces a patch of skin on an aircraft panel. Figure 25 shows the sensor locations as used on the NOAA WC-130 aircraft in recent Air Force tests. This CML-7 sensor is not optimized for

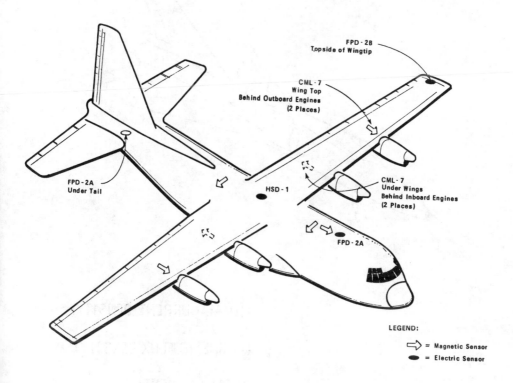

Fig. 25. Sensor locations on WC-130 NOAA aircraft.

high-frequency bandwidth, but rather for signal level vs. physical size and for aerodynamic consideration. The upper frequency
response is about 35 MHz with a 10 ns risetime.

7.2 F-106 Experiments

A program at NASA, Langley, has recently obtained some data
on the surface current density and normal total current density
on the skin of a specially instrumented F-106 under direct strike
conditions [106, 107]. A schematic of the aircraft with its
measurement locations is given in Fig. 26. For these measure-
ments, a set of $\partial \vec{B}/\partial t$, $\vec{J}_t$, and $\partial I/\partial t$ sensors were fabricated

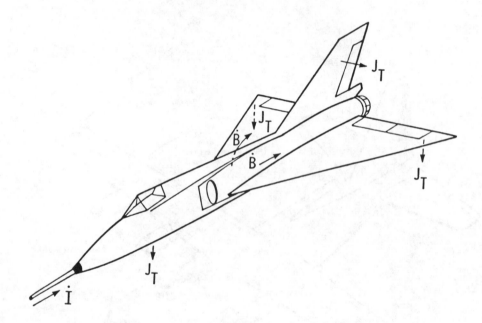

$$J_T = \qquad \text{(TOTAL CURRENT DENSITY)}$$

$$\dot{B} = \frac{\partial}{\partial t} \text{ (MAGNETIC FLUX DENSITY)}$$

$$\dot{I} = \frac{\partial}{\partial t} \text{ (TOTAL CURRENT)}$$

Fig. 26.   Sensor locations on NASA F-106 aircraft.

following the previously discussed designs at NASA. Their parameters are summarized as

Type of Sensor	Quantity Measured	Sensitivity	Risetime (10%-90%)
Multigap Loop	B-dot	5.73×10^{-3} m^2	0.85 ns[1]
Flush-plate Dipole	J_t	4.09×10^{-2} m^2	2.4 ns[1]
Outside Moebius Mutual Inductance	I-dot	2.13×10^{-9} H	0.23 ns[2]

Notes: [1]Measured
        [2]Calculated

To record the transient signals, some of the data channels
used specially modified digital recorders for long data records.
The transient recorders used in NASA Direct Strike Lightning
Research [105] are modified Biomation Model 6500 Waveform Recorders. The Biomation 6500 is a fast analog-to-digital converter (ADC) with internal storage for 1,024 six-bit words. At
the fastest rate, the Biomation 6500 samples an analog signal
every 2 nanoseconds and has an input analog bandwidth of 100 MHz.
The recorders operate in an "endless loop" mode, wherein the recorder continuously samples and stores data until a trigger event
occurs which causes the record phase to end. The recorded data
is then output for permanent storage at a slower rate. Once the
memory has been read out, it returns to the record phase to await
the next trigger event.

A memory expansion from 1,024 to 131,072 samples allows a
significantly longer data record than obtainable with the basic
unit [105]. For example, the modified recorder can store a 1300
microsecond "snapshot" of data with a time resolution of 10
nanoseconds per data sample. The major characteristics of the
modified recorder are storage of 131,072 samples as compared to
1,024, increase of memory integrated circuits from 48 to 192, and
provision for delaying the trigger by up to 100,000 sample intervals.

The major components of the transient recorder are indicated
in Fig. 27. The signal to be recorded is presented to the input
attenuators, amplifier, digitized, and then stored in the memory.
The control circuitry selects the data period to be retained.
Once a "snapshot" has been obtained, it is routed to permanent

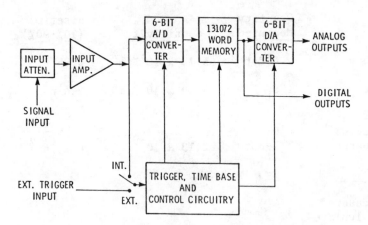

Fig. 27. Schematic of transient recorder.

storage via the analog or digital outputs. The Biomation 6500
recorder achieves its high sampling rate by employing a parallel
ADC technique. The output of the ADC is converted to a six-bit
gray code. In order to provide more time for the storage of the
data, the ADC output is double buffered into two six-bit registers on an alternating basis. The output of these two registers
is presented to the memory board on two six-bit parallel buses.
The expanded memory is partitioned into 32 sections, each employing memory devices which have a 55 nanosecond write cycle time.
The write cycles of the memory devices are overlapped through a
series of intermediate registers. The ADC, operating at its designed maximum sample rate, generates a data word every 2 nanoseconds. The first data word is routed to the first memory
location in memory section one. Two nanoseconds later the second
data word is routed to the first memory location in memory section two, and so on, until the first 32 data samples have been
routed to the 32 memory sections. The elapsed time of 64 nanoseconds allows the first memory device to complete its write
cycle (55 nanoseconds) in time for the next sequence of data
words. Through the use of memory interleaving, an effective
speed of 2 nanoseconds between samples is thus obtained using
memory devices of 55 nanosecond write-cycle time.

     While not all the results are in, the early results are
quite remarkable. Figure 28 shows the normal current density
(conduction plus displacement with positive polarity outward from
the skin) taken below and behind the nose (i.e., the chin) of the

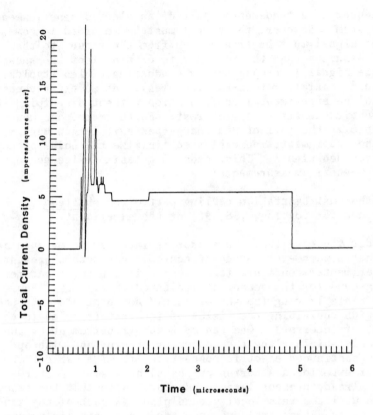

Fig. 28.   Example of total current density normal to aircraft
            skin under direct strike conditions.

F-106 at the position indicated in Fig. 26.   Note the early-time
resonant behavior with times corresponding to transit times on
the aircraft exterior.   Note also the continuing current density
of about 5 A/m^2 for about 4 μs.   Neglecting corona, this can be
integrated to give about 2 MV/m for the change in electric field.

These results point out the significance of electric parameters (as well as magnetic) and high frequency resonance effects
on aircraft and the necessity for including them in simulation of
lightning on aircraft as discussed in a previous paper [108].
Note also the risetimes in the 30 ns regime, agreeing with the
field measurements in [100].

8. TOPOLOGY OF INSTRUMENTATION CABLING

The sensor is a fundamental part of an electromagnetic measurement system. However, the signal must be recorded in some
way and the signal must be transported from the sensor to the
recorder. Assuming that the recorder is distant from the sensor
and that the signal is propagated by conducting cables (typically
well shielded coaxial or twin-axial cables), one must fit these
cables into the experimental configuration without disturbing the
electromagnetic quantities of interest. Furthermore, one would
like to minimize the current and charge-per-unit length magnitudes on the instrumentation cables to minimize the noise pickup
with the recorded signal. This leads to a basic design concept
for electromagnetic measurements:

Make the instrumentation cabling part of or shielded by
the conductor topology [98, 96] of the experiment.

Consider the case that the sensor is intended to measure an
electromagnetic parameter on one of the good conductors present
in the experimental configuration. Then this conductor becomes
the local ground for the sensor as indicated in Fig. 29. Any
conducting cable leaving the sensor should not protrude into the
(upper) region containing the fields of interest (sampled in the
measurement of interest). One can meet this requirement by running the cable shield along the conductor present and with approximately continuous electric contact to it. The cable then
behaves (for exterior scattering purposes) as a small perturbation on an already present large conductor. Note that the sensor
itself then utilizes this local ground plane as part of the sensor itself in that frequency response, accuracy, and field configuration are strongly influenced by this ground "plane."

Continuing, the cable transports the signal from the sensor
along the large metal conductor to somewhere else, where some-
thing is done to the signal. This somewhere else may have, perhaps, an oscilloscope or other recorder in a screen box (also
well grounded to the original conductor) or perhaps some modula-
tor which converts the signal to another frequency band and tele-
meters the signal location for demodulation and recording (as in
Fig. 30a). Note that one must have frequent connections to the
original conductor from sensor through screen box [93]. The
spacing between connections should be less than a half wavelength
at the highest frequencies of interest. In special places, such
as near the sensor, as well as the sensor "ground plane" itself,
it is good practice to increase the number of electrical connections to the original conductor.

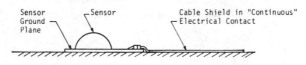

a. Field Sensor: electric (surface charge density) or magnetic
   (surface current density)

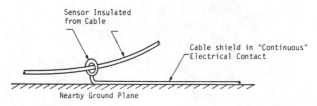

b. I or ∂I/∂t Sensor

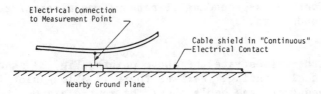

c. V or ∂V/∂t Sensor

Fig. 29. Local sensor grounding topology.

In some cases one can further improve the measurement by lo-
cating larger objects, such as a screen box, at larger distances
from the sensor to minimize the influence of the electromagnetic
scattering from such objects to the sensor. As illustrated in
Fig. 30b, one might position such objects in a place where the
scattering to the sensor is shadowed by the original conductor.
An example might be the positioning of sensor and screen box on
opposite sides of an aircraft fuselage or wing.

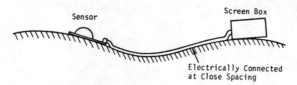

a. Conforming measurement conductors to experiment conductors

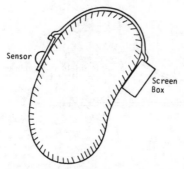

b. "Shadowing" sensor from screen box

Fig. 30.   Continuous grounding of sensor/transmission/recorder
            conductors.

To further illustrate the above points, Fig. 31 shows unacceptable cable routing. Perhaps there is a long slot in the
original conductor; the cable should be routed around it instead
of across it. Perhaps one is measuring a signal in an equipment
rack; one should avoid the temptation of jumping the cable from
the rack to another structure (e.g., a wall). One should follow
the rack conductors until (if and when) they electrically connect
to other conductors (e.g., floor, conduit, etc.).

One can go a step further in some cases by using the original conductor of interest as a shield. Figure 32 shows the case
that the original conductor of interest is locally approximately
planar and serves as a shield in that the field(s) of interest on
one side are large compared to the fields on the other side.
Then, on this other side, instrumentation cables can be routed
with minimal effect on the experiment and minimal noise pickup in
the cables. The sensors are any that are mounted on (or near)
ground planes with the cable now being fed through (with electrical connection to) the local ground plane.

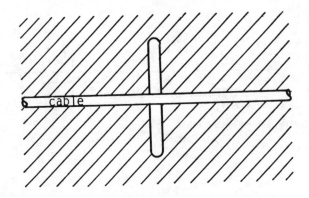

a. Crossing slot or other aperture

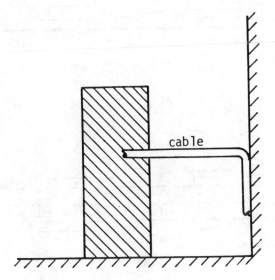

b. Jumping from one structure to another

Fig. 31.   Unacceptable cable routing.

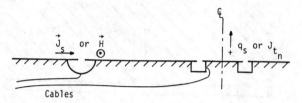

a. Cabling behind ground plane

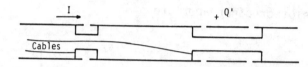

b. Cabling inside

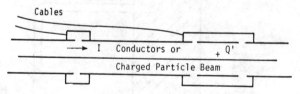

c. Cabling outside

Fig. 32. Cabling shielded by conductors of experiment topology.

Besides measuring local surface current and charge densities, one might also measure such integral quantities as current and charge per unit length on or in cylindrical conductors as indicated in Figs. 32b and 32c. Suppose, as in Fig. 32b, one wishes to measure the current and/or charge per unit length on a circular conducting cylinder (such as a pipe or tube). Then one can insert appropriate sensors which preserve the electrical continuity (and hence shiedling of the interior). By leaving a shielded passage through the center, the instrumentation cables from various such sensors can be routed through other such sensors to the recorders (or telemetering devices) without disturbing the measurements by these other devices.

Figure 32c illustrates the complementary problem in which the electromagnetic fields of interest are inside the conducting pipe. There is some current and associated charge per unit length propagating along or near the axis of the pipe. This may be via a conductor (as in the center conductor of a coaxial cable) or via energetic charged particles coming from some particle accelerator. Sensors can be built for this application which preserve the electrical continuity of the pipe, and thereby allow the instrumentation cables to leave by various routes from the pipe.

9.   SYMMETRY CONSIDERATIONS IN SENSORS AND INSTRUMENTATION
     CABLING

Suppose now that one wishes to measure electromagnetic parameters at positions removed from the conductors in the experiment. What does one do with cables from the sensor? Symmetry of the sensor, cabling, and/or electromagnetic-field configuration can be used to minimize the errors associated with instrumentation-cable scattering. This leads to another basic design concept for electromagnetic measurements:

Configure the sensor and cabling such that

a.   the cabling is orthogonal to the incident electric field (minimizes the scattering) and/or

b.   the large field components scattered by the cabling exterior are orthogonal to the sensor response characteristics (sensor symmetry with respect to cabling).

Of course one may also choose to remove this cable scattering problem by removing the cable (such as by telemetering the data from the immediate vicinity of the sensor).

Symmetry is a powerful concept in that it allows one to make useful statements concerning some of the properties of a physical system, even a very complex one, without detailed calculation (whether analytical or numerical, recognizing accuracy limitations). If one had a highly conducting circular cylinder (boom) as the outer conductor of the cable or a conduit enclosing one or more cables, one might consider the two dimensional pure-rotation group C_∞ [97]. As in Fig. 33, such a sensor boom might have

coordinate systems with origin at the sensor. Cylindrical coordinates (Ψ', ϕ', z') would have the z' axis as the axis of rotation for the circular cylindrical boom. C_∞ symmetry leads to $\cos(n\phi)$ and $\sin(n\phi)$ terms (integer n) for an infinite set of separate terms in the electromagnetic-field expansion in the presence of this boom.

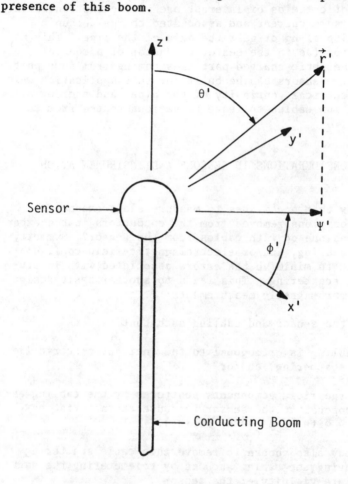

Fig. 33. Coordinates for sensor boom.

The sensor, however, may not in general possess C_∞ symmetry with respect to the z' axis because of various other requirements in its design. Fortunately, the main results of interest are achieved with a lower order symmetry, namely reflection (or planar) symmetry [97]. Consider planes containing the z' axis (e.g., the x'z' and y'z' planes) and note that these are symmetry planes for the conducting boom and that some number of such planes (typically 2) can be symmetry planes of the sensor as well.

Reflection symmetry, say group R, is illustrated by Fig. 34. The coordinate system in Figure 34a has chosen our symmetry plane of interest as the xy plane which lets us define

$$\overset{\leftrightarrow}{R} \equiv \begin{pmatrix} 1 & 0 & 0 \\ 0 & 1 & 0 \\ 0 & 0 & -1 \end{pmatrix} \equiv \text{reflection matrix}$$

$$= \overset{\leftrightarrow}{R}^{-1}$$

$$\vec{r} \equiv x\vec{1}_x + y\vec{1}_y + z\vec{1}_z \equiv \text{position or coordinates} \qquad (24)$$

$$\vec{r}_m \equiv \overset{\leftrightarrow}{R} \cdot \vec{r} \equiv \text{mirror position or mirror coordinates}$$

$$= x\vec{1}_x + y\vec{1}_y - z\vec{1}_z .$$

The function of the reflection matrix is to map each position into its "mirror image" through the symmetry plane. Having this symmetry plane apply to a sensor and boom, means that "whatever" is at $\vec{r}$ is at $\vec{r}_m$ also, this "whatever" being typically conductors and insulators, at least as far as external scattering is concerned. This applies to tensor parameters such as permittivity, conductivity, and permeability (where used); such tensors are reflected from $\vec{r}$ to $\vec{r}_m$ by a similarity transformation using $\overset{\leftrightarrow}{R}$. This type of symmetry analysis is considered in much greater detail in a previous paper [89].

With respect to such a symmetry plane, one can decompose electromagnetic fields, potentials, etc. into two uncoupled parts which we term symmetric (subscript sy) and antisymmetric (subscript as). For some of the common quantities we have:

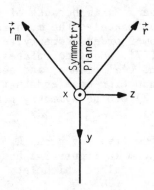

a. Coordinates and mirror coordinates

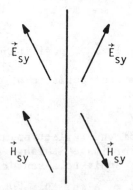

b. Symmetric fields

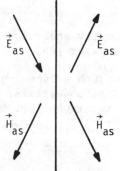

c. Antisymmetric fields

Fig. 34. **Electromagnetic symmetry with respect to a plane.**

$$\vec{E} = \vec{E}_{sy} + \vec{E}_{as} \qquad \text{(electric field)}$$

$$\vec{J} = \vec{J}_{sy} + \vec{J}_{as} \qquad \text{(current density)} \qquad (25)$$

$$\vec{\rho} = \vec{\rho}_{sy} + \vec{\rho}_{as} \qquad \text{(charge density)}$$

$$\vec{H} = \vec{H}_{sy} + \vec{H}_{as} \qquad \text{(magnetic field)} .$$

The construction of the symmetric and antisymmetric parts uses combinations of the fields at $\vec{r}$ and $\vec{r}_m$ [89].

The symmetric part reflects as indicated in Fig. 34b. Note that the electric type vectors have tangential components continuous through the plane while magnetic type vectors have the normal component continuous. Specifically we have

$$\vec{E}_{sy}(\vec{r}_m) = \overleftrightarrow{R} \cdot \vec{E}_{sy}(\vec{r})$$

$$\vec{J}_{sy}(\vec{r}_m) = \overleftrightarrow{R} \cdot \vec{J}_{sy}(\vec{r})$$

$$\rho_{sy}(\vec{r}_m) = \rho_{sy}(\vec{r}) \qquad (26)$$

$$\vec{H}_{sy}(\vec{r}_m) = -\overleftrightarrow{R} \cdot \vec{H}_{sy}(\vec{r}) .$$

The antisymmetric part reflects with exactly opposite signs as

$$\vec{E}_{as}(\vec{r}_m) = -\overleftrightarrow{R} \cdot \vec{E}_{as}(\vec{r})$$

$$\vec{J}_{as}(\vec{r}_m) = -\overleftrightarrow{R} \cdot \vec{J}_{as}(\vec{r})$$

$$\rho_{as}(\vec{r}_m) = -\rho_{as}(\vec{r}) \qquad (27)$$

$$\vec{H}_{as}(\vec{r}_m) = \overleftrightarrow{R} \cdot \vec{H}_{as}(\vec{r}) .$$

In this case, normal components of electric type vectors are continuous through the plane as are tangential components of magnetic type vectors.

Consider now the scattering of an electromagnetic wave from a conducting plane. Note that the antisymmetric part has zero tangential electric field and as such is not scattered by the conducting plane. Only the symmetric part is scattered. Considering some plane containing the z' axis in Fig. 33, the antisymmetric fields have little scattering from a cable or boom of small cross section. It is the case of an electric field parallel to the cable which has a large scattering. An incident electric field with a large z' component constitutes a large symmetric part with respect to all planes containing the z' axis.

Minimization of boom (or cable) scattering means having an antisymmetric field distribution with respect to two planes containing the z' axis (e.g., the x'z' and y'z' planes).

Enforcing the sensor to have two symmetry planes containing
the z' axis also helps. Say there is an incident electric field
with a z' component, one may still try to measure the x' and y'
components and use sensor symmetry to ideally have no response to
the z' component. However, since the scattering of a z' incident
electric field is so large near the sensor, one may still encounter signal-to-noise problems because of construction tolerances which introduce small distortions of the sensor from the
desired symmetry.

In the case of the magnetic field, the situation is somewhat
better. The incident z' electric field produces a large charge
near the end of the boom and on the sensor. However, the same
position is also a current minimum (for net axial current) which
implies a relatively small scattered magnetic field from the boom
in the vicinity of the sensor. It is thus generally possible to
measure all three components of the incident magnetic field, in
the usual dot product sense, at the end of a boom. For the electric field, only two components are possible in this sense, with
some signal-to-noise restrictions, depending on the relative
amount of z' incident electric field present and sensor precision
in its symmetry.

Note that we have been only considering sensor and boom symmetry. One should also consider other scatterers which may be
present. Figure 35 shows a sensor and boom mounted on the earth,
say to measure the fields from an airborne EMP simulator [91].
In general, this other scatterer (in this case the local earth
topography, conductivity, etc.) may have no symmetry plane in
common with the boom. However, a transient wave has its leading
edge propagate with the local speed of light in the media of concern. This allows one in many cases to have a measurement valid
for a period of time (the "clear time") before reflections from
these other scatterers can arrive at the sensor. During this
"clear time" the foregoing symmetry analysis is still applicable.

10. SUMMARY

As should be apparent by now, electromagnetic field measurements can be simple or complex depending on frequency, sensor
size, and other environmental parameters. Obtaining more signal
and bandwidth from a sensor and minimizing the influence of ad-
verse environmental conditions is quite possible if one is will-
ing to go beyond the elementary concepts of electrically small

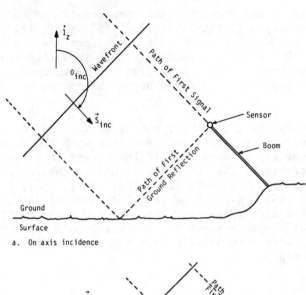

a. On axis incidence

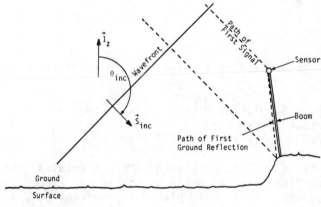

b. Off axis incidence

Fig. 35. Modes of operation of sensor boom.

electric- and magnetic-dipole antennas in linear, uniform, isotropic media.

Much is now known concerning the design of sensors for accurate measurement of electromagnetic-field parameters. These
special antennas come in many varieties and can be used to measure various types of electromagnetic environments including those
of EMP, lightning, pulse-power machinery, charged particle beams,
etc.

Various other sensor concepts have been considered and some
will likely be developed in the future [2], [5], [12], [21],
[35], [44], [87]. Special passive voltages probes are one likely
candidate. Other EMP related sensors, such as radiation sensors
for γ rays, X rays, and neutrons have also been developed for EMP
applications, but such are considered beyond the scope of this
paper.

Besides the sensors themselves, one needs to consider various electromagnetic design problems associated with the objects
to which the sensor is attached to hold the sensor in a given
position and/or transport the resulting signal to the recorder,
as well as other adjacent objects [23], [28], [29], [32], [40],
[42], [49], [62], [63], [64]. Fundamental to such an integration
are the concepts of electromagnetic topology and symmetry of the
sensor and/or fields with respect to scattering conductors such
as sensor cables. For many details the reader may consult the
references of this paper.

REFERENCES

In the interest of brevity the following abbreviations are
used:

    SSN   Sensor and Simulation Notes
    IN   Interaction Notes
    MN   Measurement Notes

These are part of the note series on EMP and related sub-
jects described in "The Complete Guide to the Notes," INDEX 1-1,
February 1973, by Baum. These have played a prominent role in
transient/broadband technology. Listings of these can also be
found in the IEEE AP-S Newsletter. Copies of these papers can be
requested from the author, from the Defense Documentation Center,
Cameron Station, Alexandria, VA 22314, or from the Editor,
Dr. Carl E. Baum, Air Force Weapons Laboratory (NTATT), Kirtland
AFB, NM 87117. In addition, these notes are available at many
universities and companies doing research in EMP and electromagnetic theory. Other report numbers used are:

    AFWL-TR   Air Force Weapons Laboratory Technical Report
    (AFWL-TRs are available from Defense Documentation Center,
    Alexandria, VA 22314.)
    AL   EG&G WASC
              Albuquerque Operation
              P.O. Box 9100
              Albuquerque, New Mexico 87119

1. R. E. Partridge, "Invisible Absolute E-field Probe," SSN 2, Feb 64.

2. R. E. Partridge, "Combined E and B-dot Sensor," SSN 3, Feb 64.

3. K. Theobald, "On the Properties of Loop Antennas," SSN 4, Feb 64.

4. C. E. Baum, "Underground Testing of Close-in EMP Sensors," SSN 5, Oct 64.

5. C. E. Baum, "Minimizing Transit Time Effects in Sensor Cables," SSN 6, Oct 64.

6. C. E. Baum, "Characteristics of the Moebius Strip Loop," SSN 7, Dec 64.

7. C. E. Baum, "Maximizing Frequency Response of a B-dot Loop," SSN 8, Dec 64.

8. R. E. Partridge, "Capacitive Probe E-field Sensors," SSN 11, Feb 65.

9. C. E. Baum, "Electric Field and Current Density Measurements in media of constant conductivity," SSN 13, Jan 65.

10. G. L. Fjetland, "Design Considerations for a Twinaxial Cable," SSN 14, Mar 65.

11. C. E. Baum, "Radiation and Conductivity Constraints on the Design of a Dipole Electric Field Sensor," SSN 15, Feb 65.

12. L. E. Orsak, A. L. Whitson, "Electric Field Sensor for EMP Simulators," SSN 18, Dec 65.

13. C. E. Baum, "Combining Voltage or Current Dividers with Sensor Cables," SSN 19, Nov 65.

14. C. E. Baum, "A Technique for the Distribution of Signal Inputs to Loops," SSN 23, Jul 66.

15. C. E. Baum, "A Technique for Measuring Electric Fields Associated with Internal EMP," SSN 24, Aug 66.

16. C. E. Baum, "The Multiple Moebius Strip Loop," SSN 25, Aug 66.

17. C. E. Baum, "The Influence of Finite Soil and Water Conductivity on Close-in Surface Electric Field Measurements," SSN 26, Sept 66.

18. C. E. Baum, "The Influence of Radiation and Conductivity on B-dot Loop Design," SSN 29, Oct 66.

19. C. E. Baum, "The Single-gap Cylindrical Loop in Nonconducting and Conducting Media," SSN 30, Jan 67.

20. C. E. Baum, "Two Types of Vertical Current Density Sensors," SSN 33, Feb 67.

21. R. S. Hebbert, L. J. Schwee, "Thin Film Magnetoresistance Magnetometer," SSB 34, Feb 67.

22. C. E. Baum, "Parameters for Some Electrically-small Electro-Magnetic Sensors," SSN 38, Mar 67.

23. C. E. Baum, "Some Electromagnetic Considerations for a Seawater-based Platform for Electromagnetic Sensors," SSN 39, Mar 67.

24. C. E. Baum, "Conducting Shields for Electrically-small Cylindrical Loops," SSN 40, May 67.

25. C. E. Baum, "The Multi-gap Cylindrical Loop in Nonconducting Media," SSN 41, May 67.

26. C. E. Baum, "A Conical-transmission-line Gap for a Cylindrical Loop," SSN 42, May 67.

27. C.E. Baum, "Some Considerations for Electrically-small Multi-turn Cylindrical Loops," SSN 43, May 67.

28. R. W. Sassman, R. W. Latham, A. G. Berger, "Electromagnetic Scattering from a Conducting Post," SSN 45, Jun 67.

29. R. W. Latham, K.S.H. Lee, "Minimization of Induced Currents by Impedance Loading," SSN 51, Apr 68.

30. C. E. Baum, "Some Electromagnetic Considerations for a Rocket Platform for Electromagnetic Sensors," SSN 56, Jun 68.

31. C. E. Baum, "Some Considerations for Inductive Current Sensors," SSN 59, Jul 69.

32. R. W. Sassman, R. W. Latham, K.S.H. Lee, "A Numerical Study on Minimization of Induced Currents by Impedance Loading," SSN 61, Aug 68.

33. C. E. Baum, "An equivalent-charge Method for Defining Ge-
     ometries of Dipole Antennas," SSN 72, Jan 69.

34. C. E. Baum, "Parameters for Electrically-small Loops and
     Dipoles Expressed in Terms of Current and Charge Distribu-
     tions," SSN 74, Jan 69.

35. C. E. Baum, "Electrically-small Cylindrical Loops for Meas-
     uring the Magnetic Field Perpendicular to the Cylinder
     Axis," SSN 78, Mar 69.

36. C. E. Baum, "The Circular Parallel-plate Dipole," SSN 80,
     Mar 69.

37. C. E. Baum, "Some Further Considerations for the Circular
     Parallel Plate Dipole," SSN 86, Jun 69.

38. C. E. Baum, "The single-gap Hollow Spherical Dipole in Non-
     conducting Media," SSN 91, Jul 69.

39. C. E. Baum, "The circular Flush-plate Dipole in a Conducting
     Plane and Located in Nonconducting Media," SSN 98, Feb 70.

40. R. W. Latham, K.S.H. Lee, "Magnetic-field Distortion by a
     Specific Axisymmetric Semi-infinite, Perfectly Conducting
     Body," SSN 102, Apr 70.

41. R. W. Latham, K.S.H. Lee, "Capacitance and Equivalent Area
     of a Disk in a Circular Aperture," SSN 106, May 70.

42. C. E. Baum, "Two Approaches to the Measurement of Pulsed
     Electromagnetic Fields Incident on the Surface of the
     Earth," SSN 109, Jun 70.

43. R. W. Latham, K.S.H. Lee, "Capacitance and Equivalent Area
     of a Spherical Dipole Sensor," SSN 113, Jul 70.

44. C. J. Hall, "The Asymmetric Dipole as a Transient Field
     Probe," SSN 115, Aug 70.

45. L. Marin, "Scattering by Two Perfectly Conducting, Circular
     Coaxial Disks," SSN 126, Mar 71.

46. C. E. Baum, "Further Considerations for Multi-turn Cylindri-
     cal Loops," SSN 127, Mar 71.

47. K.S.H. Lee, R. W. Latham, "Inductance and Current Density of
     a Cylindrical Shell," SSN 130, Run 71.

48. F. M. Tesche, "Optimum Spacing of N Loops in a B-dot Sensor," SSN 133, Jul 71.

49. S. W. Lee, V. Jamnejad, R. Mittra, "Near Field of Scattering by a Hollow Semi-infinite Cylinder and its Application to EMP Studies," SSN 149, May 72.

50. P. H. Duncan, Jr., "Analysis of the Moebius Loop Magnetic Field Sensor," SSN 183, Sept 73.

51. K.S.H. Lee, "Electrically-small Ellipsoidal Antennas," SSN 193, Feb 74.

52. C. E. Baum, "A Figure of Merit for Transit-time-limited Time-derivative Electromagnetic Field Sensors," SSN 212, Dec 75.

53. C. E. Baum, "Electromagnetic Pulse Interaction Close to Nuclear Bursts and Associated EMP Environment Specification," IN 76, Jul 71.

54. C. E. Baum, "Some Design Considerations for Signal Transmission Lines for Use with Sensors in a Nuclear Radiation Environment," MN 17, Oct 73.

55. H. Whiteside and R.W.P. King, "The Loop Antenna as a Probe," IEEE, Transactions on Antennas and Propagation, pp. 291-297, May 64.

56. A. H. Libbey, et al., "Development of Hardened Magnetic Field Sensors," AFWL-TR-69-58, (Volumes I and II), Jun 69.

57. R. Morey, et al., "Development and Production of Multi-gap Loop (MGL) Series EMP B-dot Sensors," AFWL-TR-70-153, Feb 71.

58. J. K. Travers and J. H. Kraemer, "Development and Construction of Electric Field and D-dot Sensors," AFWL-TR-70-154, Mar 71.

59. W. R. Edgel, "Hollow Spherical Dipole D-dot Sensor (HSD-4) Development," AFWL-TR-75-77, Jan 75. Also AL-1147, Jan 75.

60. S. L. Olsen, "Asymptotic Conical Dipole D-dot Sensor Development," AFWL-TR-75-263, Jan 76. Also AL-1185, Sept 75.

61. T. Summers, "I-dot Sensor Design and Fabrication," AL-516, Nov 70.

62. R. Christiansen and T. Summers, "EMP Sensor Application Guide," AL-524, Dec 70.

63. M. Bumgardner, "Noise Associated with the Differential E-field Sensors," AL-592, Jun 71.

64. W. Motil and M. Bumgarnder, "A Technique for Evaluating Measurements Obtained from HSD-sensors," AL-647, Oct 71.

65. W. Edgel, "I-dot Sensor Design and Fabrication Phase II," AL-678, Dec 71.

66. J. Harrison, "Fabrication and Testing of a Multi-turn B-dot Sensor," AL-735, Feb 72.

67. W. Edgel, "I-dot Sensor Development," AL-921, Jun 73.

68. W. Edgel, "B-dot Sensor Development," AL-952, Jul 73.

69. S. Olsen, "Flush Plate Dipole D-dot Sensor Development," AL-1100, Aug 74.

70. W. Edgel, "MGL-6 B-dot Sensor Development," AL-1101, Aug 74.

71. W. Edgel, "Radiation Hardened I-dot Sensor (OMM-1A) for Use in Baum Antenna," AL-1102, Aug 74.

72. W. Edgel, "Conforming Flat D-dot Sensor Development," AL-1103, Aug 74.

73. W. Edgel, "MGL-S7A B-dot Sensor Development," AL-1104, Aug 74.

74. W. Edgel, "MTL-2 B-dot Sensor Development," AL-1105, Aug 74.

75. W. Edgel, "A small, Radiation Hardened B-dot Sensor (CML-4A)," AL-1106, Aug 74.

76. S. Olsen, "Sensor MGL-8B Sensor DW," AL-1186, Sept 75.

77. S. Olsen, "MSL-S8 B-dot Sensor Development," AL-1187, Sept 75.

78. G. D. Sower, et al., "Cylindrical Moebius Loop Radiation Hardened B-dot Sensor (CML-6A(R) Development," AL-1224, Jun 76.

79. G. D. Sower, et al., "Flush Moebius Mutual Inductance Radiation Hardened J_n-dot Sensor (FMM-1A) Development," AL-1226, Jun 76.

142   C. E. BAUM

80. G. D. Sower, et al., "Cylindrical Moebius Loop Radiation
     Hardened B-dot Sensor (CML-X3A(R) and CML-X5A(R)) Develop-
     ment," AL-1229, Jun 76.

81. G. D. Sower, et al., "Outside Moebius Mutual Radiation
     Hardened I-dot Sensor (OMM-2A) Development," AL-1232,
     Jun 26.

82. G. D. Sower, et al., "Parallel Mesh Dipole Radiation Hard-
     ened E-field Sensors (PMD-1C) Development," AL-1233, Jun 76.

83. B. C. Tupper, et al., "EMP Instrumentation Development,"
     Final Report, SRI Project 7990, Stanford Research Institute,
     Menlo Park, CA, Jun 72.

84. C. E. Baum, ed., "Electromagnetic Pulse Sensor Handbook,"
     EMP Measurement 1-1, Jun 71 (original issue).

85. C. E. Baum, ed., "Electromagnetic Pulse Instrumentation
     Handbook," EMP Measurement 2-1, Jun 71 (original issue).

86. V. V. Liepa and T.B.A. Senior, "Measured Characteristics of
     MGL and ACD Sensors", SSN 276, Sept 82.

87. C. E. Baum, "Idealized Electric- and Magneticfield Sensors
     Based on Spherical Sheet Impedances", SSN 283, Mar 83.

88. D. V. Giri and C. E. Baum, "Airborne Platform for Measure-
     ment of Transient or Broadband CW Electromagnetic Fields",
     SSN 284, May 1984.

89. C. E. Baum, "Interaction of Electromagnetic Fields with an
     Object which has an Electromagnetic Symmetry Plane", IN 63,
     Mar 71.

90. C. E. Baum, "Electromagnetic Topology: A Formal Approach to
     the Analysis and Design of Complex Electronic Systems",
     IN 400, Sept 80, and Proc. EMC Symposium, Zurich, Mar 81,
     pp. 209-214.

91. C. E. Baum, "Two Approaches to the Measurement of Pulsed
     Electromagnetic Fields Incident on the Surface of the
     Earth", SSN 109, Jun 70.

92. C. E. Baum, "EMP Simulators for Various Types of EMP Environ-
     ments", SSN 240, Jan 78, IEEE Transactions on Antennas and
     Propagation, Jan 78, pp. 35-53, and IEEE Transactions on
     EMC, Feb 78, pp. 35-53.

93. E. L. Breen, "The Application of B-dot and D-dot Sensors to Aircraft Skin Surfaces", instrumentation development Memo 3, July 74.

94. I. D. Smith and H. Aslin, "Pulsed Power for EMP Simulators, IEEE Transactions on Antennas and Propagation", Jan 78, pp. 53-59, and IEEE Transactions on EMC, Feb 78, pp. 53-59.

95. G. D. Sower, "I-dot Probes for Pulsed Power Monitors, 3rd IEEE International Pulsed Power Conference", Albuquerque, NM, Jun 80.

96. Mini-Symposium on Electromagnetic Topology, FULMEN 4, University of New Mexico, Mar 80.

97. M. Hammermesh, Group Theory and its Application to Physical Problems, Addison Wesley, 1962.

98. C. E. Baum, "The Role of Scattering in Electromagnetic Interference Problems", in P.L.E. Uslenghi (ed.), Electromagnetic Scattering, Academic Press, 1978.

99. C. D. Weidman and E. P. Krider, "Submicrosecond Risetime in Lightning Radiation Fields", Proc. Lighting Technology, NASA conf. Pub. 2128, and FAA-RD-80-30, April 1980, pp. 29-38, also as Submicrosecond risetimes in lightning Return-Stroke Fields, Geophysical Research Letters, Vol. 7, No. 11, Nov 80, pp. 955-958.

100. C. E. Baum, E. L. Breen, D. L. Hall, C. B. Moore, and J. P. O'Neill, "Measurement of Electromagnetic Properties of Lightning with 10 Nanosecond Resolution", Proc. Lightning Technology, NASA Conf. Pub. 2128 and FAA-RD-80-30, Apr 80, pp. 39-82. Also same title, (revised with more data) as lightning phenomenology Note 3, Feb 82.

101. C. E. Baum, "Electromagnetic Characterization of Lightning in the Submicrosecond Regime", URSI XXth General Assembly, Wash., D.C. Aug 81.

102. R. K. Baum, "Airborne Lightning Characterization, Proc. Lightning Technology", NASA Conf. Pub. 2128 and FAA-RD-80-30, Apr 80, Supplement pp. 1-20.

103. G. Dubro, Private Communication.

104. T. F. Trost and K. P. Zaeptel, "Broadband Electromagnetic Sensors for Aircraft Lightning Research", NASA Conf. Pub. 2128 and FAA-RD-80-30, Apr 80, pp. 131-152.

105.   R. M. Thomas, Jr., "Expanded Interleaved Solid-state Memory for a Wide Bandwidth Transient Waveform Recorder", NASA Conf. Pub. 2128 and FAA-RD-80-30, Apr 80, pp 119-129.

106.   F. L. Pitts and M. E. Thomas, "1980 Direct Strike Lightning Data", NASA Tech. Memo 81946, Feb. 81.

107.   F. L. Pitts, "Electromagnetic Measurement of Lightning Strikes to Aircraft", AIAA Paper 81-0083, Jan 81.

108.   C. E. Baum, "Simulation of Electromagnetic Aspects of Lightning", Lightning simulation Note 1, Feb 80, and Proc. Lightning Technology, NASA Conf. Pub. 2128 and FAA-RD-80-30, Apr 80, pp. 283-299.

109.   Special Joint Issue on the Nuclear Electromagnetic Pulse, IEEE Trans. on Antennas and Propagation, Jan 78, and IEEE trans. on EMC, Feb 78.

ULTRAFAST ELECTRICAL VOLTAGE AND CURRENT MONITORS

W. Pfeiffer

Technische Hochschule, 6100 Darmstadt,
Schlobgraβen 1,
FRG

1. INTRODUCTION

In power electronics, high voltage engineering, and plasma physics, very often the measurement of rapidly changing voltages and currents is required. Therefore, it is necessary to know, not only the characteristics of different measurement devices, but also the transfer characteristics of the whole measuring circuit. Also, it is always necessary to be aware of the limiting technical and physical properties.

The measurement range, especially for small duty cycles, may be up to about 10 MV or some MA. Such high values always have to be reduced before being fed to the measuring apparatus. Therefore, precise voltage dividers or current shunts are necessary [1, 2].

In order to get some idea of the difficulty of the measuring problem, it is necessary to know the steepness of the transient phenomenon and the maximum amplitude. Therefore, a good criterion is the maximal rate of voltage or current change. In order to get a rough survey about what is possible, let us look at a rather long air gap. Thereby the mean electrical field will be in the order of 3 kV/cm. If breakdown occurs, the highest possible voltage change might be about 10^{14} V/s, on condition that the breakdown phenomena propagate with the velocity of light. Higher values are possible in compressed gases, especially in homogeneous gaps. The current changes resulting from such phenomena depend upon the characteristic impedance of the test arrangement and are in the order of 10^{12} A/s. In this range of voltage or

current, changes in the precision of the measurement will be
greatly influenced by leakage inductances and stray capacitances.
In most cases a screened coaxial arrangement will be inevitable.
Such coaxial arrangements, including a coaxial pulse generator,
shall be considered here [3].

2.   COAXIAL VOLTAGE DIVIDERS

First some methods for the construction of coaxial voltage
dividers shall be discussed. The transfer characteristics of
the purely resistive divider are determined by the influence of
the distributed earth capacitance [4]. For a homogeneous construction, the divider can be regarded as a recurrent network
with $R' = R/n$ and $C_e' = C_e/n$ (Fig. 1). The dependency on frequency of the attenuated voltage U_2 can be calculated as follows

$$\frac{U_2}{U_1} = \frac{\sinh(\sqrt{R \cdot j\omega C_e}/n)}{\sinh(\sqrt{R \cdot j\omega C_e})} = \frac{\sinh R' \cdot j\omega C_e'}{\sinh R \cdot j\omega C_e} \tag{1}$$

The quantity n is the number of components or the divider ratio,
respectively. For $n \gg 1$ the exact value of the upper cut-off-
frequency is

$$f_c = 1.46/(RC_e) \tag{2}$$

and the step-response rise time is,

$$T_a = 2.35 \cdot \tau , \qquad \tau = R \cdot C_e/\tau^2 , \qquad T_a = 0.237 \cdot RC_e . \tag{3}$$

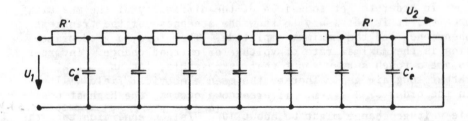

Fig. 1.   Resistive divider as a recurrent network.

The earth capacitance can be calculated from the characteristic impedance and the propagation velocity

$$v = \frac{1}{\sqrt{L'C'_e}} \ , \quad Z = \sqrt{\frac{L}{C'_e}} \quad \rightarrow \ C_e = \frac{1}{v \cdot Z} = \frac{\sqrt{\varepsilon_{r} \cdot 1}}{c \cdot Z} \ . \tag{4}$$

Let us take a real example. For measuring 100 kV, about 15 cm length of the divider will be necessary. For Z = 50 Ω, this will result in a total earth capacitance of at least 10 pF. If we could make the total divider resistance so low that it would be, for instance, the terminating resistor for a 50 Ω - system, then we would get a rise time of only 120 ps. However, the power dissipation in this resistor would be 200 MW and the technical realization of this element is rather troublesome [5]. Moreover, the loading of the high voltage circuit might not be admissible. Therefore, the upper limit for the use of ohmic dividers with very short rise times is some 10 kV.

In some cases, especially for the measurement of short pulses, the lumped resistors may be replaced by distributed resistors in form of the different characteristic impedances of two coaxial lines with low losses [6]. In this reflection type divider the characteristic impedance Z_2 is much lower than Z_1 (Fig. 2). According to this, the incoming voltage wave is nearly completely reflected

$$U_r = U_{v1} \ (Z_2 - Z_1) \ / \ (Z_1 + Z_2). \tag{5}$$

Only the small fraction

$$U_{v2} = U_{v1} \cdot 2Z_2/(Z_1 + Z_2) \tag{6}$$

is measured across Z_2. However, it has to be kept in mind that the input resistance of such reflection dividers (Z_1) is rather small. On the other hand, this resistance is not significantly influenced by voltage and temperature nonlinearities. For voltages up to 100 kV, rise times below 100 ps are possible. The greatest disadvantage is that the measuring time is limited to twice the transient time on the first line. Afterwards multiple reflections occur and the output voltage increases which may cause danger for the measuring apparatus. If the length of the first line is greatly increased, in order to prevent this, impulse attenuation and dispersion may cause severe distortion of the pulse form.

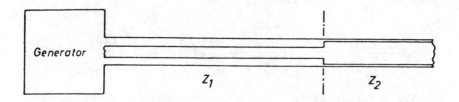

Fig. 2.   Reflection type divider.

In practice, especially for high divider ratios, there is another problem. It is very difficult to realize lines with extremely low characteristic impedance Z_2. If we try to do this by use of a ceramic dielectric [4], this will reduce the limiting hollow waveguide frequency,

$$f_c = \frac{c}{\sqrt{\varepsilon_r}\,(D+d)\,\frac{\pi}{2}} \qquad Z_2' = \frac{Z_2}{\sqrt{\varepsilon_r}} \ . \tag{7}$$

It is, therefore, more advantageous to use many equal lines with regular impedance in parallel [6]. For very low values of Z_2, it may be useful to use a lumped resistor R_2 (Fig. 3). This resistor however must be "zero" inductive (L/R_2). Regarding the terminated measuring line Z_m, the divided voltage is

$$U_2 = U_1 \cdot \frac{2R_2}{Z_1 + R_2 + Z_1 \cdot R_2 / Z_m} \ . \tag{8}$$

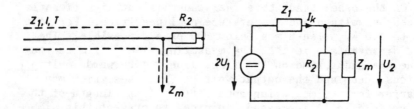

Fig. 3.   Reflection type divider with secondary resistor.

The resistor R_2 must fulfil the same requirements as coaxial shunts, which shall be discussed later. The divider ratio may be as great as 10^3 ... 10^4 and the rise time below 100 ps.

In many measuring problems, however, the very strong interaction of these dividers with the measuring circuit and, additionally, the occurence of multiple reflections cannot be permitted. If we look for a divider with low interaction, this can only be a capacitive one. However, for short rise time, we will only get useful transfer characteristics if the divider capacitances are "zero" inductive and without external connections. In coaxial systems these conditions can be fulfilled if the divider is integrated in the line (Fig. 4) [7-9]. For this reason, only a thin metal foil has to be arranged close to the outer conductor. In connection with the coupling capacitance to the inner conductor, the divider ratio may be greater than 10^3. This ratio, additionally, can be influenced by the dielectric of C_2 (linearly and not with ε_r compared with the reflection type divider). For good transfer characteristics, the axial length of this divider should be shorter than $\lambda_{min}/10$. Neglecting stray capacitances (homogeneous dielectric in C_1 and C_2), the divider ratio can be calculated in a very simple way

$$\frac{U_2}{U_1} = \frac{C_1}{C_1+C_2} = \frac{\ln \frac{D}{D-2t}}{\ln(\frac{D}{D-2t})+\ln(\frac{D-2t}{d})} = \frac{\ln \frac{D}{D-2t}}{\ln \frac{D}{d}} \ . \tag{9}$$

Therefore, the only limitation of the measuring voltage is the dielectric strength of the coaxial system. Neglecting the problem of hollow waveguide resonances, there is no connection between rise time of the divider and voltage amplitude to be measured. This divider combines the advantages of the reflection type divider with essentially no interaction with the measuring circuit. However, in practice, the measurable pulse duration is also limited. It is not possible to have only capacitive loading across C_2. Either the measuring apparatus or the measuring line has a rather low high frequency impedance. Terminating the measuring line at its end with the characteristic impedance would result in a rather low tail time constant

$$\tau_R = Z_m (C_1 + C_2). \tag{10}$$

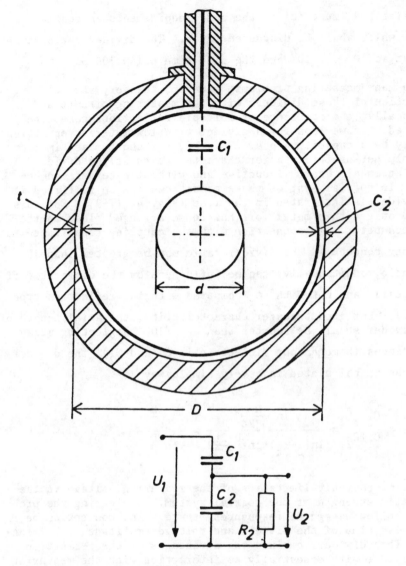

Fig. 4. Coaxial capacitive divider.

In case of a measuring apparatus with a high input resistance,
this can be prevented by a primary termination of the measuring
line (Fig. 5). However, this procedure is only useful for rather
short measuring lines where the cable capacitance can be neglected compared with C_2 [10]. Otherwise there will be a voltage

drop across C_2 during twice the transit time on the measuring

cable, reducing its initial value

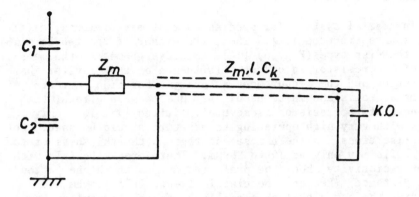

Fig. 5. Capacitive divider with primary termination of the
         measuring line.

$$U_{20} = U_1 \cdot \frac{C_1}{C_1+C_2} \tag{11}$$

to

$$U_{21} = U_1 \cdot \frac{C_1}{C_1+C_2} \cdot e - 2T/(C_1+C_2)2Z_m \tag{12}$$

If this value does not, by chance, coincide with the final voltage value,

$$U_{2\infty} = U_1 \cdot \frac{C_1}{C_1+C_2+C_k} , \tag{13}$$

multiple reflections will occur [11].

     When measuring with extremely short rise times, it has to be
regarded that measuring apparatus usually has 50 Ω input resistance which automatically causes end side termination of the
measuring line and a low tail time constant. This can only be
avoided by use of special high impedance probes. If we are aware
of the fact that in coaxial systems, with low losses, voltage and
current are connected by the ohmic characteristic impedance, then
we could avoid any direct current measurement. On the other
hand, some times it is more advantageous to make a precise current measurement and to avoid a problematic voltage measurement.

3. LOW OHMIC SHUNTS

In coaxial systems for precise current measurements, mainly ohmic shunts are used [12]. These are either of the tubular [13] or of the disk type (Fig. 6) [14]. Besides geometry, the material of the resistor is of great influence on the transfer characteristics and the measurement accuracy. For high currents, only metal foils should be used. As, due to skin effect, the thickness of the resistor is severely limited [15, 16], only alloys with very high operating temperature should be used. For pulse rise times, in the nanosecond region, the thickness of the metal foils may only be 10 to 20 μm. Therefore, materials with higher resistivity should be preferred due to the higher attenuation distance. Besides the classical materials Isaohm and Manganin, the new nickel-chrome-alloys are very advantageous. Vacromium, which shall be considered here, is available in foils of 10 μm thickness, has an attenuation distance of 5.5 μm, at 10 GHz, and only 1,1% change in resistance for a temperature rise from 20 to 200°C.

The transfer characteristic of the coaxial disk shunt is only limited by the influence of skin effect. For the calculation, the resistor is regarded as a line without series resistance and shunt capacitance (Fig. 7). Neglecting the characteristic impedance of the measuring line, compared with the de-resistance of the disk ($I_2 = 0$), we get the measured voltage for x = d.

$$\frac{U_2}{I_1} = R_o \frac{k \cdot d}{\sinh(kd)} \quad R_o = \frac{1}{2\pi kd} \cdot \ln \frac{D_2}{D_1} \; ; \; k = \frac{1+j}{\delta} \; ; \qquad (14)$$

$$\delta = \sqrt{2/(\omega\mu k)} \; ;$$

where k is conductivity

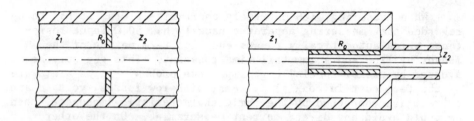

Fig. 6. Coaxial disk or tubular shunts.

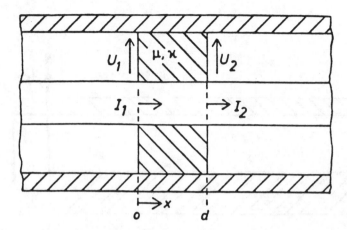

Fig. 7.   Coaxial disk shunt as a line without series resistor and
          shunt capacitance.

Therefore, U_2/I_1 is the mutual impedance (Fig. 8) which is mostly
used to characterize the screening effect [17]. The corresponding upper cut-off-frequency is

$$f_c = 1.46/(\mu\kappa d^2).\tag{15}$$

For a 10 μm Vacromium foil, this will be 12.7 GHz. The step response of such a shunt is described by

$$\frac{Z_{K(t)}}{R_o} = 1+2 \sum_{\gamma=1}^{\infty}(-1)^\gamma \cdot e^{-\gamma^2 t/\tau}; \quad \tau=\mu\kappa d^2/\pi^2; \quad T_a=2.35\tau.\tag{16}$$

This function is approximately exponential without any overshoot.
For a 10 μm Vacromium foil, the rise time T_a is 27 ps. However,
in practice, it is rather troublesome to use coaxial disk shunts.
Especially it is difficult to adapt the resistance value to the
current amplitude to be measured. As the dimensions of the coaxial system generally are fixed, only the thickness of the disk
could be varied. Here, however, we are limited by the skin effect.

     In this respect, coaxial tubular shunts have some advan-
tages, as the length of the tube is an additional parameter in-
fluencing the resistance. However, it has to be kept in mind
that the greatest value of tube length is limited by the necessary upper frequency limit. Let us first neglect the influence
of the tube length and consider the shunt as a lumped element.

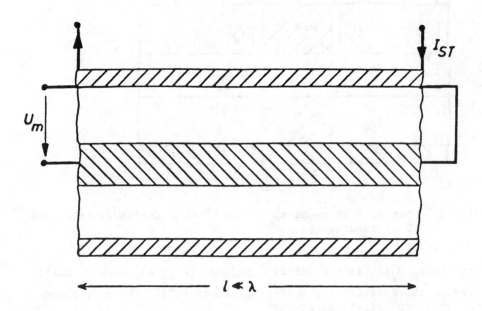

Fig. 8.   Definition of the mutual impedance.

If we further assume that the thickness of the tube d is small,
compared with the radius r_1, then we have the same plane current
displacement problem as for the disk shunt. Therefore, the
transfer function for the mutual impedance also is identical:

$$\frac{Z_k}{R_o} \doteq \frac{k \cdot d}{\sinh(kd)} .$$

(17)

The upper frequency limit can be calculated from Eq. (15)
and the step-response rise time from Eq. (16). However, this can
only be verified if a seamless tube is used. Only a small axial
gap (Fig. 9) may greatly influence the transient characteristics
and even cause overshoot [18]. This is caused by the punch-
through of the magnetic field, inducing an additional voltage in
the measuring circuit. This effect is increased with the gap
ratio b/d. Therefore, especially for thin foils which are necessary for short rise times, great care is required [13]. The

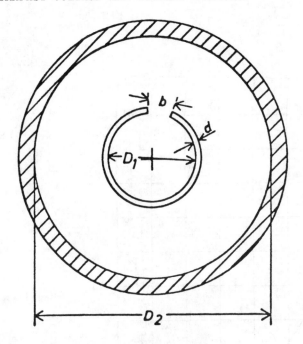

Fig. 9.   Coaxial tubular shunt with axial gap.

Nyquist plot of the normalized mutual impedance for different gap
ratios is shown in the next figure (Fig. 10). It is obvious
that, especially for thin foils, such gaps are strictly forbid-
den.

        The transfer characteristics can also be degraded by exces-
sive tube length. In some cases, skin-effect may even be negligible. For the calculation of the influence of the tube length
on the transfer characteristics, the shunt has to be regarded
as a short coaxial line (Fig. 11). We assume that the characteristic impedance Z_1 of the outer line, formed by tube and outer

conductor, is matched with the source impedance. Also, the
impedance of the inner line Z_2, formed by tube and voltage pick-

off, is matched with the impedance of the measuring circuit.
Assuming that in both lines, an air dielectric is used, the
normalized mutual impedance depends from the wavelength as
follows [19]

$$\frac{Z_k}{R_o} = \frac{\lambda}{4\pi l} \cdot \frac{\sin 2\pi \frac{1}{\lambda} + 2\pi \frac{1}{\lambda} \cos 2\pi \frac{1}{\lambda}}{(\cos 2\pi \frac{1}{\lambda} + j \cdot \sin 2\pi \frac{1}{\lambda})^2} \cdot \qquad (18)$$

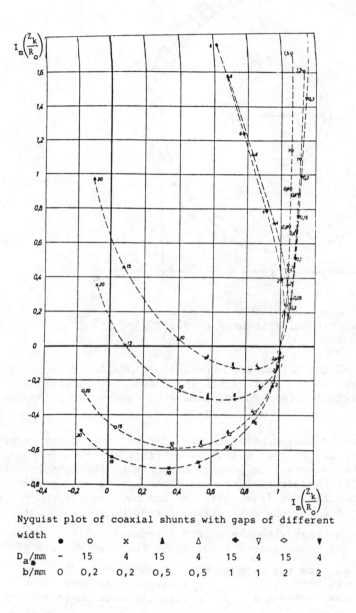

Nyquist plot of coaxial shunts with gaps of different width

width	●	○	x	▲	△	◆	▽	◇	▼
D_a/mm	–	15	4	15	4	15	4	15	4
b/mm	0	0,2	0,2	0,5	0,5	1	1	2	2

Fig. 10. Nyquist plot of coaxial shunts with gaps of different width.

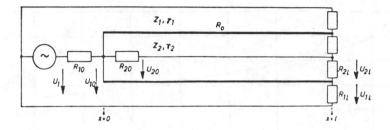

Fig. 11.   Coaxial tubular shunt as a line without losses.

The upper cut-frequency is

$$f_c(GHz) = 4.641 \ (cm).$$ (19)

If, in the voltage pick-off, a dielectric with $\varepsilon_r > 1$ is used, the results are slightly reduced (Fig. 12). If we regard the example of the 10 μm Vacromium foil and $\varepsilon_r = 3.7$ in the voltage pick-off (Fig. 13), we can see that, for tube lengths greater than 1.5 cm, the transfer characteristics are only deter- mined by the influence of tube length. On the other hand, we see that, in this case (extremely thin foil), short tubular shunts can have the same excellent transfer characteristics as disk shunts.

Another problem of shunts with thin foils is the maximal loading. In short time intervals, when we can neglect power dissipation, this is determined by the thermal capacity of the shunt. If we allow a temperature increase from 20 to 200°C the loading of a 10 μm Vacromium foil is as follows,

$$\int_0^T i^2 dt / A^2 \cdot s = 0.623 \ (D_1(mm))^2 \ .$$

Thus, with a tube diameter of 15 mm, a rectangular current pulse of 1 μs duration, an amplitude of 11.85 kA can be measured with 1 % error, due to temperature increase. For measurements in the ns-range, this maximal loading always seems to be sufficient.

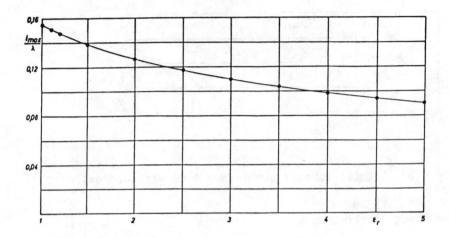

Fig. 12.   Cut-off wavelength versus dielectric constant in the
           voltage pickoff.

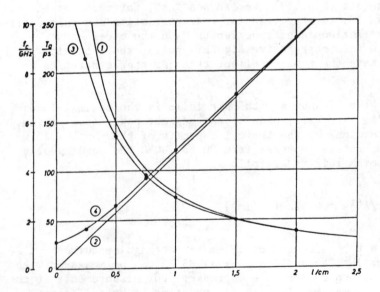

Fig. 13.   Cut-off frequency and rise time of coaxial tubular
           shunts without (1,2) and with skin effect (3,4).

4.   NANOSECOND AND SUBNANOSECOND IMPULSE MEASUREMENTS

Let us now regard the problems connected with the recording of fast transient phenomena [20]. Oscilloscopes used for this purpose both must have extremely high writing speed and short rise time (Fig. 14). Also excellent screening is necessary. In real time oscilloscopes, travelling wave cathode ray tubes are necessary together with special techniques for image storing and image enhancement. Presently, the highest cut-off-frequency is 1 GHz and the highest writing speed is 30 cm/ns for the scan converter [21] and about 3 cm/ns for a transfer storage tube.

If periodic events, with sufficient repetition rates have to be measured, sampling oscilloscopes can be used [22]. Due to their high sensitivity, this is especially useful if calibration of the measuring circuit is to be done. If a delay line is necessary, the upper frequency limit is about 2 GHz. Without a delay line, up to 14 GHz, is possible [23]. No delay line is necessary if the pulse generator used has a pretrigger output, if the random sampling technique can be used [24], or the pulse

Type	rise time	sensitivity	special features
Travelling wave tube	0,15 ns	10 V/div	high focus
same, but magnifying lense	0,15 ns	2 ... 4 V/div	medium focus
same, but with vertical amplifier	0,3 ns	10 mV/div	reduced interference suppression
sampling, with delay lne	0,15 ns	2 mV/div	repetition rate > 10 Hz noise ≈ 2 mV
same, but without delay line	25 ps	2 mV/div	pretrigger necessary noise ≈ 2 mV
random sampling	25 ps	2 mV/div	repetition rate > 10 KHz noise ≈ 5 mV

Fig. 14.   Comparison of high-speed oscilloscopes.

generator can be triggered (Fig. 15) with no jitter. The last
question mainly depends upon the switching element which is used
in the pulse generator. Unfortunately, the mercury wetted relay
[25] which, due to its high output amplitude, is very useful for
calibration measurements, can only be used in connection with
delay lines (Fig. 16). The triggerable switch, with the shortest
rise time, is the tunnel diode (T_a < 25 ps). This switch supplies rather low impulse amplitude and has significant distortion.

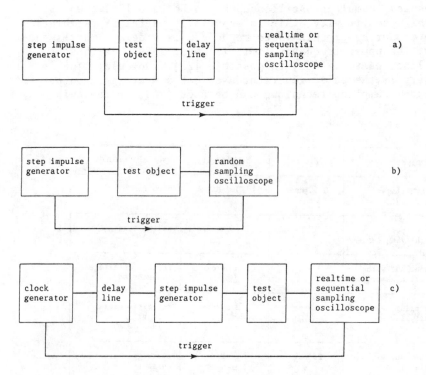

Fig. 15.   Test circuits for step response measurements;
           a) delay line in the signal path,
           b) random sampling,
           c) delay line in the trigger path.

switch	rise time	amplitude	special features
mercury wetted relais	< 100 ps	1 kV	no pretrigger available, repetition rate < 200 Hz
avalanche transistor	150 ps	50 V	pretrigger available
step recovery diode	50 ps	20 V	pretrigger available
tunnel diode	20 ps	0,5 V	pretrigger available

Fig. 16.   Comparison of step impulse generators.

In many cases, for calibration, not only the step response
has to be measured but also the matching of the measuring apparatus with the circuit must be determined. This is most simply
done in the time domain. For this reason, a pulse reflectometer
is used. Therefore, not only the reflection factor, but the
location of the mismatching can be analyzed. The amplitude
and spatial resolution greatly depend upon the circuit used
(Fig. 17). The smallest measurable reflection factors are about
0.1 %. If a lower amplitude resolution (1 %) is sufficient, a
temporal resolution of 35 ps can be obtained. Lumped disturbances with only 1 cm spatial separation can, therefore, be located separately.

Now let us give some examples for the calibration and application of coaxial voltage or current measurement systems. In
case of a coaxial, capacitive, high voltage divider, adapters for
feeding the line with the low voltage step impulse and for termination are necessary (Fig. 18). The divider was constructed
using a 50 μm mylar-foil (Polyäthylenterephtalat) with an axial
length of 1 cm. An upper cut-off frequency of at least 3 GHz
can, therefore, be expected. If we measure the step response,
using a step impulse generator with a 120 ps rise time, we can
see that the rise time of the divider must be significantly
smaller than this value (Fig. 19). There is no overshoot, the
divider ratio is ~ 900, and the tail time constant is about 60 ns
for a 50 Ω load. If we try to make an exact measurement of the
divider's specific rise time, using a step inpulse generator of
only 60 ps rise time, we can see that, even in this case, the
step impulse response has approximately the same rise time. However, due to the big diameter of the coaxial high voltage line
[26], oscillations of the lowest hollow waveguide frequency are
generated (Fig. 20). In practice this effect limits the upper

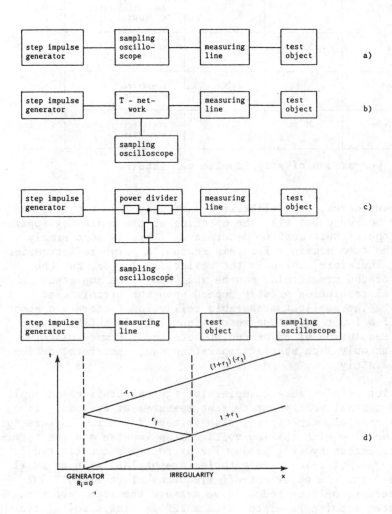

Fig. 17.   Different impulse reflectometer circuits; a) sampling
           oscilloscope with feed through input, b) T-branching,
           c) branching without reflections, d) step impulse gen-
           erator with very low output resistance (short circuit
           at the feeding end).

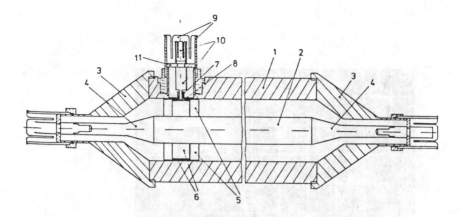

Fig. 18. Coaxial capacitive divider with adapters
    1 - outer conductor   6 - metal foil
    2 - inner conductor   7 - damping device
    3 -   8 - screening
         cone adapter
    4 -   9 - 11 - connectors
    5 - dielectric   (GR-874,50Ω)

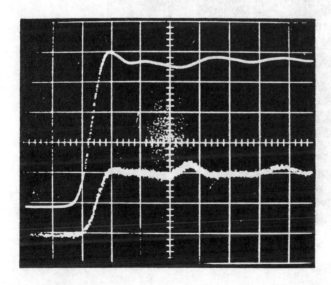

Fig. 19. Input signal and response of a coaxial capacitive
    divider; rise time of the pulse generator T_{al} - 120 ps;
    vertical scale: 2 V/div (upper), 5 mV/div (lower);
    time scale: 200 ps/div.

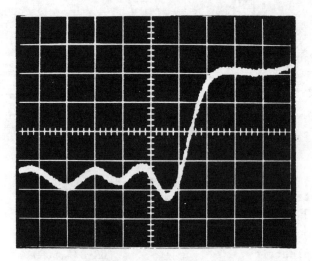

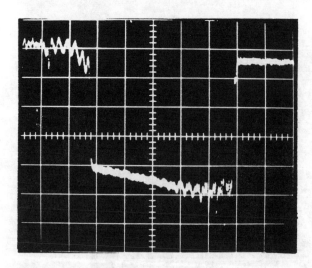

Fig. 20. Step response of a coaxial capacitive divider; rise
          time of the pulse generator T_{a1} = 60 ps;
          vertical scale: 5 mV/div;
          time scale: 2 ns/div (upper)
                       50 ps/div (lower)

frequency of such measuring devices in big coaxial systems.
Of course, if we record the step response with a rather low
frequency oscilloscope (1 GHz) it will be strictly aperiodic
(Fig. 21).

     One example for the application of such dividers is the
measurement of breakdown processes in compressed gas insulation
[27]. Especially in compressed SF_6, the voltage change is extremely fast and requires such a measuring apparatus (Fig. 22).
We can see that, in the range of parameters investigated here,
two different mechanisms of voltage breakdown occur. The higher
breakdown voltage belongs to the extraordinary steep voltage
collapse, whereas the lower breakdown voltage is accompanied by a
slower, quasi-exponential voltage collapse. Another example is
the investigation of the distribution of the total time lag of
highly overvolted gaps in air (Fig. 23). We can see both, the
full wave of the high voltage square wave generator (T = 4 ns)
and a breakdown after a very small delay time.

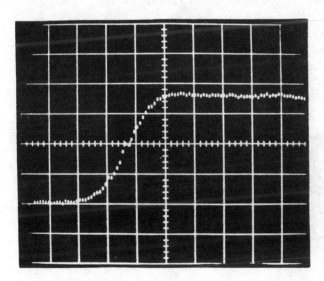

Fig. 21.   Step response of a coaxial capacitive divider; limited
           rise time of the sampling oscilloscope T_{ao} < 0.4 ns;
           vertical scale: 89 mV/div; time scale: 200 ps/div.

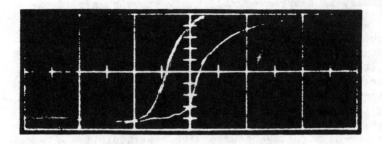

Fig. 22.   Breakdown in compressed SF_6; gap width

s = 5 mm; gas pressure = 4 bar;
vertical scale: 70.5 kV/div;
time scale: 2 ns/div.

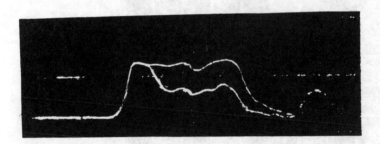

Fig. 23.   Breakdown of an extremely high overvolted air gap;
s = 1.625 mm; p = 1 bar;
vertical scale: 39.25 kV/div;
time scale: 2 ns/div; total time
lag: $t_s + t_f = 0.62$ ns;
overvoltage factor: $f_s = 5.73$

The calibration of a short-circuit current measurement system, using a tubular shunt, is shown in Figure 24. In this case, only one adapter for feeding a low amplitude current impulse, which indeed is a low voltage impulse across the ohmic characteristic impedance, is necessary. In this case, it is very difficult to get good matching of the outer line to the source impedance and, especially, of the inner line to the impedance of the measuring circuit. Both should be matched to 50 Ω. This is checked with an impulse reflectometer (Fig. 25). We can see that, in both lines, some mismatching occurs.

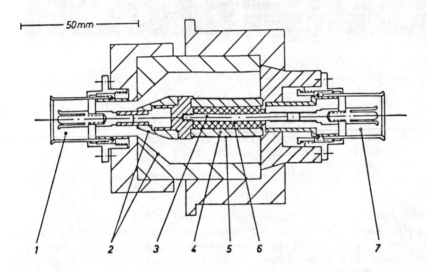

Fig. 24.   Coaxial tubular shunts: Outline of the test arrangement

1 - input plug
2 - cone adapter
3 - voltage pickoff
4 - ceramic support

5 - metal foil
6 - epoxy sealing
7 - output plug

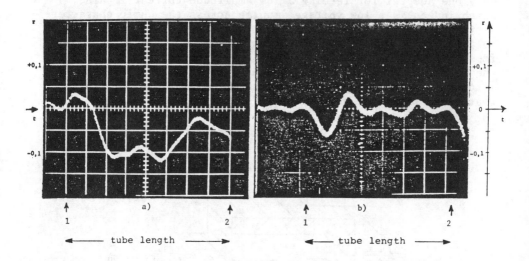

Fig. 25.   Input - (a) and output (b) reflection factors of a
           tubular coaxial shunt; R_o = 52 mΩ;

           vertical scale: 0.05 r/div; time scale: 50 ps/div.

We know that the step response rise time, for a 10 μm
Vacromium foil, is mainly influenced by the tube length.   Therefore, for an approximately 4 cm tubular shunt, we measure a
rather long rise time, however, with strictly aperiodic shape
(Fig. 26).   However, a very short tubular shunt can have extraordinary short rise time (70 ps), as it is shown in the next
oscillogram (Fig. 27).   This is also proven by the superimposed
ringing due to the lowest, hollow, waveguide frequency of the
outer line.

It is also possible to make transfer measurements of the
shunts in the frequency domain.   This is first done for the long
shunt and the result, as far as the cut-off-frequency is concerned, is in good agreement with the measured rise time (Fig.
28).   However, the frequency response alone does not say very
much about the transfer characteristics in the time domain.   Of
course, severe deviations from the regular response, as they
occur, because of the influence of a thin axial gap, especially
in tubes of small diameter (4 mm), can also be detected in the
frequency domain.   However, it requires great experience

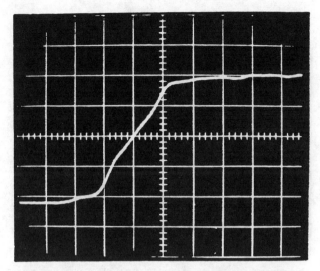

Fig. 26. Step response of a rather long coaxial tubler shunt;
          R_o = 123 mΩ; 1 = 38 mm; rise time of the pulse
          generator T_{al} = 60 ps; vertical scale: 20 mV/div;
          time scale: 200 ps/div.

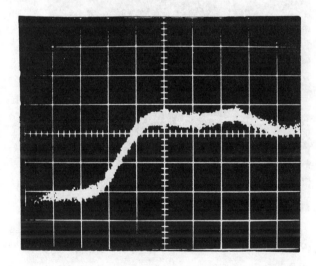

Fig. 27. Step response of a very short coaxial tubular shunt;
          R_o = 8.6 mΩ; 1 = 5 mm; rise time of the pulse generator
          T_{al} = 60 ps;
          vertical scale: 2 mV/div;
          time scale: 50 ps/div.

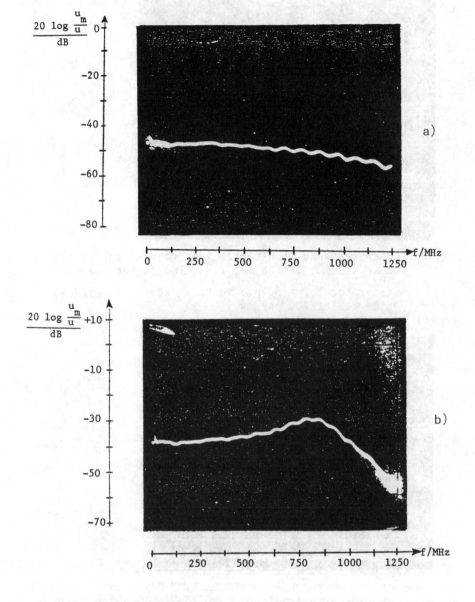

Fig. 28.   Frequency response of coaxial tubular shunts;
          a) R_o = 123 mΩ; 1 = 38 mm
          b) R_o = 295 mΩ, excessive axial gap.

to get some idea of the step response which might result from
this.

We can use such shunts in order to measure the short-circuit
current of high-voltage surge generators. From the first oscillogram we can see that, even in highly compressed nitrogen, it
is extremely difficult to get current pulses with rise times
below 1 ns. However, it is shown that the electrode capacitance
already has some effect on the rise time (Fig. 29). If we,
therefore, virtually increase this capacitance by a very small
distributed capacitance, just in front of the high-voltage electrode, we can decrease the system rise time significantly (Fig.
30).

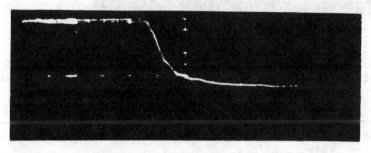

a) electrode radius
   r = 2 mm

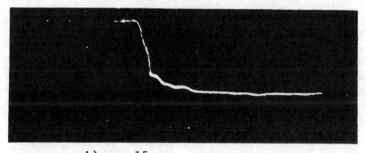

b) r = 15 mm

Fig. 29.   Discharge current in compressed N_2; s = 1.07 mm;
           p = 21 bar;
           vertical scale: 695 A/div;
           time scale: 2ns/div.

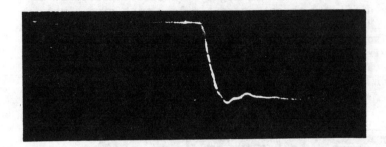

Fig. 30.   Discharge current in compressed N_2; pulse forming by a
           shunt capacitor near the high voltage electrode;
           $r = 15$ mm; $l_c = 44$ mm (dielectric insert);
           vertical scale: 695 A/div;
           time scale: 2 ns/div.

REFERENCES

1.   A. Schwab, "Hochspannungsmeßtechnik", 2. Auflage, Springer-
     Verlag, Berlin 1982.

2.   M. Di Capua, "Electrical Voltage and Current Monitors", see
     Chapter 2 this volume.

3.   W. Pfeiffer, "Impulstechnik", Carl Hanser Verlag, München
     1976.

4.   R. C. Fletcher, "Production and Measurement of Ultra-High
     Speed Impulses", Rev. Sci. Instr. Vol. 20, p. 861-869
     (1949).

5.   D. G. Pellinen, Q. Johnson, A. Mitchell, "A Picosecond Rise
     Time High Voltage Divider", Rev. Sci. Instr. Vol. 45,
     p. 944-946 (1974).

6.   R. J. Thomas, "High-voltage Pulse Reflection-type Attenua-
     tors with Subnanosecond Response", IEEE Trans. Instrum. a.
     Meas. Vol. IM-16, p. 146-154 (1967).

7. D. F. McDonald, C. J. Benning, "Subnanosecond Rise Time
    Multikilovolt Pulse Generator", Rev. Sci. Instr. Vol. 36,
    p. 504-505 (1965).

8. F. Wesner, "Kapazitiver Spannungsteiler zur Messung Rascher
    Spannungsänderungen an Koaxialen Kabelanordnungen", ETZ-A
    Vol. 92, p. 633-636 (1971).

9. W. Pfeiffer, "Aufbau and Anwendung Kapazitiver
    Spannungsteiler Extrem Kurzer Anstiegszeit fur Gasisolierte
    Koaxialsysteme", ETZ-A Vol. 94, p. 91-94 (1973).

10. W. Zaengl, "Ein Beitrag zur Schrittantwort Kapazitiver
    Spannungsteiler mit Langen Meßkabeln", ETZ-A Vol. 98,
    p. 792-795 (1977).

11. W. R. Fowkes, R. M. Rowe, "Refinements in Precision Kilovolt
    Pulse Measurements", IEEE Trans. Instr. Meas. Vol. IM-15,
    p. 284-292 (1966).

12. A. Schwab, "Low-resistance Shunts for Impulse Currents",
    IEEE Trans. Pow. App. Syst. Vol. PAS-90, p. 2251-2255
    (1971).

13. W. Pfeiffer, "Messung von Impulsströmen Extrem Kurzer
    Anstiegszeit in Koaxialen Rohrsystemen", ETZ-A Vol. 95,
    p. 297-302 (1974).

14. F. Wesner, "Koaxiale Flächenwiderstände zur Messung hoher
    Stoßströme mit Extrem Kurzer Anstiegszeit", ETZ-A Vol. 91,
    p. 521-524 (1970).

15. F. D. Bennett, J. W. Marvin, "Current Measurement and
    Transient Skin Effects in Exploding Wire Circuits",
    Rev. Sci. Instr. Vol. 33, p. 1218-1226 (1962).

16. G. Hortopan, V. Hortopan, "Eine allgemeine Theorie der
    Meßwiderstände im Übergangszustand", ETZ-A Vol. 92,
    p. 470-474 (1971).

17. I. L. Hoeft, "Shielding of Cables and cable trays", pre-
    sented at NATO Aug, 1983, Fast Electrical and Optical Diag-
    nostics.

18. H. Kaden, "Wirbelströme und Schirmung in der Nach-
    richtentechnik", Springer Verlag, Berlin 1959.

19. H. Jungfer, "Die Messung des Kopplungswiderstandes
    von Kabelabschirmungen bei hohen Frequensen", NTZ Vol. 9,
    p. 553-560 (1956).

20. N. S. Nahman, "The Measurement of Baseband Pulse Rise Times
     of Less Than 10^{-9} Second", Proc. of the IEEE Vol. 55,
     p. 855-864 (1967).

21. R. Hayes, "A Silicon Diode Array Scan Converter for High-
     speed Transient Recording", IEEE Trans. El. Dev. Vol. ED-22,
     p. 930-938 (1975).

22. R. Sugarman, "Sampling Oscilloscope for Statistically
     Varying Pulses", Rev. Sci. Instr. Vol. 28, p. 933-938
     (1957).

23. W. M. Grove, "Sampling for Oscilloscopes and Other RF Sys-
     tems: DC Through X-band", IEEE Trans. Micro. Th. Techn.
     Vol. MTT-14, p. 629-635 (1966).

24. G. J. Frye, N. S. Nahman, "Random Sampling Oscillography",
     IEEE Trans. Instr. Meas. Vol. IM-13, p. 8-13 (1964).

25. F. Meyer, "Analyse von Impulsflanken mit einer Zeitauflösung
     von 12 Picosekunden", AEÜ Vol. 27, p. 19-24 (1973).

26. E. Flad, "Die Verformung extrem kurzer Impulsflanken an
     Störstellen in Leitungen mit übergropem Querschnitt",
     Arch. Elektrotechnik Vol. 55, p. 320-329 (1973).

27. W. Pfeiffer, "Ultra-high-speed Methods of Measurement for
     the Investigation of Breakdown Development in Gases", IEEE
     Trans. Instr. Meas. Vol. IM-26, p. 367-372 (1977).

HIGH SPEED ELECTRIC FIELD AND VOLTAGE MEASUREMENTS

Marco S. DiCapua

Lawrence Livermore National Laboratory
Livermore, CA
USA

1.   INTRODUCTION

Electric field and voltage measurements are necessary, in conjunction with magnetic field and current measurements, to complete the specification of the electromagnetic field environment in pulsed power apparatus.

In this section, we introduce the concept of displacement vector ($\vec{D}$) measurements with floating electrodes, we discuss measurements with resistively grounded electrodes, we explore how measurements of voltage and electric field are related, and we introduce voltage measurements with resistive dividers. We then address voltage measurements with resistive attenuators in 50 Ω systems.

We then discuss the concept of voltage measurements with probes and measurements in the presence of time varying magnetic fields and how the voltage measured depends on the path of the measuring circuit. Our next task is to discuss the response of resistive dividers to time varying fields.

Finally, we discuss measurements in pulsed power apparatus in the presence of propagating waves and corrections that are required to compensate for transit time delays. In the last section we explore the application of carbon composition resistors to pulse power measurements.

Several useful articles, of a more general nature, on high speed field and voltage measurements have appeared in the

literature. Baum [1] discusses methods and definitions for field
measurements in a lightning environment. Thomas [2] takes a
practicing engineer viewpoint of voltage and current measurements. Craggs [3] provides an excellent review of classical
voltage measuring techniques, current to this day, including
voltage calibration scales using the threshold of nuclear reactions. A National Bureau of Standards publication provides a
broad overview of measurements in pulsed power systems [4]. We
will reference this last publication often in this article.

1.1 Voltage Measurements in Transmission Lines

The subject of voltage measurements in transmission lines is
critically important for pulse power applications, since transmission lines are used in pulsed power apparatus to form the
electromagnetic energy pulse, deliver it to the load, as well as
to transmit signals from diagnostic devices to metering apparatus. In this chapter, we will consider measurements on transmission lines that propagate TEM waves, where both the electric
field and magnetic field vectors lie in a plane perpendicular to
the axis of propagation, normally the longitudinal axis of the
pulsed power apparatus. For such waves, Ref. [5] shows that the
transverse electric field distribution at a given axial position
is the gradient of a scalar potential that is independent of frequency and is identical to the static field configuration for the
same conductor geometry.

As a consequence of this very important result, we will
treat pulse power field and potential measurements as if they
were performed under electrostatic conditions. Therefore, the
transmission line and static field measurements are identical.
At the end of the chapter, however, we will discuss the impact of
time dependence on the interpretation of the measured quantities
and the conditions leading to the presence of non-TEM waves.

2.   POTENTIAL AND DISPLACEMENT VECTOR MEASUREMENTS WITH FLOATING
      ELECTRODES

2.1 Some General Considerations

The concept of potential measurements with a floating electrode may be introduced by considering an arbitrary disposition
of conductors as shown in Fig. 2.1, where conductors 0 and 1 give
rise to the field and conductor A represents the sensing element.
The size of A, of course, is grossly exaggerated since the most
desirable configuration is that which disturbs the field the
least.

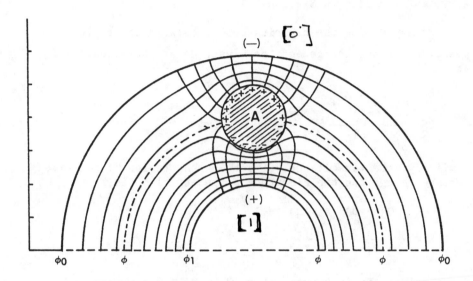

Fig. 2.1.   Electric field measurement with a floating electrode.
A is isolated from conductor 0 and 1 and attains potential ϕ. Net charge on A is zero.

We notice that A will float to a potential ϕ that lies in
between ϕ_0 and ϕ_1. Since there is no physical connection between
A and either electrode, and we assume A does not emit or collect
charge from the surrounding medium, the net charge on A is zero.
As the figure shows, however, there are regions of positive and
negative surface change on A and, in the case of the two dimensional geometry shown, these regions are divided by the equipotential surface corresponding to the floating potential of A.

Our discussion in Section 1.1 revealed that the local electric field in a transmission line can be derived as a scalar potential that is independent of frequency and identical to the
static field configuration for the same conductor geometry. When
the floating electrode does not <u>disturb</u> the equipotentials originally present in the transmission line, it reaches instantaneously a fraction of the potential difference across the transmission
line corresponding to the floating potential of A. Its response
time, theoretically, is the time required to redistribute the
charges on A.

This form of divider forms part of a more general class of dividers called geometric dividers since its division ratio is ideally independent of frequency and depends only on the geometry, as discussed in Section 1.1.

We may define the capacitance C_b, between the region of positive surface charge and the electrode at the potential ϕ_0, as

$$C_b = \frac{\int\limits_{A_+} \vec{\sigma}_+ \cdot d\vec{A}_+}{\phi - \phi_0} = \frac{Q_+}{\phi - \phi_0}.$$

Similarly, the capacitance, C, between the region of negative surface charge and the electrode at potential ϕ_1 is,

$$C = \frac{\int\limits_{A_-} \vec{\sigma}_- \cdot d\vec{A}}{\phi - \phi_1} = \frac{Q_-}{\phi - \phi_1}.$$

Since A is an isolated conductor, $Q_+ + Q_- = 0$ and,

$$\int\limits_{A_+} \vec{\sigma}_+ \cdot d\vec{A}_+ + \int\limits_{A_-} \vec{\sigma}_- \cdot d\vec{A}_- = 0 .$$

Consequently, $C_b(\phi - \phi_0) + C(\phi - \phi_1) = 0$.

The quantity ϕ then becomes,

$$\phi = \frac{C_b\phi_0 + C\phi_1}{C + C_b} .$$

Since we are interested only in differences in potential, we may arbitrarily set $\phi_0 = 0$ and obtain the desired result

$$\phi = \frac{\phi_1}{1 + \dfrac{C_b}{C}} .$$

This result, which is very well known from electrostatics, forms the basis of voltage measurements with capacitive dividers.

2.2 Measurements With A Floating Electrode Closely Coupled To Ground

When $C_b \gg C$, the floating electrode is closely coupled to ground. In this case, of great practical importance, we note that

$$\phi = \frac{C}{C_b} \phi_1 \ .$$

Subsequently $\phi \ll \phi_1$, $\phi-\phi_1 \simeq \phi_1$, and

$$C = \frac{\int_{A_-} \vec{\sigma}_- \cdot d\vec{A}_-}{-\phi_1} \ .$$

By Gauss's law, $\vec{\sigma}_- = -\vec{D}_{inc}$ and, by assuming $\vec{D}$ to be uniform over the surface A_-, we may approximate

$$\int_{A_-} \vec{\sigma}_- \cdot d\vec{A} \simeq -\vec{D}_{inc} \cdot \vec{A}_{eq} \ .$$

If we define the potential ϕ_1 to be equal to the product of an average electric field, $\vec{E}_{inc}$, times the distance ℓ_{eq} separating A and conductor 1,

$$\phi_1 = -\vec{E}_{inc} \cdot \vec{\ell}_{eq}$$

and, assuming there is a linear relationship between $\vec{D}$ and $\vec{E}$, we define C as [1],

$$C = \frac{\vec{D}_{inc} \cdot \vec{A}_{eq}}{\vec{E}_{inc} \cdot \vec{\ell}_{eq}} = \frac{\varepsilon \vec{A}_{eq}}{\vec{\ell}_{eq}} \ .$$

Hence,

$$\phi_1 = \frac{-\vec{D}_{inc} \cdot \vec{A}_{eq}}{C}$$

$$\phi = \frac{-\vec{D}_{inc} \cdot \vec{A}_{eq}}{C_b} \ .$$

In general, we construct the sensor in such a way that the quantity $\vec{A}_{eq}/C_b$ is fixed and obtainable by calibration. Hence, the potential the sensor attains is directly proportional to the dot product of its equivalent area, $\vec{A}_{eq}$, and the incident displacement vector, $\vec{D}_{inc}$.

Since

$$\phi_1 = \frac{C_b}{C} \, \phi \, ,$$

the measurement of ϕ, in principle, should be sufficient to determine ϕ_1. There are some practical difficulties, however, because the quantity C depends on how the sensor is installed and the field configuration in its vicinity. Consequently, to obtain ϕ_1, we require a calibration of the sensor in its environment or some accurate determination of the field geometry in the region between the sensor element and the high voltage conductor.

2.3 Measurements with a Floating Electrode Closely Coupled to The High Voltage Electrode ($C_b \ll C$)

In this case $\phi \sim \phi_1$, since the floating electrode is closely coupled to the high voltage electrode. Consequently

$$C_b = \frac{\int\limits_A \vec{\sigma}_+ \cdot d\vec{A}}{\phi_1}$$

and since,

$$\int\limits_{A+} \vec{\sigma}_+ \cdot d\vec{A}_+ + \int\limits_{A-} \vec{\sigma} \cdot d\vec{A}_- = 0,$$

$$\phi_1 = \frac{\vec{D}_{inc} \cdot \vec{A}_{eq}}{C_b} \, .$$

Is is evident that this arrangement is not very practical for measurement purposes since the potential of the sensing element reaches the potential of the high voltage electrode. Practical

designs, therefore, require a resistor to hold the sensing electrode close to ground potential. We describe the role of these
resistors, that perform also a metering function, in the following section.

2.4 Metering of Floating Sensors with Grounding Resistors

2.4.1 Self integration sensors. The sensors we have described
in the previous sections require, for practical purposes, a
metering resistor to measure the potential ϕ of the conductor A.
If we connect the conductor A to ground, with a metering resistor
R_b, we obtain the circuit shown in Fig. 2.2. For this circuit,

$$\phi_1 = \frac{1}{C} \int i \, dt + \phi, \quad i = \int \dot{D}_{inc} \cdot d\vec{A},$$

$$i = C_b \frac{d\phi}{dt} + \frac{\phi}{R_b} \,,$$

$$\text{then } \phi_1 = \frac{1}{R_b C} \int_o^t \phi \, dt' + \left(\frac{C_B}{C} + 1\right) \phi \,.$$

We can easily distinguish two cases as before. In one, $\frac{C_b}{C} \gg 1$
and

$$\frac{d\phi_1}{dt} = \frac{\phi}{R_b C} + \frac{C_b}{C} \frac{d\phi}{dt} \,.$$

The value of $R_b C$, relative to $\frac{C}{C_b} t_c$, where t_c is a characteristic time associated with changes in ϕ_1, affords two options.

If $R_b C \gg \frac{C}{C_b} t_c$, then $R_b C_b \gg t_c$ and

$$\frac{d\phi_1}{dt} = \frac{C_b}{C} \frac{d\phi}{dt}$$

or:

$$\phi_1 = \frac{C_b}{C} \phi \,.$$

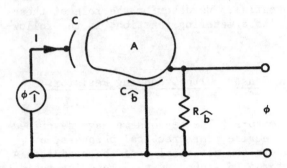

Fig. 2.2.   Floating sensor circuit diagram. C is capacitance
            coupling sensor element to high voltage electrode
            while C_b is capacitance coupling element to ground.

            R_b is the metering resistor.

The monitor, in this case, is called either self integrating or
a capacitive divider since it delivers an output voltage directly
proportional to the input voltage. Moreover, for $R_b C_b \gg t_c$,
the divider is geometric since its division ratio is independent
of frequency.

     As an example, consider the monitor that appears in Fig.
2.3. The sensor element is a prismatic cylinder that is nested
in a grounded cylindrical well. The sensor element connects
to ground through the resistor R_b which may also be part of a
secondary divider. The sensor element faces the high voltage
electrode and we assume that the dielectric is water. For the
approximate dimensions shown in the figure

$$C = \frac{\pi r^2 k \varepsilon}{d} \simeq 9 \times 10^{-12} \text{ F} ,$$

where:

$r = 2 \times 10^{-2} \text{m}$

$k = 81$ (relative dielectric constant)

$\varepsilon = 8.86 \times 10^{-12} \text{ F m}^{-1}$

$d = 10$ cm .

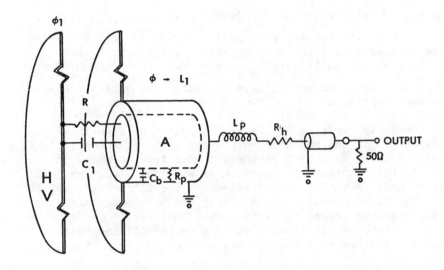

Fig. 2.3. Capacitive monitor schematic. High voltage (HV) conductor is at potential ϕ_1. C_1 is the capacitance
coupling the sensing element A and HV. C_2 is the
burden capacitance between the sensing element and
ground. R and R_p are the parasitic resistances
arising from the finite resistivity of the dielectric.
L_p is the parasitic inductance associated with the
flux in the annular space between A and ground. It
may be neglected for most applications. The burden
resistence $R_b = R_h + 50\ \Omega$ also forms a secondary
divider. Section 2.6 discusses the effects of the
parasitic resistances.

The value of C_b can be calculated as

$$C_b \sim 2 \times \frac{2\pi r h}{d_{mon}} k_\varepsilon = 1.8 \times 10^{-9}\ F$$

where $h = 5 \times 10^{-2}$m, $d_{mon} = 5 \times 10^{-3}$m and the factor of 2 takes
into account the capacitance between the prismatic cylinder and
the bottom of the well.

For this example, $C_b/C \sim 200$ and a voltage of $\phi_1 = 2.5$ MV, on the high voltage electrode, delivers a voltage $\phi = \frac{C}{C_b} \phi_1 = 1.25 \times 10^4$ V at the sensing electrode. It is, therefore, necessary to further reduce the voltage with a secondary divider that can be easily built into R_b. If we take $R_b = 2500$ Ω, $R_b C_b = 4.5$ μs is a very convenient decay constant for water insulated pulsed power apparatus.

A commercial version of the self-integrating capacitive monitor described in the above example appears in Reference [6]. It is designed to measure high potentials of either polarity in a high dielectric constant medium such as water or glycerine. The sensor element is the top-most cylinder. The flange attaches the monitor to the accelerator. The box at the bottom contains the metering resistor/secondary divider that connects to the recording oscillograph or digitizer through a cable attached to the BNC connector. Smaller versions of this monitor are also available [6].

Reference [7] describes a self-integrating sensor for free field measurements. A cross section of the sensor and its schematic diagram appears in Figs. 2.4 (a) and (b), respectively. The sensor element behaves as a current source $I = -A_{eq} D = -A_{eq} \omega \varepsilon_o E$. C_p represents the capacitance between the sensor element and the ground plane. C_L is a loading capacitor that improves the low frequency response while R_L and the 50 Ω signal lead form a resistive divider. C_c compensates for the stray capacitance C_s across R_L. The resistive divider matches the high impedance metering resistor required for good low frequency response to the 50 Ω signal lead.

Capacitive dividers have also been developed to measure fast, high voltage pulses in 50 Ω systems. Pfeiffer describes [8] a subnanosecond rise time divider and a nanosecond rise time divider [9] for coaxial cables which is available commercially. Fujimoto [10] discusses the application of coaxial dividers to detect partial discharges in gas insulated switch gear.

2.4.2 Differentiating sensors. A second possibility is to have

$$R_b C \ll t_c \frac{C}{C_b} \, .$$

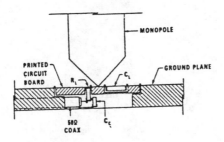

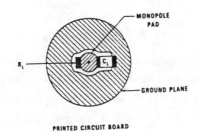

(a) Cross section of sensor.

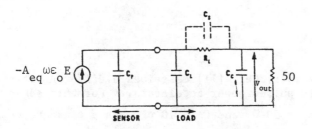

(b) Circuit diagram.

Fig. 2.4. Self integrating sensor for free field measurements
[7]. The cross section in (a) shows the disposition of
the circuit elements in the sensor. The circuit diagram appears in (b). The sensor element behaves as a
current source $-A_{eg} D = -A_{eq} \omega \varepsilon_o E$. C_p represents the
capacitance of the sensor while C_L is a loading capaci-
tor to improve the low frequency response. R_c and 50 Ω
form a resistive divider and C_c is used to compensate
for the stray capacitance, C_s across R_L. Sensor response is flat between 10 MHz and 1 GHz.

In this case

$$R_b C_b \ll t_c$$

and the equation $\dfrac{d\phi_1}{dt} = \dfrac{\phi}{R_b C} + \dfrac{C_b}{C} \dfrac{d\phi}{dt}$ becomes

$$\frac{d\phi_1}{dt} = \frac{\phi}{R_b C} \ .$$

Consequently, the monitor is differentiating.

The monitoring arrangement of Fig. 2.3 could be used in the differentiating mode by making R_b very small. As an example, $R_b = 1\ \Omega$ would provide $\tau_D = 1.8$ ns. The upper frequency cutoff of the monitor is $f_{co} = (2\pi R_b C_b)^{-1} = 88$ MHz. If, for example, the monitor was subject to a ramp voltage with a derivative $\dfrac{d\phi_1}{dt} = 5 \times 10^{13}$ Vs^{-1}, the monitor would deliver a voltage

$$\phi = R_b C \frac{d\phi_1}{dt} = 450 \text{ V}.$$

As another example, Ref. [11] describes a differentiating sensor for a liquid pulsed power accelerator. For this sensor, $C = 5 \times 10^{-13}$ F, $R_b = 50\ \Omega$, and $C_b \sim 10$ pF. As a consequence, the cutoff frequency (-3 dB) is $f_{co} = (2\pi t_c)^{-1} = 320$ MHz. The reference also describes two independent calibration procedures for the sensor and shows the frequency response of the monitor coupled to a passive RC integrator.

Differentiating sensors for free field applications [12] can take several forms. One, commercially available, consists of a hemisphere slotted around the equator. The signal current in this sensor flows to ground through the 50 Ω coaxial cable. Another version has two such hemispheres that drive a 100 Ω twin axial signal cable. A third version consists of an asymptotic conical dipole. In this sensor, the surface of the sensor corresponds to the equipotential surface that is asymptotic to a 100 Ω bicone at its apex. The advantage of this design is that it has the least stray capacitance to ground for a given sensor area. Consequently, its figure of merit is quite high.

A fourth version that minimizes field enhancement is the flush plate dipole consisting of a conducting disk centered in a circular aperture in a conducting ground plane. Since the capacitance to ground for the sensing element is the largest, this sensor has the lowest cutoff frequency of all the designs discussed so far.

2.5   Displacement Vector Measurements With Grounded Electrodes

We can utilize the results of the previous sections to obtain the relationship between sensor output and applied displacement vector.

We recall that in the case of $C_b/C \gg 1$, the differential equation describing the sensor circuit is:

$$R_b C \frac{d\phi_1}{dt} = \phi + R_b C_b \frac{d\phi}{dt} \, ,$$

where $\tau_D = R_b C_b$.

Therefore, for

$$\phi_1(t) = \begin{cases} 0 & t < 0 \\ \phi_1(t) & t > 0 \end{cases}$$

the result is,

$$\phi(t) = \begin{cases} 0 & t < 0 \\ \frac{C}{C_b} e^{-t/\tau_D} \int_0^t \frac{d\phi_1}{dt'} e^{t'/\tau_D} dt' & t > 0 \, . \end{cases}$$

Since,

$$\phi_1 = \frac{-DA_{eq}}{C} \, ,$$

where we have replaced the dot product $\vec{D} \cdot \vec{A}_{eq}$ by DA_{eq},

$$\phi(t) = - \frac{A_{eq}}{C_b} e^{-t/\tau_D} \int_0^t \dot{D} \, e^{t'/\tau_D} dt' \, .$$

For $t_C \ll \tau_D$, then

$$\phi(t) = - \frac{A_{eq} D}{C_b} .$$

For $t_C \gg \tau_D$,

$$\phi(t) = - \frac{A_{eq}}{C_b} \tau_D \dot{D} = - A_{eq} R_b \dot{D} .$$

This expression, of course, is also valid for $C_b/C \ll 1$.

2.6 Sources of Error in Measurements With Floating Electrodes

It is appropriate at this point to describe the errors arising in voltage and field measurements with floating electrodes. While this description is by no means exhaustive, it covers the most frequent occurrences in pulsed power practice.

The most frequent error, common to measurements with floating electrodes and resistive dividers, arises from magnetic flux linking the measuring circuit and wave propagation delays between the location of the sensor and the position where the measurement is required. These errors are of practical importance at the load end of pulsed power drivers, where an accurate knowledge of the magnitude of the accelerating voltage and its phasing with respect to the current is required for energy and power calculations. Due to their importance, these errors deserve the separate discussion of Section 4.

One error, particular to measurements with floating electrodes, arises from changes in the configuration of the electric fields in the vicinity of the sensor due to switching of energy storage and pulse forming elements [14]. As a consequence, the field configuration, during accelerator operation, may be different than the field configuration when the sensor was calibrated. This error can be eliminated by keeping floating electrode sensors away from switch regions.

Another error arises from charge transfer to the sensor by a conduction current density, $\vec{j}$. Since $\vec{j}$ and $\dot{D}$ are indistinguishable, charge reaching the sensor from electron/ion currents in electrolytes and vacuum, or local breakdowns will introduce an error in the measurements. Capacitive sensors in vacuum are particularly sensitive to field emitted stray electrons.

We can estimate the magnitude of this error by considering the circuit of a sensor (Fig. 2.3), in a water dielectric with a finite resistivity. The resistor, R, connects the high voltage electrode to the metering point and represents a shunt conductance across the dielectric of capacitor C. The shunt resistance, R_s, across C_b, is large compared to R_b and can be neglected.

The equivalent circuit for a self integrating sensor appears in Fig. 2.4. According to our analysis, the transient response of the sensor is given by $\phi = \frac{C}{C_b}\phi_1$. However, at lower frequencies, by examining the current balance at node A, we obtain

$$\frac{d\phi}{dt} = \frac{\phi_1}{C_b R} - \frac{\phi}{C_b R_b} .$$

Substituting the relationship for ϕ above we obtain

$$\frac{\Delta\phi}{\phi} = \left(\frac{1}{RC} - \frac{1}{R_b C_b}\right) \tau_{pulse}$$

where τ_{pulse} is the pulse duration.

The result is quite illuminating. When $RC \gg R_b C_b$, the sensor output exhibits the usual decay associated with $R_b C_b$. When $RC = R_b C_b$, the monitor is compensated and produces an output that is independent of frequency. When $RC > R_b C_b$, the monitor integrates the input pulse exponentially. The output waveforms for such a monitor; for different values of R_b, appear in Fig. 2.5.

For a differentiating sensor, $\phi = R_b C \frac{d\phi_1}{dt}$. At the metering point the error is $\Delta\phi = \frac{\phi_1 R_b}{R}$.

After integration, the error becomes

$$\Delta = \frac{\int \Delta\phi dt}{\int \phi dt} = \frac{\int \phi_1 dt}{RC\phi_1} \simeq \frac{\tau_{pulse}}{RC} .$$

However, signals from these monitors are commonly integrated with a passive integrator of time constant τ_{int}.

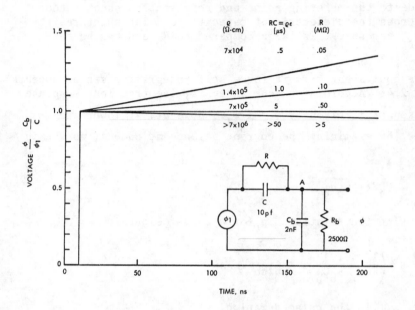

Fig. 2.5.   Calculated step response of a capacitance divider for several values of RC. R_bC_k = 5 μs. When RC > R_bC_b the output increases with time. When RC = R_bC_b the monitor is frequency compensated. For RC >> R_bC_b monitor output decreases exponentially.

The above analysis shows that the error, unless compensated by integrator decay, increases with increasing pulse length and decreasing resistivity. Due to the longer pulse lengths, it is bound to be larger in the slower stages of pulsed power apparatus.

The ideal way to eliminate this error in water dielectric apparatus is to arrange the integration time constant τ_{int} = R_bC_b = RC. Since R and C share common electrodes RC = ρε, when ρ is the resistivity of the medium and ε is the dielectric constant. For 1 MΩ-cm water, ρε = 7 μs. This time constant can be easily accommodated with conventional components and nominal sensor dimensions in the case of self integrating sensors.

2.7 Concluding Remarks On Measurements With Floating Electrodes

The biggest advantage of floating electrode voltage measurements, by measuring the displacement vector field in the vicinity of the ground conductor, is due to the minimal impact of the sensor on the mechanical and the electrical design of the apparatus. The low power dissipation of floating electrode sensors is another big advantage for repetitive applications.

Their biggest disadvantage is that they require in-place calibration to convert the local sample of field into an integral quantity such as the voltage. Secondly, being inherently AC devices, they have a built-in, low frequency cutoff. Thirdly, their sensitivity to stray charges excludes them from voltage measurements in vacuum where electric fields exceed the field emission threshold. They also exhibit a problem with charge collection when they are subject to repetitive pulses in oil dielectrics where it is possible to inject charges into the liquid. Also, they cannot be used in locations where the field configuration changes after calibration. Resistive dividers described in the next action circumvent some of these difficulties.

3. POTENTIAL MEASUREMENTS WITH RESISTIVE DIVIDERS IN TRANSMISSION LINES AND STATIC ENVIRONMENTS

3.1 Some General Considerations

The concept of potential measurements with resistive dividers is quite simple. Consider, as before, two conductors, 0 and 1 at ground and at potential ϕ_1, respectively. A resistor R, with a tap at R_ℓ, connects between 0 to 1, such that $R = R_u + R_\ell$. (See Fig. 3.1).

By Ohm's Law, the current in the resistor, R, is $i = \phi_1/(R_u + R_\ell)$ and $\phi = iR_\ell = \phi_1 R_\ell/(R_u + R_\ell)$. Consequently, the measuring principle is to measure the voltage ϕ, developed across the metering resistor R_ℓ, by the current i flowing in the resistor. The goal of the design is reduction of i to the lowest possible value, thereby minimizing the loading of the circuit and reducing the ohmic heating of R.

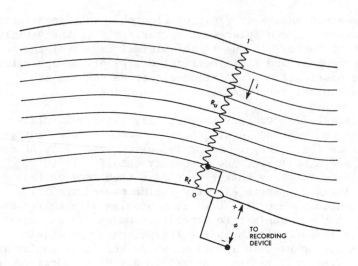

Fig. 3.1.   Voltage measurement with a resistive divider. Potential difference between 0 and 1 drives current i in
the divider. Current i develops a voltage ϕ across
R_ℓ. In ideal case $\phi = \phi_1 R_\ell (R_\ell + R_u)^{-1}$.

The discussion of Section 1.1 revealed that the local electric field in a transmission line structure may be derived as a
gradient of a scalar potential that is independent of frequency
and identical to the static field configuration for the same conductor geometry. Consequently, by disposing the resistor such
that its equipotentials correspond to the equipotentials between
the transmission line conductors, the voltage at the tap-off will
be a fraction of the total voltage, independent of frequency, re-
gradless of the length of the divider.

The instantaneous response of a resistor placed in a transmission line with a TEM wave such that the electric field lines
are parallel to the resistor (and perpendicular to the direction
of propagation) deserves a special emphasis [15]. In this case,
the wave front establishes a voltage gradient along the resistor,
instantaneously driving a current flow from one end to the other,
with no transit time. This would not be the case if the resistor
were not parallel to the undisturbed electric field in which case
a finite rise time of the monitor would arise.

From the circuit viewpoint [16], the performance of the
divider arises from the lack of stray capacitance resulting from
congruent equipotentials (divider and transmission line). Moreover, if the resistance of the monitor is sufficiently high
($R \gg Z_o$, where Z_o is the characteristic impedance of the transmission line) the energy of the fields surrounding the monitor is
negligible with respect to the energy of the fields in the transmission line. Consequently, the divider response is instantaneous.

The divider arrangement appears schematically in Figure 3.1,
where the resistor R bridges conductors 1 and 2. Since most resistors have a uniform resistance per unit length, the ideal location for such a monitor is along an electric field line, in a
location where the field is uniform, like, for example, where the
transmission line conductors are uniform.

In diverging fields, one could change the resistance per
unit length of the resistor to accommodate the equipotentials.
This elaboration, however, is not necessary in practice.

As an example, Barth [15] describes the rise time response
of a 1-m long voltage monitor that tracked accurately a 200 ps
rise time pulse in a TEM test cell.

The presence of residual capacitances C_u and C_r in R_u and
R_ℓ causes a current flow in the divider when a dV/dt appears
across the divider. Consider as an example the resistive divider
appearing in Fig. 3.2. As before, the divider has an upper
branch and a lower branch with resistances R_u and R_ℓ and residual
capacitances C_u and C_ℓ, respectively. The ratio of the voltages
in the lower and upper branches is

$$\frac{V_\ell}{V_u} = \frac{R_\ell}{R_u} \frac{1 + j\omega R_u C_u}{1 + j\omega R_\ell C_\ell} .$$

The division ratio is independent of frequency only when $R_u C_u =
R_\ell C_\ell$. At low frequencies ($\omega R_u C_u = \omega R_\ell C_\ell \ll 1$), the voltage di-
vision takes place across the resistors such that $V_\ell/V_u = R_\ell/R_u$.
At high frequencies ($\omega R_u C_u \gg 1$), the voltage division is capacitive and $V_\ell/V_u = C_u/C_\ell = R_\ell/R_u$.

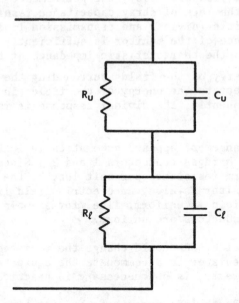

Fig. 3.2. Resistive divider with residual capacitances asso-
           ciated with R_u and R_ℓ. For $\tau \gg R_\ell C_\ell$ and $R_u C_u$
           division is resistive. For $t \ll R_\ell C_\ell$ and $R_u C_u$
           division is capacitive.

As we discussed before, the measurement of very high voltages requires large values of resistance. The goal $\omega R_u C_u \ll 1$
is, therefore very difficult to achieve. As an example, for
typical pulsed power applications $R_u = 10^4$ Ω, $C_u = 10^{-11}$ F, and
pure resistive division requires $\omega < 10^{+7} s^{-1}$.

At this point, we notice that a geometric resistive divider,
with residual capacitance, is conceptually equivalent to a compensated geometric capacitive divider with a parasitic resistance
(Section 2.6). The distinction between capacitive and resistive
division then takes place in the frequency domain with the crossover frequency between capacitive and resistive division being
established by the RC product of the divider.

3.2 Liquid Electrolyte Resistive Dividers

An elegant way to implement a geometric divider, with a uniform resistance and capacitances per unit length, is to use a
homogeneous column of resistive dielectric such as an aqueous
solution of electrolyte or a homogeneous string of resistors. In

general, a metal electrode, in a cross sectional plane of the
fluid, senses the geometric fraction of the voltage applied to
the divider. When the divider is placed in a uniform field
region, the homogeneity of the column guarantees the fraction is
constant, independent of frequency.

Other advantages of electrolytic geometric dividers are:

a. Aqueous solutions of electrolytes operate reliably at
fields as high as 300 kV cm^{-1}.

b. The division ratio is independent of nonlinearities in
the electrolyte behavior (field and temperature) as long as the
whole column is homogeneous, i.e., cross sectional areas and
electrolyte concentrations are constant.

c. Resistance of the column may be changed at will be adjusting the concentration of the electrolyte.

d. There is considerable freedom in the design of the outside shape and material of the containment vessel. As long as it
withstands the applied field on its external surface, it can consist of flexible tubing, rigid nylon tubing, or a vacuum insula-
tor structure.

e. The divider can be shaped to follow an electric field
line.

f. The heat capacity of the electrolyte (water solution) is
quite large (1.75 gm^{-1})

The electrolytic system of choice is generally copper sul-
phate aqueous solution and copper electrodes [17]. The advantages of this system are that it is readily available, nonhazard-
ous, it does not produce gas bubbles during operation unless
the resistor is polarized [17], and its periodic maintenance
intervals (6 mo. to 1 yr.) are no shorter than those of other
components in pulsed power systems. These useful range of
resistivity of solutions (1 - 1000 Ω-cm) allows significant freedom in divider design. Resistivity, as a function of concentration, appears in Fig. 3.3. Over long periods, a green insoluble
deposit appears on the surface anode (Brochantite, $CuSO_4(OH)_6$),

affecting operation at low fields. Easy removal of this deposit,
by mechanical means, restores the divider to original specifications.

According to Martin [18], reliable operation of Cu_2SO_4 dividers requires a current density above 0.3 A cm^{-2}. Below this current density, the output from the divider tap off becomes noisy. Steady densities are limited by the depletion of copper ions at the cathode so the resistor should be operating, preferrably with the cathode down.

Another electrolyte system consists of an aqueous solution of sodium thiosulphate ($Na_2S_2O_3$) with aluminum electrodes. The electrolyte does not grossly attack aluminum, does not produce gases by electrolysis, and has a negligible voltage coefficient in the 20 V cm^{-1} to 120 kV cm^{-1} range [19]. Its resistivity, for a given concentration, is lower than that of $CuSO_4$ solutions, as shown in Fig. 3.3. The only defect of this system is that the $S_2O_3Na_2$ decomposes, producing gaseous SH_2, that precipitates as bubbles. Consequently, these dividers need frequent maintenance to remove bubbles.

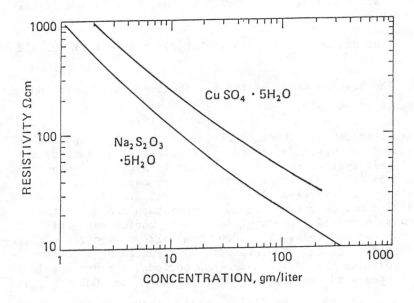

Fig. 3.3 Resistivity at 21° C of aqueous solutions for resistive dividers.

Geometric dividers with flexible tubing [20] can be obtained
in arbitrary lengths with a division ratio depending only on the
length (Fig. 3.4(a)). Geometric dividers, with a rigid wall and
adjustable length [14], that withstand fields as high as 250 kV
cm^{-1}, are part of the voltage monitoring system of a large water
dielectric accelerator. Voltage dividers that withstand fields
as high as 200 kV/cm have been used in vacuum [20] (Fig. 3.4(b))
and are available commercially [21].

The breakdown along the outer envelope of the divider sets
the operational limits for resistive dividers operating in vacuum
or at atmosphere. For oil or water immersion, the limit arises
from breakdown of the liquid or bulk breakdown of the envelope of
the divider. The risk is greater in water and glycol mixtures
when the orientation of the column is oblique to the prevalent
electric fields.

For reliable operation, it is very important to remove dis-
solved gases by filling the divider under vacuum with solution
that has been boiled. Otherwise field enhancement will break
down the bubbles.

3.3 Attenuators

Coaxial attenuators are widely used geometric dividers to
reduce the input signal to a recording device such as digitizer
or cathode ray oscillograph. Attenuators, developed originally
for microwave measurements, have found application in pulsed
power measurements due to their large bandwidths.

When used with caution, microwave-type attenuators are extremely useful for pulsed power measurements. Their design is
aimed to provide a compact, wide band, nominal (~ 2W) power dissipation device. Resistive elements of small linear dimensions
and compact disposition satisfy these requirements. Consequently, ordinary microwave, nominal power dissipation, large
bandwidth units, cannot withstand high voltages.

Some wide band attenuators have a thin metal or carbon film
resistor deposited on an insulating substrate. Others employ
conventional carbon composition resistors. The thickness of thin
film resistors is so much smaller than the skin depth that, for
all practical purposes, the resistance of the film is independent
of frequency. These resistors can be manufactured to high tolerances and materials, such as Nickel-Chromium alloys [22], yield
small temperature coefficients of resistance. The temperature
coefficient of resistance of the film is important for pulsed
power applications. Since the time scale for heat transfer into

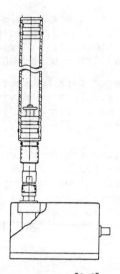

(a) Geometric voltage divider with flexible tubing.[3.6]

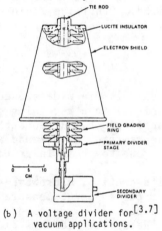

(b) A voltage divider for[3.7]
vacuum applications.

Fig. 3.4. Voltage dividers with liquid electrolyte.

the substrate can be orders of magnitude larger than the length
of the pulse, the film can heat up considerably during a pulse.

Glass [22] and ceramic substrates are currently in use.
Glass substrates allow deposition of a very smooth and uniform
metal film. This smoothness minimizes high field arcing in the
resistor that initiates normal to the current flow direction,
across film imperfections, arising from an uneven substrate surface. Attenuators [23] using the glass substrate technique consequently have the highest voltage rating as shown in Table 1.

Table 1.   Ratings of some wide-band attenuators for pulsed
           applications.

Manufacturer	Type	Construction	Single Pulse Peak Power into 50 Ω (kW)	Peak Voltage (KV)	Peak Energy (J)	Average Power at 70°C (W)
Barth [23]	202	Film on glass	500	5	0.2 1)	2
Gilbert [24]	GR-874	Film on cermanic	2	0.32	2)	1/2
Elcom [25]	AT-52	Carbon Composition	1	0.23	3)	1/2
PI [26]	PIM-230	Carbon Composition (20:1 50:1 100:1)	80	2	4)	10
PI [26]	PIM-230	Carbon Composition (2:1 5:1 10:1)	20	1	5)	1

1) 5 kV, 400 ns pulse into 50 Ω.
2) T pad ceramic
3) π pad construction, 1/2 W Allen Bradley, EB Resistors.
4) π pad construction, 2 x 2W Allen Bradley, HB resistors in series.
5) π pad construction, 1/2 W Allen Bradley, EB resistors.

Another type of thin-film, resistor attenuator commonly applied in pulsed power measurements has smaller resistors [24] deposited on a ceramic substrate. These support lower fields and,
due to their small size, their voltage rating is considerably
lower even though, from a heat transfer viewpoint, the ceramic
is a better material. The ratings of these attenuators appear in
Table 1 as well. Thicker films are much better for high voltage
applications.

The second construction technique incorporates carbon composition resistive elements [25, 26, 27]. Carbon composition resistor attenuators have a higher pulsed energy threshold for
permanent damage. However, their pulse response, while adequate
for pulsed power applications, is not as good as the one for film
units. Compensation of stray capacitances by shielding and partitioning components within the attenuator can equalize, to some
extent, their pulse response. Ratings of these units also appear
in Table 1.

The response of the Barth, Gilbert and Elcom units of Table
1 to a 500 ps rise time pulse is the same. The PI units exhibit
a resonance at 1.3 GHz. This disadvantage is offset, however, by

a connector limited maximum voltage and a very large energy
absorption capacity that generously forgives operator errors.

Besides outright destruction by pulses that severely exceed
their maximum ratings, resistors in attenuators exhibit positive
temperature coefficients of resistance above 25° C, generally
negative voltage coefficients of resistance at high voltages, and
permanent changes of resistance after repetitive pulsing at high
voltage.

A periodic measurement, with 0.1% accuracy of the DC resistance and attenuation ratio at a fixed temperature, will reveal
permanent changes in attenuators caused by high voltage pulsing.
Increases larger than 0.1% for metal thin film units and 0.2-0.4%
for carbon composition units are a sufficient reason to remove
attenuators from service in critical applications such as single
shot experiments where an attenuator failure could cause loss of
significant data.

3.4 Voltage Measurements with Resistive Probes

3.4.1 Some general considerations. The emphasis of the discussion, up to this point, has been on dividers (capacitive and resistive) that reside in the transmission line, causing a minimum
of disturbance. Under many circumstances, it is not possible to
place the divider within the transmission line. A probe is then
required to make point measurements with minimum circuit perturbation.

One of the main difficulties in making such measurements
arises from the presence of stray capacitances [3] originating
from the distortion of the equipotentials by the presence of the
probe. These stray capacitances cause a D current flow along the
probe resistor, yielding a signal proportional to the time rate
of change of the electric field.

Distributed, stray capacitances, between the resistive element and ground, slow down the rise time of the voltage reaching
the lower stage. To examine this effect, we consider a divider
of length ℓ, total resistance R, stray capacitance C_s, and residual capacitance C, distributed uniformly over the length of the
divider.

The difference equations that govern the differential section of the divider appearing in Fig. 3.5.

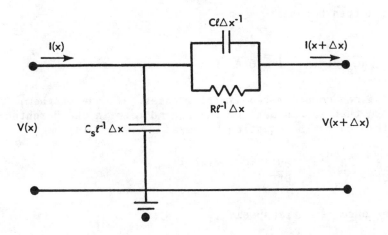

Fig. 3.5. Differential section of resistive divider with a stray
           capacitance to ground.

$$I = - \frac{C\ell}{\Delta x} \frac{d}{dt} (V(x+\Delta x)-V(x)) - \frac{\ell}{R\Delta x} (V(x+\Delta x)-V(x))$$

$$I = (x+\Delta x) - I(x) = - \frac{C_s}{\ell} \Delta x \frac{dV_x}{dt} \ .$$

In the limit as $\Delta x \to 0$,

$$I = - C\ell \frac{\partial^2 V}{\partial x \partial t} - \frac{\ell}{R} \frac{\partial V}{\partial x}$$

$$\frac{\partial I}{\partial x} = - \frac{C_s}{\ell} \frac{\partial V}{\partial t} \ .$$

Combining both equations, we obtain,

$$RC \frac{\partial}{\partial x^2} \left(\frac{\partial V}{\partial t} + \frac{V}{RC} \right) = \frac{RC_s}{\ell^2} \frac{\partial V}{\partial t} \ .$$

We can now introduce the dimensionless variables t^*, x^*, and V^*
defined as:

$$t = RC_s t^*, \ x = \ell x^*, \ \text{and} \ V = V_o V^* \ .$$

The equation then becomes:

$$\frac{\partial^2}{\partial x^{*2}} \left(\frac{RC}{RC_s} \frac{\partial V^*}{\partial t^*} + V^* \right) = \frac{\partial V^*}{\partial t^*} .$$

As usual, we can consider two limiting cases. In one of them, $RC \gg RC_s$ so that we can neglect the second term in the parenthesis. In this case, the equation becomes

$$\frac{RC}{RC_s} \frac{\partial}{\partial t^*} \frac{\partial^2 V^*}{\partial x^2} = \frac{\partial V^*}{\partial t^*} .$$

Integrating once, and transposing,

$$\frac{\partial^2 V^*}{\partial x^{*2}} = \frac{RC_s}{RC} V^*(x) \simeq 0 ,$$

such that

$$\frac{\partial V^*}{\partial x^*} = K = 1 ,$$

and:

$$V = \frac{V_o x}{\ell} .$$

Hence, we conclude that a capacitively ballasted divider where $C \gg C_s$, has a uniform, time independent, voltage per unit length. Consequently, its division is geometric.

In the other limiting case $(RC \gg RC_s)$, we can neglect the first term in the parenthesis. Hence:

$$\frac{\partial^2 V^*}{\partial x^{*2}} = \frac{\partial V^*}{\partial t^*} .$$

This is the diffusion equation. As a consequence of diffusion, the rise time is smeared. The same equation governs the behavior of shunts [28].

To demonstrate the effect of the ratio C_s/C on divider response, we have performed some calculations on a simple $N = 10$ section divider network with $R = 1$, $C_s = 1$, and $C = 1$, .1, .01,

0. Figure 3.6 shows the response of the N = 10 section divider, as a function of t*, for a step function input. The ordinate is NV_1^*, where V_1^* is the output of the first stage. The response is a ramping square wave for $C_s/C = 1$ and begins to show a rise time deterioration for $C_s/C = 10$. For $C_s/C = 100$, the pulse shape is indistinguishable from the diffusion dominated case $C_s/C = \infty$.

The very fast rise times associated with $C_s/C = 1$ and $C_s/C = 10$ arise from the capacitive division that takes place across the residual capacitance of the divider.

Since 10^{-12}F stray capacitances to ground and $10^4 \, \Omega$ resistances are common for pulsed power applications, t* is typically 10^{-8}s. Therefore, for a good frequency response, large residual capacitances are desirable in the divider. Dividers with aqueous solutions have a large advantage in this respect because the dielectric constant of water ($\varepsilon = 81$) is so large.

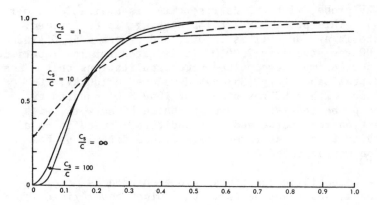

Fig. 3.6.   Response of a divider with residual capacitance, C, and stray capacitance to ground curves corresponding to $\frac{C_s}{C}$ = 1, 10, 100, and ∞.

3.4.2 Some specific probe designs. The Tektronix P6015 [29],
1000x, 40 kV, passive probe has been the workhorse of pulsed
power laboratories for many years. With a 100 MΩ input resistance, a 4 ns rise time, and a required loading of 12 to 47 pf and
1 MΩ at the metering input, it is compatible with most commercial
oscillographs and digitizers. Caution should be used not to exceed voltage (power dissipation) rating of probe at high frequencies [30]. A 3.3-m long resistive cable [31] links the probe
to the compensation box. This requires closed proximity of the
apparatus under test and the recording apparatus, since extending
this cable will degrade probe performance. From an interference
and safety standpoint, in pulsed power laboratory work, this is
sometimes unacceptable.

A fast buffer amplifier has been developed [32] to couple
the compensator box to a 50-Ω data recording system. A 4 ns rise
time is claimed for the probe and amplifier combination. Also,
probe operation to 60 kV has been extended by the addition of a
corona shield [32].

Reference [33] describes a more compact probe to measure
peak voltages of 20 kV, with a rise time of 50 ps, driving a 50-Ω
recording system. To achieve the frequency response, ferrite
beads, configured as a lossy transmission line, yield a surge impedance of 300 Ω at very high frequency.

A 60 kV probe, with a 2 ns rise time, is available commercially [34]. This probe has a nylon bushing, is DC coupled,
will operate in air as well as in oil, and drives a 50 Ω system.
Since its DC resistance is 2500 Ω, it is rated for low repetition
rate, low source impedance pulsed power applications only. Los
Alamos National Laboratory [35] has developed a 100 kV, DC-
coupled probe, with a DC resistance of 1000 M Ω, shunted by
13 pF. The probe consists of a distributed DC network, with carbon composition resistors, and RF capacitors, housed in an oil
filled acrylic tube. The reference provides details on how
compensation was achieved.

Carbon composition resistors have been used as metering
strings, loads, and shunt belts in the pulsed power field for a
long time. Their simple and solid construction results in a reliable, durable, and rugged product. Thin film resistors instead
are more fragile, and do not have the ability to withstand large,
transient overloads. Microscopic cracks in substrates, contami-
nants in the current path and thin or bare spots in the film itself do not affect resistor performance for conventional applications but can lead to early failures in high voltage pulsed power
apparatus.

The biggest single advantage of carbon composition resistors is their ability to withstand transient overloads and to perform well under unexpected conditions. Their performance is legendary, as attested by the pulsed power pioneer J. "Charlie" Martin [16].

"In general, 1W carbon resistors will stand pulse volts up to about 15 kV per resistor, for short pulses, and will absorb one or two joules with only a modest long term drift. At 5J per pulse per resistor, this change becomes larger, but can be cancelled to a first order by tapping off across a representative resistor. At the level of about 10J per resistor, they blow up with a satisfying bang!"

Reference [36] summarizes results of capacitor discharge testing on Allen Bradley (AB) carbon composition resistors. The results of Table 2, reproduced from the reference, confirm J. C. Martin's assessment. The reference provides no data, however, on the long term drift of resistance, resulting from high voltage repetitive operation or the voltage coefficient of resistance arising at high voltage. Table 3 reproduces limited data from Ref. [27], showing that, even operating at energies as high as half of that necessary for a 10% rupture probability, the permanent $\Delta R/R$ for 100 and 150 Ω resistors does not exceed 4%. These data resulted from a capacitive discharge with an RC in the 3.2 - 4.8 msec range. The references do not document the peak instantaneous temperature of the resistor at the end of the pulse.

For shorter pulses, the energy limitation no longer applies. A field limitation on high value resistors and a current density limitation for low value resistors applies instead. This area is a subject of current research [38]. There are published data showing that 3600, 100ns, FWHM pulses, with V < 10 kV, caused < 0.1% permanent resistance changes on 3-kΩ, 2W A-B resistors. A larger (0.5%) permanent change appeared at 20 kV [23]. These changes may result from arcing development at the carbon granule surfaces during high field operation.

More complete data [39] for a 56 Ω, 2W A-B resistor, with 100 ns and 260 ns FWHM pulses, showed a maximum deviation from linearly of 1 and 2% respectively in the 0 to 6 kV range. Large error bars in the data suggest that 1 and 2% may be a very pessimistic upper bound of the voltage coefficient of resistance.

Data also exist [39] for the fractional change of resistance as a function of three pulse voltages (2.5 kV, 5.5 kV, 15.3 kV) of 260 ns width, applied N = 1, 10, 100, and 1000 times, to different, 10 unit, samples of the 56Ω, 2W, A-B resistors. The results were % $\frac{\Delta R}{R}$ = -0.5 - 0.22 ln N, at 2.5 kV peak,

Table 2. Pulse characteristics of Allen Bradley (A-B) resistors
          (1 Watt second = 1 joule) [36].

Rated Watts	A-B Type	Many Pulse Watt-Secs.	Single Pulse Watt-Secs.			Thermal Time Constant (Seconds)
		Rupture Probability				
		Withstand Millions	10%	50%	90%	
1/8	BB	.14	.72	.9	1.08	4
1/4	CB	.56	2.8	3.5	4.2	8
1/2	EB	2.24	11.2	14.0	16.8	16
1	GB	8.9	44.0	55.0	66.0	32
2	HB	12.8	64.0	80.0	96.0	64

Table 3. Test matrix for Allen Bradley carbon composition
          resistors. $\frac{\Delta R}{R}$ must not exceed 4% after one pulse [37].

Rated Wattage	A-B Type	Energy Watt-Sec.	RC (ms)	Approx. Applied DC Volts	Capacitance of Capacitor	Resistance (Ohms)
1/8	BB	0.45	3.2	670	2 μf ±10%	100
1/4	CB	1.8	4.8	600	10 μf ±10%	150
1/2	EB	6.4	4.8	630	32 μf ±10%	150
1	GB	16.0	3.2	1000	32 μf ±10%	100
2	HB	44.0	3.2	1650	32 μf ±10%	100

-1.0 - 0.22 ln N, at 5.5 kV peak, and -0.5 -0.58 ln N, at 15.3 kV
peak. Consequently, one would expect less than a 2% change in
resistance, for 10^6 pulses, at a peak voltage of 5.5 kV.

Other authors have considered the performance of carbon
composition resistors as well. Extensive tests at Sandia National Laboratories, Albuquerque (SNLA) [40] produced data on the

maximum, safe, voltage for operation of A-B carbon composition
resistors. The criterion for the test was that the change of
resistance should not exceed a predetermined value (not stated
in the reference). The ranges of voltages, for 20 μs wide pulses
are tabulated below in Table 4:

Table 4. Performance of Allen Bradley, carbon resistors [37].

A-B Resistor EB (1/2 W)		A-B Resistor GB (1W)	
R (k Ω)	V_{safe} (kV)	R (k Ω)	V_{safe} (kV)
0.051 < R < 0.5	< 0.75	0.051 < R < 0.2	$1 < V_{safe} < 2$
1 < R < 2	< 2	0.24 < R < 1	< 8
20 < R < 1000	< 12	2.2 < R < 1	< 14
		20 < R < 1000	< 18

Other tests have been performed at SNLA [18] on carbon
composition resistors suggesting that a construction similar
to the Allen-Bradley construction yields the smallest $\frac{\Delta R}{R}$, as a
function of voltage. For this construction, $\frac{\Delta R}{R} < 1\%$ for 2 <
V < 20 kV, for a single-pulse, on a 75-W, 2-Ω, resistor. One
hundred pulses at 20 kV changed the value by 3%, over 100 pulses,
with the largest changes occurring during the first 30 pulses.

Finally, a program at Lawrence Livermore National Laboratory, that used carbon composition resistors as fuses on a large
capacitor, established the empirical relation between the energy
required for thermal rupture in terms of the wattage rating, W,
of the resistor [2]. For A-B (1/2 W) EB type resistors
$E_{rupture} = 56.2 \, W^{1.3}$. For GB (1 W) and HB (2 W) types,
$E_{rupture} = 63.7 \, W^{1.32}$. These equations yield 23, 64, and 160
joules for the EB, GB, and HB types, respectively. The reference
recommends energy absorption below two thirds of the rupture
thermal capacity to avoid permanent changes in carbon composition
resistive dividers.

4.   SPECIAL CONSIDERATIONS ON VOLTAGE MEASUREMENTS IN THE
      TRANSIENT REGIME

4.1 Electric Field and Voltage Measurements in the Presence of
     Time Varying Magnetic Fields

     Our discussion so far has not considered the effect of time
varying magnetic fields. In their presence, the electric field
is no longer irrotational because the circuital voltage $\oint \vec{E} \cdot d\ell$ is no longer vanishing.

     We can define the electric field by Faraday's Law as:

$$\vec{E} = - \vec{\nabla}\phi - \frac{\partial \vec{A}}{\partial t} ,$$

when ϕ is the scalar electrostatic potential and $\vec{A}$ is the magnetic vector potential such that,

$$\vec{B} = \vec{\nabla} \times \vec{A}, \quad \nabla \cdot \vec{A} = 0 .$$

We now consider again the two arbitrary conductors of Fig. 2.1,
this time with a time-varying magnetic field. We perform a line
integral of E (Fig. 4.1) along the path 0, 1, 1', 0', 0. In this
case,

$$- \int_0^1 \vec{E} \cdot d\ell = U_1, \quad - \int_1^1 \vec{E} \cdot d\vec{\ell} = 0, \quad - \int_{1'}^{0'} E \cdot d\ell = -U_2,$$

$$- \int_{0'}^0 E \cdot d\ell = 0 .$$

Consequently,

$$\oint E \cdot d\ell = U_1 - U_2 = - \frac{\partial}{\partial t} \int \vec{A} \cdot d\vec{\ell} = - \frac{\partial}{\partial t} \iint_s (\vec{\nabla} \times \vec{A}) \cdot d\vec{s}$$

$$= - \frac{\partial}{\partial t} \int_s \vec{B} \cdot d\vec{s} = - \frac{d\phi}{dt} ,$$

where ϕ is the magnetic flux threading the circuit.

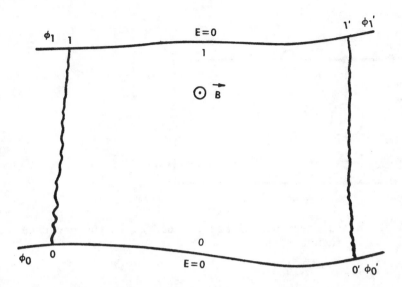

Fig. 4.1. Voltage measurements in the presence of time varying
           fields. Electric field is zero on conductors 1 and
           2. $\phi_1 - \phi_0 = U_1$ and $\phi_1' - \phi_0' = U_2$.

This description, which is valid when the propagation delay
of an electromagnetic wave between a and b can be neglected with
respect to the inverse of a characteristic frequency describing
the pulse, is useful to "correct" the voltage for time varying
fluxes in pulse power apparatus. As an example, U_1 could be the
voltage at the accelerating tube of a relativistic electron beam
accelerator and U_2 could be the accelerating voltage appearing
across the gap between the emitting cathode and the target anode.
Practical ways to implement such a voltage correction have been
described by Moses [41], Thomas, [42], and DiCapua [43].

To explore the issue of voltage measurements in the presence
of time varying magnetic fields in come more detail, we may consider the circuit of Fig. 4.2, where we have a monitor V_m connected by an inductor L, to a load Z_L. We are interested in
deducing the voltage on Z_L, in the presence of a current with a
harmonic dependence $I = I_0 \sin \omega t$.

In this case,

$$V_m = I_0 Z_L \left(\sin \omega t + \frac{\omega L}{Z_L} \cos \omega t \right) .$$

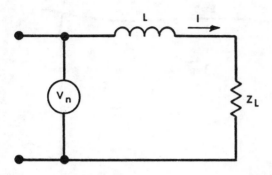

Fig. 4.2. Measurement of voltage on load Z_L in the presence of
           an inductor, L.

When $\dfrac{\omega L}{Z_L} \ll 1$, $V_M = I_o Z_L \sin\omega(t + \dfrac{L}{Z_L}) = I_o Z_L \sin\omega t'$,

where $t' = t + \dfrac{L}{Z_L}$

and $V_L = I_o Z_L \sin\omega t$.

Consequently, the presence of the inductor introduces a "delay",
$\Delta t = \dfrac{L}{Z_L}$, on the voltage at V_L, since $V_L(t+\Delta t) = V_M(t)$. This
result can be generalized for pulsed excitation as long as
$\omega L/Z_L \ll 1$ holds for all the frequency components of the pulse.

4.2 Voltage Measurements in Pulsed Power Systems in the Presence of Propagating Waves

Interpretation of voltage measurements, in the presence of
propagating waves, can become a problem when:

a. Reflections are present that introduce ambiguities in
the interpretation of the data.

b. Comparison of monitor signals from different locations
in the apparatus is required.

To outline some of the difficulties associated with 1 and 2, let us consider a transmission line apparatus operating in the principal, TEM Mode, as outlined in Section 1.1. Assuming a lossless dielectric, the well known transmission line equations are [44]:

$$\frac{\partial V(x,t)}{\partial x} = -L \frac{\partial I(x,t)}{\partial t}$$

$$\frac{\partial I(x,t)}{\partial x} = -C \frac{\partial V(x,t)}{\partial t} ,$$

where L and C are an inductance and a capacitance per unit length respectively. We choose a time dependence $e^{j\omega t}$ and obtain the wave propagation equation:

$$\frac{d^2}{dx^2} + k^2 \begin{cases} V(x,\omega) \\ I(x,\omega) \end{cases} = 0 ,$$

where

$$k^2 = \omega^2 LC$$

and

$$v = \frac{\omega}{L} = (LC)^{-1/2} .$$

The solutions to the wave equation are:

$$V(x,\omega) = V(\ell,\omega)\cos k(x-\ell) - jZ_o I(\ell,\omega)\sin k(x-\ell)$$

$$I(x,\omega) = I(\ell,\omega)\cos k(x-\ell) - jY_o(\ell,\omega)\sin k(x-\ell)$$

where:

$$Z_o = \frac{L}{C} = \frac{1}{Y_o} ,$$

so that

$$\omega L = k Z_o$$

$$\omega C = k Y_o .$$

The above solutions show that, once the voltage <u>and</u> the current are known at any point $x = \ell$, the voltage and current at any other point can be obtained. If the load is at $x = \ell$, then the current and voltage are related by $V(\ell,\omega) = Z_L I(\ell,\omega)$, where Z_L is the load impedance.

For pulsed excitation, we resolve the pulse into its harmonic components:

$$V(x,\omega) = \int_0^\infty dt\ e^{-j\omega t}\ V(x,t)\ .$$

The inverse transformation brings us to $V(x,t)$ as,

$$V(x,t) = \frac{1}{2\pi} \int_0^\infty d\omega\ e^{j\omega t}\ V(x,\omega)\ .$$

Using $k = \omega v^{-1}$,

$$V(x,t) = \frac{1}{2\pi} \int_{-\infty}^\infty d\omega\ e^{j\omega t}\ V(\ell,\omega)\cos\frac{\omega}{v}(x-\ell)$$

$$- jZ_o I(\ell,\omega)\sin\frac{\omega}{v}(x-\ell)$$

$$I(x,t) = \frac{1}{2\pi} \int_{-\infty}^\infty d\omega\ e^{j\omega t}\ I(\ell,\omega)\cos\frac{\omega}{v}(x-\ell)$$

$$- jY_o V(\ell,\omega)\sin\frac{\omega}{v}(x-\ell)\ .$$

Let us look at a couple of specific examples:

a. $\frac{\omega}{v}\left|x-\ell\right| \ll 1$

In this case, the points are considered to be close together. An equivalent statement is $\frac{t_{x,\ell}}{T} \ll 1$, where the transit time between the two points is $t_{x,\ell} = \left|x-\ell\right|/v$ and the pulse rise time is T ($\sim 1/\omega$). Under these conditions, we have

$$V(x,t) = \frac{1}{2\pi} \int_{-\infty}^{\infty} d\omega e^{j\omega t} \, V(\ell,\omega) - jZ_o \frac{\omega}{v} (x-\ell)I(\ell,\omega)$$

$$\simeq V(\ell,t) - (x-\ell)L \frac{\partial I(\ell,t)}{\partial t} \ .$$

Similarly, for $I(x,t)$,

$$I(x,t) = I(\ell,t) - (x-\ell)C \frac{\partial V(\ell,t)}{\partial t} \ .$$

Of course, we could have derived the last two expressions using the finite difference form of the partial differential equations at the beginning of this section.

    b.   Short circuit at $x = \ell$ ($V(\ell,\omega)=0$). Does the transmission line behave like an inductor when one looks at the relationship between voltage and current at $x = 0$?

We solve for $I(\ell,\omega)$ and find:

$$V(o,t) = \frac{1}{2\pi} \int_{-\infty}^{\infty} d\omega e^{j\omega t} jZ_o \frac{\omega\ell}{v} \frac{\tan \dfrac{\omega\ell}{v}}{\dfrac{\omega\ell}{v}} \, I(o,\omega) \ .$$

Since $Z_o \omega\ell/v = \omega\ell L$, this shorted transmission line behaves like an inductor only if the bracket on the right is equal to 1, i.e., when $\frac{\omega\ell}{v} \ll 1$.

4.3 Higher Order Modes in Transmission Lines

The presence of higher order modes in transmission lines can also introduce difficulties in the interpretation of voltage measurements since TM waves and TE waves can propagate in these structures as well [45]. The cutoff wavelength for TM waves in coaxial lines with large radii of curvature is:

$$\lambda_c \simeq \frac{2}{p} (r_o - r_i); \ p = 1, 2, 3 \ . \ . \ .$$

These modes are akin to waves in a parallel plane guide when planes are p wavelengths apart.

TE waves are of great importance in pulsed power apparatus and have established a practical limit to coaxial, low impedance,

water dielectric accelerators. The cutoff wavelength for TE
waves is:

$$\lambda_c = \frac{2\pi}{n} \frac{r_o + r_i}{2} \; ; \; n = 1, 2, 3 \ldots ,$$

i.e., a wavelength equal to the average perimeter of the apparatus. Since mean perimeters of pulsed power apparatus can be
of the same order as the length of pulse forming lines [46]
($\ell = 3m$, $r \sim 1m$), TE waves can be easily excited by the introduction of penetrations, local breakdowns, or time delays between
multiple switching sites.

The effect of these waves is to produce signals, from monitors at the same axial location, that depend on the azimuthal
location of the monitor. Such waves can seriously overstress
water vacuum interfaces, reduce the electrical efficiency of the
apparatus, and complicate enormously the interpretation of the
measurements.

5.   GENERATION OF IMPULSE VOLTAGES FOR SENSOR CALIBRATION

Establishing the step response of voltage and $\dot{D}$ sensors,
for pulse applications, requires test pulses that: a) have risetimes and durations similar to the pulse under measurement; b)
provide a measurable output signal from the sensor; c) stresses
the components of the sensor to operational level; d) are triggerable, repeatable, and provide pulses with a known magnitude.
Moreover, the pulsers must be reliable and transportable, permitting in-situ calibrations.

All of these conditions cannot be satisfied simultaneously.
Pulsed wave generators that fulfill some of these requirements
consist of a (Fig. 5.1) a 50-Ω coaxial or strip transmission line
charged to a known voltage. A switch, that maintains a 50 Ω wave
impedance geometry, connects the charged line to the output upon
application of a command trigger. On some occasions, a pulse
sharpening, self breaking switch (gas or magnetic cove) is inserted between the triggered switch and the output.

The performance of the switch determines, to a very large
extent, the performance of the pulser. The switch incorporates
several desirable, but conflicting properties: a) high hold-off
voltage, b) low current drain in the open state, c) short transition time, d) low inductance, e) current independent voltage drop
in the closed state, f) triggerability, and g) low jitter.

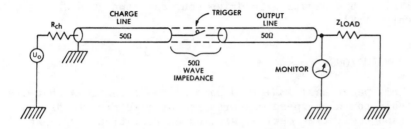

Fig. 5.1. Calibration Pulser Schematic. To avoid distortions,
           pulser must maintain a 50 Ω wave impedance throughout.
           Transition time and unavoidable residual inductance of
           the switch set practical limitations to output rise-
           time.

     For charge voltages up to 2 kV, a popular design that de-
livers pulses with 0.5 ns risetime or less, utilizes a mechanical
switch (mercury wetted reed relay in high pressure nitrogen) [47,
48]. Pulsers, with this type of switch, that deliver lower voltages (50v output), are no longer manufactured but still exist at
many laboratories [49].

     Other pulsers in the 200V range use avalanche transistors
[50]. These form the basis of commercially available fiducial
generators [51] and fast pulsers [51b]. Transition times in the
300 ps range are common for these switches. Avalanche transistor
pulse shaping with a step recovery can decrease the pulse risetime at some expense in amplitude.

     Schemes for generation of higher voltages employ triggered
spark gaps [52-56] and Krytrons. The jitter and limited life-
time precludes their use for all but laboratory applications.

     Impulse generators with 1/2 ns risetime, or less, and low
jitter are a subject of present research, since user driver
generator requirements call for larger outputs and shorter risetimes. In turn, their outputs drive a need for improved diagnostic devices. The 1/2 ns transition time performance limits are
mainly established by the formation times of carriers in the gap.
Faster, high voltage switches (picosecond regime) require optoelectronic techniques that rely on high power lasers for carrier
injection and formation such as Auston switches [57-61].

A promising technique, that does not rely on optical triggers, uses power, metal oxide semiconductors (MOS), operating in parallel with separate gate drive sources, connected directly to the chips [62].

6. CONCLUSIONS

This paper has summarized some of the techniques that are available for high speed measurements of electric fields, voltages, and their time derivatives, using resistive and capacitive dividers. We have not discussed particle [63, 64] and electrooptical techniques [65]. As in the case of magnetic field and current measurements, there is no single sensor configuration that will fulfill all the instrumentation requirements for electric field and voltage measurements. The purpose of this paper has been to survey the available approaches and objectively present their merits and limitations.

REFERENCES

1.   Carl E. Baum, Edward L. Breen, Felix L. Pitts, Gary D.
     Sower, and Mitchel E. Thomas, "The Measurement of Lightning
     Environmental Parameters Related to Interaction with Elec-
     tronics Systems," IEEE Transactions on Electromagnetic Com-
     patibility, EMC-24, pp. 123-137 (1982).

2.   Robert J. Thomas, "High Impulse Current and Voltage Measure-
     ment," IEEE Transactions on Instrumentation and Measurement,
     IM-19, pp. 103-117 (1970).

3.   J. D. Craggs and J. M. Meek, High Voltage Laboratory Tech-
     nique, Butterworths, London (1954).

4.   R. H. McNight and R. E. Hebner, eds., Measurement of Elec-
     trical Quantities in Pulse Power Systems, NBS SP-628, USGPO,
     Washington, D.C. (1982).

5.   Robert Plonsey and Robert E. Collin, Principles and Applica-
     tions of Electromagnetic Fields, McGraw Hill Book Co., New
     York (1961).

6.   Monitor PIM 197-AE, Physics International Company, San
     Leandro, CA, 94577; also Monitor PIM 197-V.

7.   Ronald J. Spiegel, Charles A. Booth, and Edwin Bronaugh, "A
     Radiation Measuring System with Potential Automotive Under-
     Hood Application," IEEE Transactions on Electromagnetic
     Compatibility, EMC-25, pp. 61-69 (1983).

8. Wolfgang Pfeiffer, "Ultra High Speed Methods of Measurement for the Investigation of Breakdown Development in Gases," IEEE Transactions on Instrumentation and Measurement, IM-26, pp. 367-373 (1977).

9. Monitor PIM 197-TE, same as [6].

10. N. Fujimoto, S. A. Boggs, and R. C. Madge, "Measurement of Transient Potentials in Coaxial Transmission Lines Using Co-axial Dividers," in Ref. [4].

11. Mark Wilkinson and Edmond Chu, "Calibration of Capacitive Voltage Probes in Water Dielectric, High Power Pulse Genera-tors," pp. 59-68, same as [4].

12. Carl Baum, et al., "Sensors for Electromagnetic Pulse Mea-surements Both Inside and Away from Nuclear Source Regions," IEEE Transactions on Antennas and Propagation," AP-26, pp. 22-35 (1978).

13. EG&G WASC Inc., Albuquerque, NM, 87119, also: Prodyr Ser-vices, Albuquerque, NM, 87119.

14. Steven R. Ashby, "An Adjustable Length Voltage Divider for Highly Stressed Transmission Lines," Proc. 4th IEEE Pulsed Power Conference, Albuquerque, NM, pp. 393-395, IEEE 83CH1908-3 (1983).

15. J. E. Barth and W. J. Sarjeant, "Direct Subnanosecond Volt-age Monitors," Proc. 3rd IEEE International Pulsed Power Conference, Albuquerque, NM, pp. 171-174, IEEE 81CH1662-6 (1981).

16. J. C. Martin, "Nanosecond Pulse Techniques," AWRE SSWA/JCM/ 704/49 (1970).

17. R. V. Whiteley and J. M. Wilson, "Electrochemical Character-ization of Liquid Resistors," pp. 654-657, same as [14].

18. T. H. Martin, "Effects of High Voltage Pulses on Various Resistors," SNLA RS-5245/013 (1974).

19. Donald A. Pellinen and Marco S. DiCapua, "Two MV Divider for Pulsed High Voltages in Vacuum," Rev. Sci. Instr., 51, pp. 70-73 (1980).

20. Monitor PIM 197-A, same as [6].

21. Monitor PIM 197-B, same as [6].

22. D. Woods, "Improvements in Precision Coaxial Resistor Design," IRE Transactions on Instrumentation, pp. 305-309 (1962).

23. John Barth, Barth Electronics, Boulder City, NV, 89005, private communication (1984).

24. Gilbert Engineering, Phoenix, AZ, 85063.

25. Elcom Systems, Inc., Boca Raton, FL, 33431.

26. Attenuator PIM 230-NN, same as [6].

27. Walter Sarjeant and Richard Dollinger, "A 50 ohm High Voltage Attenuator with 100 Picosecond Rise Time," pp. 403-406, same as (14).

28. Marco S. DiCapua, "Rogowski Coils, Fluxmeters, and Resistors for Pulsed Current Measurements," pp. 175-193, same as [4].

29. Tektronix, Inc., Beaverton, OR, 97005.

30. Walter E. McAbel, Probe Measurements, Measurement Concepts Series, 062-1120-00, same as [29] (1969).

31. Patrick McGovern, "Nanosecond Passive Probes," IEEE Transactions on Instrumentation and Measurement, IM-26, pp. 46-52 (1977).

32. Jim Sargeant, "A High Voltage Probe Line Driver," pp. 170-173, same as [4].

33. W. J. Sarjeant, "High-Voltage Probe System with Subnanosecond Rise Time," Rev. Sci. Instr. 47 pp. 1283-1287 (1976); also, U. S. Patent 4,051,432 (1977); also, same as Ref. [15].

34. Monitor PIM 220-DB, same as [6].

35. J. Power, W. Nunnally, and D. Yound, "A 100-kV, 2-ns Risetime, dc-Coupled Probe," pp. 46-53, same as [4].

36. Jack R. Polakowski, "Pulse Withstanding Capability of Allen Bradley Carbon Composition Resistors," Allen Bradley, El Paso, TX, 79935 (1975).

37. Jack R. Polakowski, "Some Thoughts About Film Resistors," same as [36].

38. Richard Dollinger, SUNY-Buffalo, private communication (1984).

39. Jim Sarjeant and Richard Dollinger, "A 50-ohm Coaxial High Voltage Attenuator with 100 Picosecond Risetime," same as [14], pp. 403-406.

40. C. R. Lennox, "Experimental Results of Testing Resistors under Pulse Conditions," SC-TM-67-559, SNLA (1967).

41. K. G. Moses and T. Korneff, "Voltage Measurements in the Presence of Strong Fields," Rev. Sci. Instr., 34 pp. 849-853 (1963).

42. Robert J. Thomas and Joseph R. Hearst, "An Electronic Scheme to Measure Exploding Wire Energy," IEEE Transactions of Instrumentation and Measurement, IM-16, pp. 51-62 (1967).

43. Marco DiCapua, John Creedon, and Robert Huff, "Experimental Investigation of High Current Relativistic Electron Flow in Diodes," J. Appl. Phys., 47, pp. 1887-1897 (1976).

44. H. S. Carslaw and J. C. Jaeger, Operational Methods in Applied Mathematics, Dover, New York (1963).

45. Simon Ramo, John R. Whinnery and Theodore Van Duzer, Fields and Waves in Communication Electronics, pp. 446-448, John Wiley and Sons, New York (1965).

46. A. Richard Miller, "Sub-Ohm Coaxial Pulse Generators," pp. 200-205, same as (15); also, "Power Flow Enhancement in the Blackjack 5 Pulser (Blackjack 5')," pp. 594-597, same as [14].

47. R. L. Garwin, Rev. Sci. Instr., 21, 903 (1950).

48. SPI Pulse 25, Spire Co., Bedford, MA 01730.

49. Pulser Types 109.

50. Werner Herden, "Application of Avalanche Transistors to Circuits with a Long Mean Time to Failure," IEEE Transactions on Instrumentation and Measurement, 25, pp. 152-160 (1976).

51. Same as [6]. Also, SF-1090, Pacific Atlantic Electronics, El Cerrito, CA, 94530. b) Picosecond Pulse Labs, Boulder, CO, 80306; also Autech Electro Systems, Ottaura, Canada

52. R. C. Fletcher, "Impulse Breakdown in the 10^{-9}-sec Range at Atmospheric Pressure," Phys. Rev., 76, pp. 1501-1511 (1949).

53. R. C. Fletcher, "Production and Measurement of Ultra-High
     Speed Impulses," Rev. Sci. Instr., 20, pp. 861-69 (1949).
     (This reference is still up to date and is required reading
     for anybody in the high speed measurements area.)

54. Daniel McDonald, Carl Benning, and S. J. Brient, "Subnano-
     second Risetime Multikilovolt Pulse Generator," Rev. Sci.
     Instr., 36, pp. 504-506 (1965).

55. Wolfgang Pfeiffer, "Aufbau und Anwendung Kapazitive
     Spannungsteiler extrem Kurzer Anstiegszeit fur Gasisolierte
     Koaxialsysteme," Elektrotechnische Zeitschrift, Ausgabe A,
     94, pp. 91-94 (1975).

56. D. L. Pulfrey, "A Generation of High Voltage Pulses with
     Subnanosecond Risetime and Adjustable Duration," J. Sci.
     Instr., 2, pp. 503-505 (1969).

57. D. H. Auston, "Picosecond Optoelectronic Switching and
     Gating in Silicon," Appl. Phys. Lett., 26, pp. 101-103
     (1975).

58. D. H. Auston, "Impulse Response of Photoconductors in Trans-
     mission Lines," IEEE Transactions, JQE-19, pp. 639-648
     (1983).

59. C. S. Chang et al, "Ultrafast Optoelectronic Switching in a
     Blumlein Pulse Generator," Appl. Phys. Lett., 41, pp. 392-
     394 (1983).

60. P. T. Ho, et al, "Diamond Switches and Blumlein Pulse Gen-
     erators for Kilovolt Optoelectronics," Picosecond Optoelec-
     tronics, Gerard Mourou, Ed., Proc. SPIE 439, pp. 95-100.

61. Russell B. Wilcox, "Silicon Switch Development for Optical
     Pulse Generation in Fusion Lasers at Lawrence Livermore
     National Laboratory," same as [60], pp. 112-115; also Gerard
     Mourou, "High Power Switching with Picosecond Precision,"
     Appl. Phys. Lett., 35, pp. 492-495 (1979).

62. Jeff Oicles and G. Krausse, "Power MOS Fast Switching Tran-
     sistors," 16th Power Modulator Symposium, Arlington, VA,
     18-20 June 1984, UCRL-91738 (1984).

63. Frank Young, "Ion Current and Voltage Determinations by
     Nuclear Techniques," pp. 104-117, same as [4].

64. Richard Leeper, et al., "Proton Current Measurements Using the Prompt Gamma Ray Diagnostic Technique," pp. 267-276, same as [4].

65. James Thompson, "Electro-optical Pulsed Voltage Measurements," pp. 1-19, same as [4].

HIGH SPEED MAGNETIC FIELD AND CURRENT MEASUREMENTS

Marco S. Di Capua

Lawrence Livermore National Laboratory
Livermore, CA
USA

1.   ELECTROMAGNETIC SENSOR BASICS [1]

An electromagnetic sensor is a special antenna with the following properties:

a. It is an analog device that converts the electromagnetic quantity of interest to a voltage or current (in the circuit sense), at a terminal pair, to drive a current through a load impedance, usually the characteristic termination of the transmission line signal lead.

b. It is passive, i.e., it does not contain any energy source other than the field under measurement.

c. It is a primary standard in the sense that its sensitivity for conversion of fields into volts and amperes is well known in terms of its geometry, i.e., it is calibratable by a ruler. The impedances of the loading elements may be measured and trimmed. In principle, the sensor is as accurate as the length and resistance standards in a calibration facility.

d. It is designed to have a specific constant sensitivity.

e. Its transfer function is devised to be simple across a wide frequency band. This means flat in terms of volts per unit of field or of time derivative of field or it may mean another simple mathematical form that can be specified with a few constants in which case more than one number defines the sensitivity and frequency response.

f. It does not perturb the quantity being measured, i.e.,
its insertion in the environment does not change the environment
configuration. This may be understood in terms of a minimum or
negligible insertion capacitance, inductance or resistance in the
circuit sense; or minimum or negligible field perturbation.

2.   MAGNETIC FIELD SENSORS

2.1 Basic Considerations

A magnetic field sensor responds to an applied magnetic
field by delivering an output voltage proportional to the applied
field at one frequency extreme, or its derivative at the other.
While the coupling between the field and the sensor may be described by a mutual inductance between the sensor and circuit, it
is more convenient to describe the sensor in terms of an equivalent area or calibration constant as described below.

When transit time effects in the sensor can be neglected
(i.e., the sensor is electrically small) the sensor may be described by its voltage source equivalent circuit as shown in
Figure 2.1.

The voltage source is the negative of the dot product of the
equivalent area of the sensor $\vec{A}_{eq}$ and the time rate of change of
the magnetic field $\partial\vec{B}/\partial t$ at the sensor. We define this dot product as

$$\frac{\partial\vec{B}}{\partial t} \cdot \vec{A}_{eq} \equiv \int_A \frac{\partial\vec{B}}{\partial t} \cdot d\vec{A} = \frac{d\phi}{dt} .$$

This definition assumes there is no relative motion between the
field and the sensor and is exact when the variation of the field
over the dimensions of the sensor is small. We define the quantity $R = R_c + R_b$, where R_c is the sensor (coil) resistance and R_b
is the burden resistance. The burden resistance R_b consists of
the parallel combination of the current viewing resistor R_{CVR}
and the termination (metering) resistor R_r at the recording device.

The circuit equation for this sensor is [2]:

$$-\frac{1}{R}\frac{d\phi}{dt} = \frac{L}{R}\frac{di}{dt} + i$$

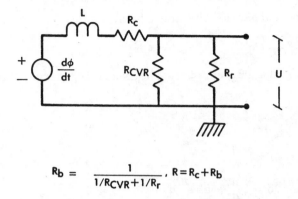

$$R_b = \frac{1}{1/R_{CVR}+1/R_r} , \quad R = R_c + R_b$$

Fig. 2.1. Equivalent circuit of magnetic field sensor. The voltage source is - dϕ/dt. L and R_c represent the inductance and resistance of the winding. R_{CVR} and R_r are
the resistance of the current viewing resistor and recording aparatus. The burden resistor R_b is the parallel combination of R_{CVR} and R_r.

The three terms in this equation suggest two time domains: in
one, the characteristic time, τ_c, for changes of current in the
primary circuit is much smaller than the characteristic time
τ_d = L/R. Therefore, (L/R) di/dt $\sim$ (τ_d/τ_c) i >> i and hence
i $\sim$ = ϕ/L. In the other time domain τ_c >> τ_d. Hence, τ_d/τ_c << 1
and - (1/R) dϕ/dt $\sim$ i. In the first time domain the sensor is
called integrating. In the second time domain the sensor is
called differentiating. It is important to note that, very early
in the pulse, the condition τ_d/τ_c << 1 cannot be satisfied so
that a differentiating sensor has an upper frequency cutoff equal
to $\sim$ 2.2 R/L. As an example, a sensor with L = 50 nH, R_c = 0,
R_{CVR} = ∞, and R_r = 50 Ω, has a 10-90% risetime of 2.2 x 1ns =
2.2 ns. In this case, for τ>> τ_d, the voltage of the measuring
device is

$$U = R_r i = - \dot{\phi} .$$

In the case $\tau_c \ll \tau_d$ it is standard practice to make $R_{CVR} \ll R_r$. One may not be able, however, to neglect R_c with respect to R_{CVR}. The output of the sensor is:

$$U = iR_{CVR} = - \phi R_{CVR}/L .$$

The solution for the current i in the measuring circuit is, for

$$\phi(t) = \begin{cases} 0 & t < 0 \\ \phi(t) & t > 0 \end{cases}$$

$$i(t) = \begin{cases} 0 & t < 0 \\ - \dfrac{1}{L} e^{-t/\tau_d} \displaystyle\int_0^t \dfrac{d\phi}{dt'} e^{t'/\tau_d} dt' & t > 0 . \end{cases}$$

Two useful solutions for pulsed applications arise. For a ramp function input:

$$\phi = \begin{cases} 0, & t < 0 \\ \phi_o \dfrac{t}{\tau_r}, & t > 0 \end{cases}$$

the solution is:

$$i = \begin{cases} 0 & t < 0 \\ - \dfrac{\dot\phi_o}{L} \dfrac{\tau_d}{\tau_r} \left[1 - e^{-t/\tau_d} \right] & t > 0 \end{cases}$$

For a step function input with an amplitude ϕ_o a duration equal to t_p;

$$\phi(t) = \begin{cases} 0 & t < 0 \\ \phi_o & 0 < t < t_p \\ 0 & t > t_p \end{cases}$$

such that:

$$\frac{d\phi}{dt} = \begin{cases} \phi_o \delta(t - 0) \\ -\phi_o \delta(t - t_p), \end{cases}$$

$$i = \begin{cases} 0 & t < 0 \\ -\dfrac{\phi_o}{L} e^{-t/\tau_d} & 0 < t < t_p \\ -\dfrac{\phi_o}{L} e^{-(t - t_p)/\tau_p} \left[e^{-t_p/\tau_d} - 1 \right] & t > t_p \end{cases}.$$

The solutions are very useful for interpreting the output of a sensor in pulsed power applications.

We can now examine how the solutions behave in the time extremes. In the time domain $\tau_c \ll \tau_d$, the quantity e^{τ'/τ_d} does not change in a timescale in which ϕ undergoes rapid change so it can be removed from under the integral sign. The general solution then becomes:

$$i = -\frac{1}{L} e^{-t/\tau_d} e^{t/\tau_d} \int_o^t \frac{d\phi}{dt'} dt' = (\phi(o) - \phi(t))/L .$$

In the time domain $\tau_c \gg \tau_d$ the quantity $d\phi/dt$ does not change in a time scale which is many times τ_d, so $d\phi/dt$ can be removed from under the integral sign. The general solution then becomes:

$$i = -\frac{1}{L} e^{-t/\tau_d} \frac{d\phi}{dt} \int_o^t e^{t'/\tau_d} dt'$$

$$= -\frac{1}{R} \frac{d\phi}{dt} \left[1 - e^{-t/\tau_d} \right] = -\frac{1}{R} \frac{d\phi}{dt} .$$

Since this solution is valid only for $\tau \gg \tau_d$, we neglect the second term in the bracket because it is much less than 1. Of course we reached the same conclusions by examining the differential equation directly.

In the frequency domain [3] the response of the sensor may be characterized by:

$$-\frac{1}{R} j\omega\phi_o = \frac{L}{R} j\omega i + i ,$$

where the $e^{j\omega t}$ dependence has been factored out of the equation.

Then:

$$\frac{i}{\phi_o} = -\frac{1}{L} \frac{jw\tau_d + \tau_d^2 w^2}{1 + w^2 \tau_d^2} \, .$$

If $w\tau_d \ll 1$ then

$$\frac{Li}{\phi_o} = -jw\tau_d \, .$$

In this case, the sensor output is proportional to the frequency and the sensor is considered differentiating.

If $w\tau_d \gg 1$ then

$$\frac{Li}{\phi_o} = -1 \, ,$$

so the output of the sensor becomes independent of frequency and is usually called self-integrating.

The output of the sensor may be displayed as a function of frequency in the plot of Fig. 2.2 that shows the differentiating domain, the self-integrating domain and the - 3 db cutoff points.

2.2 Discrete Magnetic Field Sensors--Differentiating Sensors

Discrete magnetic field sensors span a wide range of complexity, from a single loop at the end of a coaxial cable, to a multigap loop (MGL) [3] in a Moebius geometry or to miniature sensors with accurate and repeatable geometries [4] produced with printed circuit techniques. The advantage of discrete magnetic field sensors is that their construction is generally quite simple and they provide a large output signal per unit surface area. Among their disadvantages, they are high impedance devices and the currents induced by capacitive coupling with the environments, and space charge currents collected by the sensor may be comparable with the currents flowing in the sensor. Consequently, both these currents may introduce large errors in the measurement unless special precautions are taken. Another disadvantage is that the large voltages induced by rapidly varying fields may overvolt components in the measuring chain.

Regardless of the design complexity, however, the sensors may still be described by an equivalent area and the upper frequency response, for differentiating sensors, is still bound by

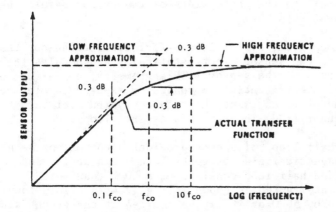

Fig. 2.2. Frequency response of an electrically small sensor
[3]. For low frequencies the sensor output is proportional to frequency (differentiating sensor). For
high frequencies the sensor output is independent of
frequency (integrating sensor). The 3dB point corresponds to the inverse of the time constant of the
sensor.

the characteristic time $\tau_d = LR^{-1}$. We emphasize that, while the
equivalent area is directly proportional to N (the total number
of turns), the inductance of the sensor is in general proportional to N^2. So, an increase in sensitivity of the sensor is
directly accompanied by a reduction in bandwidth at the upper
frequencies due to the increase of τ_d.

Construction techniques for magnetic field sensors that operate in the presence of large electrostatic fields or space
charge currents deserve special attention. Under these conditions the sensor will respond to, in addition to the magnetic
field, electrostatic fields and will act as a charge collector
as well. These two phenomena determine the accuracy of the
measurement. These difficulties may be circumvented by electrostatic and charge shielding or by a differential measuring technique. For the differential measuring technique, the sensor consists of two identical loops wound in opposite directions.

The difference between the two signals, A and B, at the recording device, adds the magnetic field contribution from the two
loops and zeroes out the common mode noise picked up by the two
loops. Of course, for good rejection of common mode noise, the

sensor must be small so that both sides of the sensor sample the same environment.

A Moebius loop geometry [5] is a differential technique that completely shields the sensor electrostatically and doubles the sensitivity per turn of the sensor. This geometry, appearing in Fig. 2.3, reduces significantly the conductor exposed to electro-static fields and provides twice the area equivalent per turn. The MGL [6,7] sensor (EGG) incorporates several loops.

A shielded half loop [3] sensor provides a convenient method to measure the magnetic field, tangential to a conducting sur-face. The shielded half loop consists of a 50 Ω coax which shield has been interrupted as shown in Fig. 2.4(a). This loop may be terminated by 50 Ω or it may be shorted at the ground side if it is used in a self-integrating configuration. The advantage of this arrangement over the simple half loop of Figure 2.4(b) is

that it discriminates against the predominantly normal $\vec{E}$ fields over a ground plane. Such a sensor is also quite simple to build.

A B

Fig. 2.3.   A Moebius loop field sensor [5]. In this geometry the conductor has a complete electrostatic shield. Output is recorded differentially as A-B.

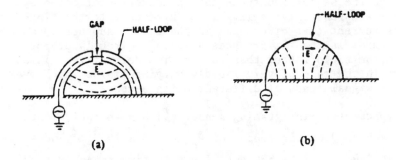

Fig. 2.4.   (a) A shielded half loop to measure magnetic fields
            tangential to a conducting plane. This arrangement
            shields the loop from electric fields which are mostly
            normal to the plane. The effect of these fields in an
            unshielded sensor appears in part (b) of the figure
            [3].

For such a loop the inductance is

$$L_p = \frac{R\mu_o}{2} \ln\left(\frac{8R}{a} - 2\right) ,$$

where R is the loop radius and a is the wire thickness. Following the example of Ref. [3], a 0.023-in.-diameter coax, bent on a
0.35-in. loop radius, yields an inductance of $\sim$ 20 nH. Consequently, risetimes of $\tau \sim L/R \sim$ 400 ps should be attainable with
this device.

    Another field sensor configuration is the cavity sensor. A
cavity, or depression on the surface of a current carrying conductor, form a loop completed by the signal lead. The cavity
sensor, widely used in pulsed power apparatus, will be discussed
in detail in the section on current measurements.

2.3 Self-integrating Discrete Field Sensors

    Self-integrating field sensors [8] have a current viewing
resistor of $10\text{-}20 \times 10^{-3}$ Ω incorporated directly across the loop
terminals, at the connection between the loop and the signal

cable. The small value of the current viewing resistor provides some distinct advantages: currents induced by capacitive
coupling to the sensor and space charge currents collected by
the sensor are generally negligible with respect to the flux-
excluding current that circulates in the loop. Another advantage
of this sensor is that it produces a voltage directly proportional to the field rather than to its time derivative. This is
important in the case of very rapidly varying fields which overstress the signal leads of differentiating sensors.

A typical self-integrating sensor [9] has an equivalent
area for a 5 cm-diameter sensor of A_{eq} = 1.8 x 10^{-3} m^2 and a time
constant $L^{-1} R_{CVR} = \tau_d$ = 550 ns. The resulting detector sensitivity is 3.3 x 10^3 V T^{-1}.

Since the output of this sensor is

$$U = -\phi L^{-1} R_{CVR} = -\vec{B} \cdot \vec{A}_{eq}/\tau_{d'} ,$$

incorporating more turns in the sensor (N turns) results in an
increase in inductance by N^2, so that the net result is a decrease in sensitivity by N. Similarly an increase in the value
of R_{CVR} is accompanied by a decrease in τ_d, resulting in a decrease in the bandwidth at lower frequencies.

3.   CURRENT MEASUREMENTS

3.1 Basic Considerations

The measurement of current passing through a surface is of
great interest in pulse power technology. Current measurements
rely upon the integral form of one of Maxwell's equations as:

$$\int_c \vec{H} \cdot d\vec{\ell} = \int_s \vec{J}_t \cdot d\vec{A} = I_t ,$$

where $\vec{J}_t$ is the total current density passing through the surface
element $d\vec{A}$. This current density, in the most general case, is
composed of three parts:

$$\vec{J}_t = \vec{J}_c + \vec{J}_\sigma + \vec{J}_\varepsilon .$$

$\vec{J}_c$ is the source current density (i.e., space charge current flow
associated with Compton electrons in a nuclear burst environment,

sheath electron currents in a magnetically insulated vacuum
transmission line or beam currents in a particle accelerator),
$\vec{J}_\sigma$ is the conduction current described by Ohm's law and $\vec{J}_\varepsilon$ is the
displacement current density. In linear, time invariant, iso-
tropic media:

$$J_t = \vec{J}_c + \sigma\vec{E} + \frac{\partial}{\partial t}\vec{D} \ .$$

Hence, the measurement of current is reduced to the measurement
of the line integral of $\vec{H}$ around a closed path, namely the curve
c bounding the surface $\vec{A}$ threaded by the current of interest.

The measurement of $\oint_c \vec{H} \cdot d\vec{\ell}$ is performed by sampling the
field around the curve c with a number of equally spaced loops.
In the case the loops form a helix, the sensor takes a special
name, Rogowski coil [10] or belt. We will discuss Rogowski coils
in detail at the end of this section.

3.2 Measurement of $\oint \vec{H} \cdot d\vec{\ell}$ with Discrete Loops

One approach to the measurement of $\oint \vec{H} \cdot d\vec{\ell}$ is to sample the
magnetic field, along the contour c, with a number of equally
spaced discrete loops [6,11] while adding the contributions of
each loop. In this case the output of each loop is

$$\frac{d\phi_i}{dt} = A_i \frac{dB_i}{dt} \ .$$

If the loops are identical and spaced a distance $\Delta\ell$ apart, then:

$$\frac{d\phi_i}{dt} = \frac{A}{\Delta\ell} \frac{dB_i}{dt} \Delta\ell \ .$$

The total flux then becomes, for N loops:

$$\frac{d\phi}{dt} = F \frac{\mu_o A}{\Delta\ell} \sum_{i=1}^{N} H_i \ \Delta\ell \equiv F \frac{\mu_o A}{\Delta\ell} I \ .$$

The factor F, of course, depends on the method of addition of the
signals from the individual loops.

One method to add the signals is to terminate each loop with
a series resistor with a value equal to the impedance of the

signal cable [11]. The signal cables, of identical length, connect to a common tie point. In this case the output of each loop
is:

$$\dot{\phi}_i = \frac{1}{2} A \frac{dB_i}{dt} .$$

The signal ϕ_{iT}, launched in the other N cables, at the common tie
point (N-1 connected to the remaining loops plus one more cable
which is the signal lead), is

$$\dot{\phi}_{iT} = \frac{2 \cdot \frac{1}{2} A \frac{dB_1}{dt}}{NZ_o + Z_o} = \frac{A \frac{dB_i}{dt}}{N + 1}$$

By superposition,

$$\dot{\phi}_T = \frac{A}{N + 1} \sum_{i = 1}^{N} \frac{dB_i}{dt} ,$$

so that the constant $F = (N + 1)^{-1}$.

The terminating resistors, connected in series with the loops,
absorb the reflections originating at the common tie point and
absorb the signals transmitted from the i_{th} sensor to all the
other sensors. Moreover, the presence of the terminating resistors can take care of minor differences in the length of the
signal leads.

Another approach [6] is to use N sensors, each connected to
a cable of impedance Z_o. The sensor end does not incorporate a
terminating resistor. The N cables feed a signal cable of impedance Z_o/N. In this case

$$\phi_{iT} = \frac{2A}{2N + 1} \frac{dB_i}{dt}$$

and using superposition:

$$\dot{\phi}_T = \frac{A}{1 + \frac{1}{2N}} \sum_{i = 1}^{N} \frac{dB_i}{dt} .$$

Since the signal lead from each sensor looks at the parallel connection of N-1 cables of impedance Z_o and one cable of impedance Z_o/N, the reflected signal from the junction is:

$$\phi_{iR} = \frac{-2\ (N-1)}{2\ (N+1)}\ \phi_i \ .$$

This signal reflects again at the sensor (shorted end) as

$$\phi_{iRR} = \frac{2\ (N-1)}{2N+1}\ \phi_i \ .$$

The transmitted signal, along the remaining N-1 cables, is

$$\phi_{iT} = \frac{2}{2N+1}\ \phi_i \ ,$$

which reflects off the sensors' shorted ends as

$$\phi_{iTR} = \frac{-2}{2N+1}\ \phi_i \ .$$

These reflected signals add up at the common tie point as:

$$\phi_{reflected} = \phi_{iRR} + (N-1)\ \phi_{iTR}$$

$$= \phi_i\ \frac{2\ (N-1)}{2N+1} - \frac{2\ (N-1)}{2N+1}\ = 0 \ .$$

Hence, after one reflection at the sensor, all the reflected signals cancel at the tie point.

This approach requires all sensors and their signal cables to be identical. Normally, signals of one pair of sensors, each driving a 100 Ω cable, are added into a 50 Ω signal lead. The resultant signal from another pair of sensors, wound in the opposite direction, is then subtracted at the recording device.

3.3 Current Measurements with Cavity Sensors

Cavity sensors are used widely for current measurements in pulsed power generators. Their geometry appears in Figs. 3.1(a) [12], (b) [11], and (c) [13]. As discussed before, a cavity on

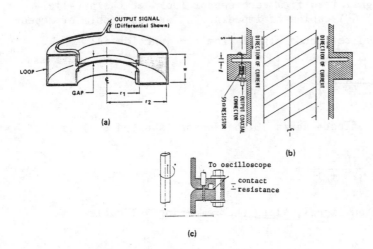

Fig. 3.1. Cavity sensors: (a) this geometry [12] compensates
           errors arising from diffusion of magnetic field in a
           cavity; (b) is a form commonly used in pulsed power
           accelerators [11]. Several pickoffs around the peri-

           meter of the cavity allow a measurement of $\oint H \cdot d\ell$;
           (c) the least desirable geometry of cavity sensor.
           Contact resistance at bolted joint can introduce serious
           random errors.

the surface of the return conductor forms part of a loop completed by the signal lead of the sensor. The area of the sensor
is then the shaded region of the figure.

    This monitor arrangement is very effective because it can
be easily incorporated in pulsed power apparatus. It does have
built in, however, two closely connected sources of systematic
error, due to the diffusion of magnetic field into the walls of
the cavity. One source is the voltage drop along the loop caused
by the current in the circuit. This voltage drop adds to $d\phi/dt$
in the source term of the equivalent circuit of Fig. 2.1. The
other source of error is the area increase of the loop, associated with the diffusion of magnetic field into the walls of the
cavity.

    We can easily estimate the contributions of these terms by
assuming a square cavity with side b, located at a radius a from
the center of a current I. We further assume that the same current I returns along the wall with a uniform linear current density equal to $I/2\pi a$. Assuming the field penetrates a distance

$\delta = \sqrt{2t^{-1}}\mu\rho$ in the wall, the error introduced by the resistive drop in the cavity is:

$$\Delta_R = \frac{IR}{\dot{\phi}} = \frac{I}{\dot{I}t}\frac{\delta}{b} \; .$$

By assuming that $\dot{I}t \sim I$, the error then becomes:

$$\Delta_R = \frac{\delta}{b} \; .$$

The error introduced by the increase in cavity area, due to diffusion of magnetic field in the cavity is:

$$\Delta_D = \frac{4b\delta}{b^2}\frac{\dot{B}}{\dot{B}} = \frac{4\delta}{b^2} \; .$$

A more rigorous treatment [11] takes both effects into account at once and yields, for a square wave input,

$$\Delta = \Delta_D + \Delta_R = \frac{4\sqrt{2}}{\pi}\frac{\delta}{b} \; .$$

The error is larger for higher resistivity wall materials and grows with a square root dependence on time.

As an example, for aluminum $\rho = 4 \times 10^{-8}$ Ω m. So, for $\tau = 0.1$ μs, $\delta = 8 \times 10^{-3}$ cm and for b = 0.5 cm, Δ = 3%. For stainless steel, on the other hand, under the same conditions the error is Δ = 9%.

In the construction technique of Figure 3.1(c), a source of systematic error is the contact resistance of the joint of the two flanges. In this setup, the magnitude of the error will change every time the apparatus is reassembled.

The errors due to diffusion of fields in the walls of the cavity may be compensated with the geometry of Fig. 3.1(a) [12], where two loops sample the field in a differential arrangement. The common mode error, due to the voltage drop in the cavity, is zeroed out in this arrangement by taking the difference of the two signals at the recording instrument. The compensation is effective only for uniform excitation of the sensor.

Since the inductance of a typical monitor is of the order of 2×10^{-9} H, the L/R risetime of the monitor, driving a 50 Ω impedance, is about 40 ps. A limitation on bandwidth arises from

the capacity that exists between the monitor lips [12]. This
capacity is of the order of 50 pF. Therefore, the monitor risetime is limited by the intrinsic risetime of the cavity itself
and the inductances of the small wire stub completing the loop.

This intrinsic risetime is characterized by $\tau = 1/f = 2\pi \sqrt{LC}$, of
the order of 2 ns. The gain/bandwidth tradeoff is also evident
for this monitor since increasing L, hence M, to increase the
sensitivity, results in an increase in cavity risetime.

The techniques we described at the end of the last section
for addition of signals apply to this monitor as well. Meas-
urements of $\oint \vec{H} \cdot d\vec{\ell}$ with these sensors are normally taken by
sampling the magnetic field at 4 or 8 azimuthal locations.

A very important limitation on the maximum value of $d\phi/dt$
the monitor can measure arises from the breakdown threshold of
the electrical hardware linking the monitor to the attenuators.
This limitation is about 5 kV for TCC and N type connectors, 1 kV
for BNC connectors and about 500 V for OSM connectors. Since
$d\phi/dt = MdI/dt$, to measure current risetime of $\sim 10^{13}$ A s^{-1} with
a monitor with a 1 cm^2 monitor area, the monitor cannot be any
closer than 5 cm to the centerline of the current distribution.

Another limitation is of course the breakdown field of the
gap of the monitor. The presence of space charge flow or radiation can substantially reduce this breakdown field. There is
evidence that this field cannot exceed 25 kV cm^{-1} [15] for reliable and consistent operation.

3.4 Measurement of $\oint \vec{H} \cdot d\vec{\ell}$ With Continuous Loops

All the current measurements we have described so far
utilize discrete loops to sample the time derivative of the magnetic field around a contour. Electrical addition of the signals
provides the line integral of the magnetic induction around the
contour. A Rogowski sensor on the other hand will yield a direct
measurement of $\oint \vec{H} \cdot d\vec{\ell}$, or its time derivative, without requiring
any further elaboration of the signals. A Rogowski sensor [10],
as shown in Fig. 3.2, consists of a coil: a) whose centerline
follows a contour c of arbitrary shape; b) whose turns are
equally spaced with a density n per unit length; c) the area $\vec{A}$ of
each turn is everywhere the same; d) the area $\vec{A}$ is locally normal
to the contour c throughout the coil, i.e., the unit vectors
associated with $\vec{A}$ and $\vec{c}$ are everywhere collinear; and e) the

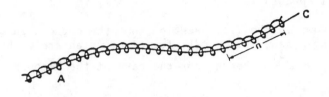

1. Crossection of Coil <u>A</u> is Everywhere the Same

2. Number of Turns <u>n</u> per Unit Length is Constant

3. B Field is Homogeneous Over the Area of One Turn

4. Area of Each Turn is Normal to <u>C</u> (locally)

$$\therefore \; \phi = nA \int_c \vec{B} \cdot \vec{d\ell}$$

Fig. 3.2. The essential features and basic assumptions for a Rogowski geometry $\int_c \vec{H} \cdot d\ell$ sensor.

distance scale for the variation of magnetic field is much smaller than the dimension of one turn.

When the above conditions are met, the flux, ϕ, linked by the turns of the sensor:

$$\phi = \int_c \int_A \int n \, (\vec{B} \cdot d\vec{A}) \, d\ell$$

is equivalent to

$$\phi = nA \int_c \vec{B} \cdot \vec{d\ell} \; .$$

Assuming no relative motion between the sensor and the field, $\dot\phi = -nA \int \dot{\vec{B}} \cdot \vec{d\ell}$ becomes the source in the circuit of Fig. 2.1. In the case the core is air or plastic, as is usually the case:

$$\phi = nA \, \mu_o \int_c \vec{H} \cdot \vec{d\ell} = nA\mu_o I_{enc} \; .$$

The RHS is valid when: (a) c forms a closed curve and (b) the curve c itself does not enclose any flux. To guarantee that c closes no flux, the coil is normally surrounded by a metal shield [2] which also acts as an electrostatic shield for the

windings. This shield, of course, has a slit to allow the magnetic field to penetrate to the coil. A typical arrangement for
a Rogowski coil appears in [2].

We have shown in the first section that the current in the
self-integrating sensor is t = ϕ/L, for times which are much
shorter than τ_d = LR^{-1}. Hence, for the case of the Rogowski coil

$$i = nA\mu_o I_{enc} L^{-1} .$$

If there is no flux leakage between the coil and the primary
circuit, L = μ_oAnN, where N is the total number of turns. Consequently, i = -I$_{enc}$/N.

In general, some flux leakage is present and, consequently,
more detailed formulas are required to calculate the self-inductance of the coil [2] and the mutual inductance between the coil
and the circuit. Another effect at very high frequencies is the
frequency dependent resistance of the winding conductor. This
frequency dependent resistance must also be taken into account
in calculating the sensitivity of the coil. For a self-integrating coil, the current i in the coil flows through the parallel connection R$_b$ of the current viewing resistor and terminating
resistance at the metering device. Hence, the voltage developed
at the metering device becomes

$$U = - \frac{\phi R_b}{L} \simeq - \frac{I_{enc}}{N} R_{CVR} .$$

The RHS is valid in the case of: no flux leakage, a current
viewing resistor whose value is much less than the terminating
resistor at the oscilloscope, and a pulse duration much less than
τ_d = L/R.

The response of a typical Rogowski coil [2], under calibration, appears in Figs. 3.3(a) through (c). Figs. 3.4(a) and
3.4(b) show representative measurements in the 100 kA and 3 MA
ranges.

3.5   Current Transformers for Pulsed Applications

The coil we have described so far is a special case of a
current transformer, constructed with a dielectric core. In many
instances it is very desirable to improve the low frequency response without sacrificing the sensitivity of the coil. One way
to accomplish this is to replace the dielectric core of the coil

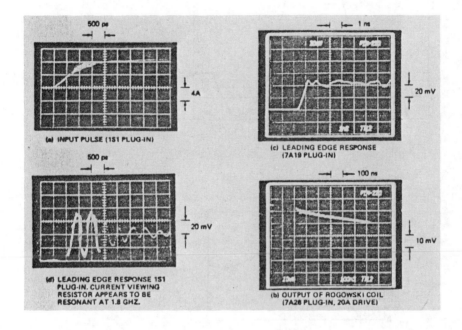

Fig. 3.3.   Calibration response of a typical Physics International PIM-220 [2] Rogowski coil to a 500 ps risetime
pulse (a). Notice resonance of current viewing resistor (d) at 1.8 GHz. Resonance is not visible with a
Tektronix 7A19 preamplifier (c). Decay of output is
visible in (b).

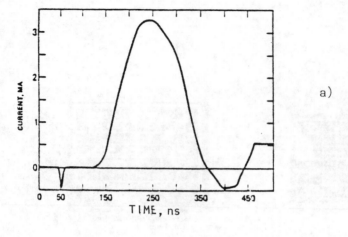

a)

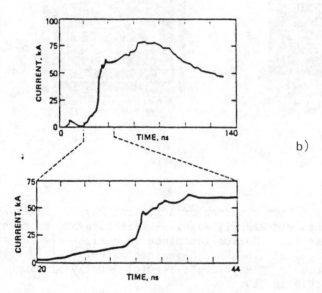

b)

Fig. 3.4. (a) Current waveforms obtained with a Physics International PIM-220 coil on a 10 m-long magnetically
insulated vacuum transmission line [35]. Notice sharp
leading edge due to erosion of the pulse.

(b) A 3 MA pulse delivered by the PITHON accelerator [2] as measured with a PIM-220 coil.

with a ferromagnetic core of a larger permeability [16,17]. From a very elementary viewpoint, the core does not change the ratio of ϕ/L in the core, since μ enters in both quantities to the same power. However, the magnetic material greatly increased the L/R decay time.

For an ideal Rogowski geometry coil, we have shown that:

$$i = - \frac{I_{enc}}{N} .$$

Consequently, there would be no magnetic induction in the core if the resistance of the secondary windings, plus burden resistance were zero. However, in a practical transformer, there is a winding resistance and a metering resistance that cause the current in the secondary to decay with a characteristic time $\tau_d = L/R$. A magnetic material can increase L and hence τ_d significantly by providing a large change in B for a small charge in H. However there is a limit set by saturation to the amount of change in B a magnetic material can accommodate. Therefore an important source of error in ferromagnetic core transformer measurements is the saturation of the core, arising from the decay of the induced current in the secondary windings of the transformer.

In general, ferromagnetic cores will tolerate a limited swing ΔB, in the magnetic field, in the core, before the core saturates. The swing in magnetic field ΔB is a property of the core material itself. To calculate ΔB [16], for a Rogowski coil, with a ferrite core, we observe that the voltage U across the resistance R, for times t $\ll$ L/R, is:

$$U = - \frac{\partial B_{core}}{\partial t} AN = - d\phi/dt .$$

We have also shown that

$$i = \frac{I}{N}, \ U = Ri .$$

Therefore:

$$I = \frac{N^2 A}{R} \frac{\partial B_{core}}{\partial t} .$$

Consequently:

$$\Delta B = \frac{R}{N^2 A} \int_0^t I dt \ .$$

ΔB must be kept below saturation at all times. The expression above is also derived heuristically in Reference [18].

For a given core material (fixed ΔB), the current time product can be increased by increasing the core area, the number of turns and by decreasing the burden resistance.

At the low frequencies, the magnetic core provides the required secondary inductance and allows a tight coupling of the primary and secondary through the large mutual flux. As the frequency increases, the effective permeability of the core decreases, reducing the coupling between portions of the winding so that uniform excitation becomes important [16]. At the highest frequencies ~ 1 GHz, the core plays very little role and the structure behaves like an air core Rogowski coil.

The ringing of a self-integrating Rogowski coil at a frequency given by the inverse of the two-way transit time of an EM wave in the windings is the main disadvantage these devices have. This ringing generally arises from a non-uniform excitation of the windings produced by a nonsymmetric current distribution. From the lumped circuit veiwpoint, these oscillations appear due to the stray capacitance between the windows and the core, the windings and the shield, and between adjacent turns in the winding.

Oscillations can be damped successfully by a network of resistors. These resistors may bridge several turns, as in the Pearson [18] current transformer, or connect every nth turn to a common but independent annular ring, as in the GE design [16,19]. This second damping method is preferable because it will attenuate propagating modes without affecting the desired response. Specifically, under uniform excitation, there is no ϕ gradient in the azimuthal direction of the coil and hence no current will flow in the damping network. Oscillations arising due to a nonuniform excitation place a limit (~ 20 cm) on the outer diameter of self-integrating coils.

Ferrite loaded current transformers are available commercially from other sources as well [20].

Current transformers have found applications in biology, too [37]. This transformer has a major diameter of 1 mm and measures the cellular currents associated with the action potential in isolated nerve bundles.

3.6 Transmission Line Description of Continuous Loop-- $\oint \vec{H} \cdot d\vec{\ell}$ Sensors

The lumped parameter circuit description is no longer valid
when the sensor undergoes nonuniform excitation by pulses whose
characteristic, time τ_c is smaller than the transit time for an
electromagnetic wave in the coil. The reason, of course, is that
currents induced at different locations in the sensor reach the
metering resistor R_b with different delays. Since these coils
are high Q devices, these waves will propagate with several reflections before damping out. These effects can be properly
accounted for by a transmission line description of the coil.

In 1963 J. Cooper [21] developed a transmission line model
for field and $\oint \vec{H} \cdot d\vec{\ell}$ sensors. The model includes the distributed capacitance per unit length between the coil and the
shields/core C, the inductance per unit length , and the excitation electric field as voltage per unit length u(X). The partial differential equations describing the coil behavior may be
solved using a Laplace transform technique. The boundary conditions applied at the sensor terminals may be short circuit,
resistive, inductive or capacitive. A differential section of a
transmission line coil appears in Fig. 3.5. A schematic of
its construction appears in Fig. 3.6 [22]. A commercially available transmission line coil is described in [23].

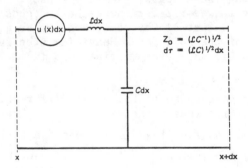

Fig. 3.5. Differential section of a transmission line coil.
D(x), $\mathcal{L}$, and C are the excitation, inductance, and
capacitance per unit length of the transmission line.

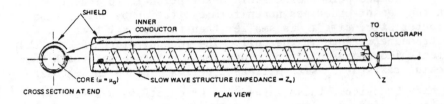

Fig. 3.6. A schematic of the construction of a transmission line
           coil. Outer shield has a slit to allow magnetic flux
           to penetrate in the coil. Terminating impedance at
           right will determine behavior of the coil.

To illustrate the effect of transit time we use a Heaviside step current, with a delta function dI/dt. We assume $I_p = I_o H(t-0)$ and $\dot{I}_p = I_o \delta (t - 0)$. Furthermore, we assume that $u(x)$ is a constant along the Rogowski geometry sensor. A short circuit and a metering resistor Z terminate the two ends of the coil respectively. The output of the coil, appears as a voltage across Z. The output of the coil for a Heaviside step function excitation, appears in Fig. 3.7, for terminations (a) $Z \gg Z_o$', (b) $Z = Z_o$, and (c) $Z \ll Z_o$.

Cases (a) and (b) correspond to the usual differentiating sensor. Indeed, for a differentiating sensor, $\oint U_{coil} \, dt = \phi = \mu_o InA$. If we assume no flux leakage between the coil and the primary circuit, $L = \mu_o AnN$ so taht $\phi = ILN^{-1}$. If we integrate the waveforms (a) and (b) of Fig. 3.7 and use $\ell (\mathcal{L}C)^{1/2} = \tau$ and $(\mathcal{L}C^{-1})^{1/2} = Z_o$, we obtain waveforms (a) and (b) of Fig. 3.8. The mean value of $\int U_{coil} \, dt = ILN^{-1}$, for $Z = Z_o$, and the final value ILN^{-1}, for $Z = Z_o$ at $t = \tau$, agree with the lumped parameter description. We notice, however, that the electrical length of the

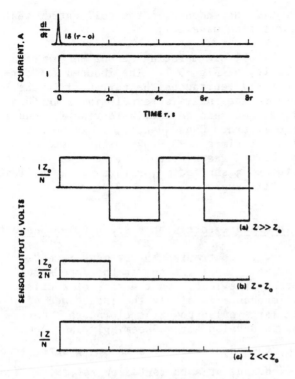

Fig. 3.7.   Response of transmission line coil to a step function
excitation (top trace). Ratio of terminating impedance Z to transmission line impedance Z_o of coil determines response. When $Z<<Z_o$ coil is self-integrating and exhibits a response comparable to the transit
time in one turn.

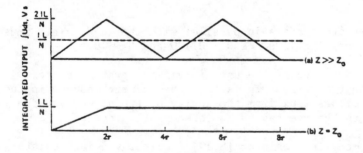

Fig. 3.8.   Integrated waveforms for a transmission line coil.
For $Z = Z_o$ the intrinsic risetime of the measurement
is the transit time in the coil.

sensor has introduced a time dependence in the coil output that
did not exist in the excitation waveform.

The advantages of the self-integrating sensor become clear
when we observe waveform (c) of Fig. 3.7. The induced I/N current in the secondary exactly reproduces the current I in the
primary. The transmission line analysis reveals why a uniformly
excited, self-integrating, coil has an intrinsic risetime equal
to the transit time in one turn. Consequently, it is possible
to build coils, with transit times $\tau > 10$ ns, with risetimes of
1/2 ns or less. Of course the restriction that we have indicated
about uniformly wound turns, a shorted major turn and a non-reac-
tive burden resistor Z still apply.

3.7   Sources of Error in Measurements with Field Coupled Sensors

There are several errors that arise in the measurement of
currents with field sensors. The first one is the impedance
that the sensor presents to the load. In the case of a differentiating sensor, this impedance is simply the inductance of the
cavity. In the case of large pulse power accelerators, substantial voltage drops can develop across seemingly innocuous
grooves. As an example, the inductance of a 25 mm^2 groove, at a
radius of 5 cm, is 10^{-10} H and, at a current slew rate of 10^{13} A
s^{-1}, the voltage drop across the groove is 1 kV.

In the case of a self-integrating sensor, the insertion
impedance Z_{ins} is simple equal to

$$Z_{ins} = \frac{R}{N^2} .$$

This impedance appears in parallel with the load. This impedance
is always negligible since $R \sim 0.1 \ \Omega$ and $N \sim 10$.

One advantage of the use of self-integrating sensors in
pulsed power accelerators is that the current induced in the
sensor excludes flux from the cavity of the sensor and the inductance of the groove cavity is consequently reduced.

The metering resistor [8,19] (current viewing resistor) in
a self-integrating sensor holds a key to the performance of the
sensor. Hence the success or failure of the measurement with
such a sensor hinges upon a reactanceless resistor of a relatively small value.

The simplest error to deal with arises from the decay of the current induced in the sensor. This decay can be compensated for using the relation:

$$I_s = I + \frac{1}{\tau_d} \int_o^t I dt \; ,$$

where I is the current as "measured" by the sensor and I_s is the compensated value of the measurement. Other errors associated with the performance of current viewing resistors are deferred to the next section.

The low frequency cutoff of a self-integrating sensor arises from the full penetration of the magnetic induction into the core. The penetration is accounted for with the compensation equation displayed above.

A high permeability core "soaks" up the magnetic induction penetrating the core and determines a much larger $\partial B/\partial t$ per unit $\partial H/\partial t$ than a vacuum core does. A high permeability core therefore can substantially increase the time constant of the coil. Two precuations are necessary, however. One is associated with the frequency dependent permeability of the core material. It is important to know what this permeability is because the coil will exhibit a frequency dependent decay time. The second precaution is associated with the saturation of the core. When the current-time product of the core, as described before, is exceeded, the transformer output drops abruptly even though current may still be flowing in the primary. This abrupt drop will occur with the characteristic time associated with the windings alone. Since the risk of saturation always exists in a coil with a ferromagnetic core, an estimate of the magnitude and duration of the current pulse must be obtained and compared with the I x t product of the transformer before a measurement is made.

As mentioned before, non-uniform excitation (u = u(x,t)) is the Achilles heel of a self-integrating coil. It produces oscillations in the output whose period is the two-way transit time for an EM wave in the coil. Due to the high "Q" of the sensor, these oscillations damp out very slowly. Resistors between turns or between turns and an isolated collector, can be used to reduce this oscillations at the expense of increased complexity and some loss of risetime in the system.

Another approach to current measurements when ringing in the sensor becomes a problem is to use a terminated differentiating sensor. In this case the risetime information is no better than the two-way transit time in the sensor as shown in Fig. 3.7.

Self-integrating coils with radii greater than ~ 10 cm become impractical due to transit time effects. Measurements in
systems with large radii are better carried out with a number of
discrete sensors located symetrically about the azimuth of the
current distribution to be measured, as discussed in Section 3.2.

The errors associated with the time response and heating of
the current viewing resistor of a self-integrating coil will be
discussed in the section on current viewing resistors.

4.   CURRENT VIEWING RESISTORS

4.1 Basic Considerations

Current viewing resistors (CVRs), as a current diagnostic,
base their operation on Ohm's law. Therefore, the measurement of
current relies upon the voltage developed when the circuit current flows across the resistor. CVRs may be used to measure the
current directly by sampling all the current in the circuit or
by sampling only a fraction of it. The fraction may be sampled
either by inductive coupling as in the case of a Rogowski coil,
or directly, as in the case of segmented CVRs.

From the viewpoint of pulsed power measurements, CVRs should
be used with caution since their direct connection to apparatus
operating at high voltages and carrying large currents may introduce troublesome electrical interference problems. Moreover,
their apparent advantage of being calibratable at zero frequency
(DC) is negated by the presence of diffusion (skin effect), parasitic inductances, and heating which can all introduce sizeable
errors in high speed current measurements.

4.2 Response of CVR

The frequency response of a CVR is limited by: (1) diffusion of the magnetic field into the CVR material; (2) the insertion inductance of the CVR in the circuit; and (3) the transit
time of electromagnetic waves in the CVR itself. Magnetic diffusion is the most important limitation; thin CVRs of small
dimensions have little insertion inductance. A small dimension
also circumvents the problem of electromagnetic wave transit
time.

The response of a CVR has been treated in detail in the
literature [24]. Consider a tubular CVR with internal radius r,
external radius a, length ℓ, CVR material thickness r-a = δ,

resistivity ρ, permeability μ, and ratio of radii $\varepsilon = ra^{-1}$. We introduce a characteristic time $t_c = \mu\rho^{-1}\delta^2$ and characteristic voltage $U_c = I_p\rho\ell(2\pi a\delta)^{-1}$, where I_p is the peak current in the circuit. We further assume a CVR with a thickness $\delta \ll a$. Then the output voltage U of such a CVR, to a step function input of magnitude I_p, is approximately

$$U^* = 2(\pi t^*)^{-1/2} e^{-\frac{1}{4t^*}} ,$$

where $t^* = tt_c^{-1} = t\rho(\mu\delta^2)^{-1}$ and $U^* = UU_c^{-1} = 2U\pi a\delta(I_p\rho\ell)^{-1}$. This equation describes a CVR response that is indistinguishable from the exact solution in the interval $0 < U^* < 0.95$. It appears as the solid line in Fig. 4.1.

The diffusion process may also be modeled with an equivalent network of inductors and resistors [25]. The circuit diagram of such a 10-section network appears in Fig. 4.2. The inductance of each section L_n is normalized with respect to the total inductance of the foil: $L = \mu\ell\delta w^{-1}$ such that $L_n^* = 0.1 = L_n L^{-1}$ where L_n is the thickness of each section and $w = 2\pi a$. The resistance of each section is normalized with respect to the total resistance of the foil: $R = \rho\ell(W\delta)^{-1}$, such that $R^* = 10 = LL_n^{-1}$. The time is normalized with respect to $t_c = LR^{-1} = \mu\delta^2\rho^{-1}$. As before, the output is normalized with respect to U_c.

The result of a simulation, using a current waveform with a risetime of $5.2 \times 10^{-2} t_c$ ($t^* = 5.2 \times 10^{-2}$), appears as the dotted waveform in Fig. 4.1. This simulation waveform is indistinguishable from the analytic approximation. Both the analytic approximation and the circuit analysis simulation show a 10 to 90 percent risetime $t^* = 0.25$.

There are several techniques to compensate for the diffusion of current in the CVR. One technique [24,26] relies on a loop that links some of the flux within the resistive material. Another taps the output midway in the resistive material [27] by using a shunt made with two parallel foils. A third technique relies on compensation with a passive network [25,28]

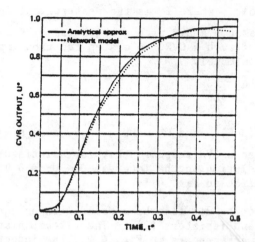

Fig. 4.1. The response of a current viewing resistor to a step
           function input. The voltage U*, time t* are non--
           dimensional as described in the text. t* is linked to
           the skin depth of the material and U* is linked to the
           DC resistance of the material.

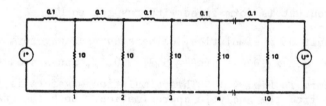

Fig. 4.2. Equivalent network model for a current viewing resis-
           tor. Each section represents a slice of the resistor
           in the direction of the magnetic field gradient. The
           series inductors and shunt resistors add up to the
           inductance and resistance of the whole resistor. In
           this case the normalized values for both add up to
           one.

4.3   Sources of Measurement Error in CVRs

To reduce error, a CVR should be constructed as a four-terminal network so that the metering contacts are independent of the current-carrying contacts. The resistive element also requires a sturdy support. Figure 4.3 [22] schematically illustrates the construction for a continuous CVR that accomplishes both objectives. The foil (a) is stretched over a metal yoke (b). The metering contacts are connected to the foil (c).

The yoke carries the current for pulse durations, which are long compared to $t_d = L_y R^{-1}$, where L_y is the inductance of the region between the foil and the yoke, which is crosshatched in Figure 4.3. A decay time of $t_d = 10^{-6}$ s is easily obtained since foil resistances of typically 10^{-3} Ω, and inductances of 10^{-9} H result from reasonable geometries. The decay associated with t_d is compensated with the algorithm

$$U_m = Y + \frac{1}{t_d} \int_o^t U \, dt \, ,$$

where U_m is the compensated value of the measurement.

There are three other possible sources of error when measuring currents with CVRs. The first source is spurious signals that arise from the direct coupling of the measuring instrument (oscilloscope or digitizer) and the circuit carrying the current. We reduce the error by introducing either a decoupling transformer or a large inductance on the ground connection between the CVR and the oscillograph.

The second source of error is timevarying flux coupled in the connection to the CVR. In the CVR construction illustrated in Fig. 4.3, it is easy to show that the ratio of resistive voltage, U_c to inductive voltage ϕ_m is:

$$\frac{U_c}{\dot{\phi}_m} = \frac{A_y}{A_m} \, ,$$

where A_y is the area between the yoke and the foil and A_m is the area subtended by the metering connections. Therefore, we must construct the CVR, minimizing the area subtended by the connections to the foil. A Park shunt minimizes this error [29].

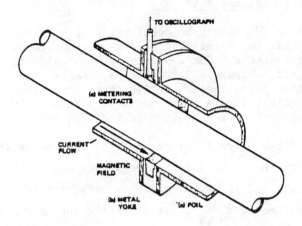

Fig. 4.3. A current viewing resistor (CVR) for high speed cur-
           rent measurements. The foil is supported on a metal
           yoke. Contacts are welded to the foil. The signal
           from the resistor exhibits a characteristic decay time
           given by the ratio of the inductance of the volume be-
           tween the foil and the yoke to the resistance of the
           foil.

Joule heating of the foil is the third source of error.
This affects the CVR response by changing foil resistance during
the pulse. Heating takes place in two phases. In the first
phase, heating occurs in the foil as the magnetic field diffuses
in the conductor. In the second phase, the current has fully
diffused and heating takes place throughout the foil.

We now discuss foil heating during the diffusion phase.
Assume a step current I_p impressed on the foil at time t = 0.
The penetration of the current in the foil at time t is
$\delta(t) = (2\rho t \mu^{-1})^{1/2}$. The factor 2 arises from full penetration
at $tt_c^{-1} = t*/2$. The resistance of the foil is $R(t) = \rho \ell (w \delta)^{-1}$.
The energy input at time t is

$$E(t) = I_p^2 \int R(t) \, dt = w I_0^2 \int \frac{\mu}{2\rho} \, t^{1/2}.$$

This heat input occurs in a volume v = $\delta \ell w$. Equating the thermal
energy of the foil with the energy dissipated in it, we obtain

$$C_p D \Delta T = \mu_0 I_0^2 w^{-2} = 2B^2/2\mu_0 \, ,$$

where ΔT is the temperature rise of the foil, $B = \mu_o Iw^{-1}$, and D is the density of the material. This result is well known and originated from work with pulsed magnetic fields [30]. It was found during the diffusion phase that the thermal energy per unit volume of the material is equal to the energy density of the magnetic field.

Once current diffusion is complete, the thickness of the foil carrying the current is constant. A parallel analysis to the one above yields.

$$C_p D\Delta t = \frac{2t}{t_c} \frac{B^2}{2\mu_o}.$$

Both equations have a similar form. Making the foil thicker does not affect the temperature rise in the diffusion phase. However, when diffusion is complete, the thicker foil will have a smaller temperature rise. Unfortunately, this is achieved at the cost of CVR response.

4.4 A Practical CVR Design

Current viewing resistors have found applications for high speed current measurements in magnetically insulated vacuum transmission lines. In these lines, the electrical stresses associated with large power flows, fast risetimes, and the presence of space charge clouds, break down magnetic field sensors and their shielding grooves. A CVR for such an application is described in [32]. Its dimensions are: $a = 1.34 \times 10^{-2}$ m; $\ell = 7.6 \times 10^{-3}$ m; $\delta = 7.6 \times 10^{-5}$ m. Therefore, $t_c = 9.6 \times 10^{-9}$ s, $U_c/I_p = 9.01 \times 10^{-4}$ V A^{-1} and the risetime ($t/t_c = 0.25$) is 2.5 ns.

We measured the response of this CVR to a 500 ps risetime current drive of 20 A, with a sampling scope at a 500 ps/div sweep speed. The response, normalized with respect to t_c and U_c, appears as the solid line in Fig. 4.4. If we compare the response to the simulation results in Fig. 4.1, there is significant oscillation ($\sim$ 1 GHz) in the leading edge of the CVR output, not visible on a 250 MHz bandwidth oscilloscope. Otherwise, the CVR output risetime of 2.5 ns compares quite favorably with the numerical simulation.

Figure 4.5 shows a measurement obtained with this CVR on a 0.95-m-long MITL. The solid line is the MITL input current,

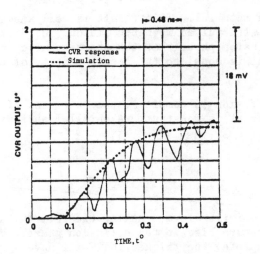

Fig. 4.4. CVR calibration waveforms. This calibration was obtained on the CVR of Ref [31]. Dotted line is the theoretical response as displayed in Fig. 4.1. Drive has a 500 ps risetime. Response has a ~ 500 ps structure.

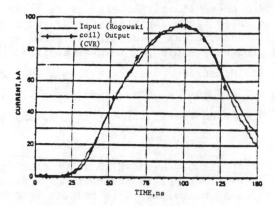

Fig. 4.5. Comparison of CVR and Rogowski sensor response on a magnetically insulated vacuum transmission line.

while the curve annotated with diamonds is the MITL output current. The input current was measured with a conventional Rogowski coil; the output current was measured with the CVR described above. The measurements were obtained with a short circuit at the output end of the MITL. The two waveforms have been displaced by the one-way transit time of light in the 0.95-m-long structure. The current measured by the CVR faithfully reproduces the input current measured by the Rogowski coil, except for a slight expected erosion of the leading edge of the pulse. The waveforms are almost identical in the vicinity of peak current.

The maximum linear current density in this CVR has been as high as 200 kA/8.4 cm = 24 kA cm^{-1}, which implies a peak magnetic field of 3.0 $Wb\text{-}m^{-2}$ at the surface of the foil. A temperature rise of about 5°C, accompanied by less than 0.5 percent change in resistance, is implied for this particular CVR at peak current.

4.5 Segmented CVRs

This section describes a segmented CVR for current measurements in the wall along the surface of a pulse power accelerator [32]. Figure 4.6 shows a schematic of a segmented CVR and Fig. 4.7 shows typical data obtained on the PITHON accelerator.

5. CALIBRATION SOURCES FOR ROGOWSKI COILS, FLUXMETERS, AND CVRs

The calibration of current sensors is very difficult in the > 100 kA current range. The difficulty arises from lack of sources with a precisely known current. In this section, we discuss the relative merits of various calibration methods.

From the standpoint of high frequency response, a cable pulser with a reed switch [33] is the best source. Its advantages are: exactly known output level, < 500 ps risetime, and reproducible repetitive output. Its main disadvantage is its low level output (~ 20 A into a short circuit), which limits its application to bench calibration of high sensitivity sensors (> 100 V/MA). In situ calibration with this pulser is even more difficult due to its low level output. A schematic of a current sensor calibration set-up appears in Fig. 4.8.

Another current source is a high voltage cable pulser. A spark gap (air or oil, self-breaking, triggered, or mechanically actuated) connects the charge line to the load under calibration. This kind of pulser can drive as much as 400 A per cable. Unless it operates in the relaxation mode or with a triggered spark gap, it is nonrepetitive. The switching losses, which are somewhat difficult to quantify, and the switch inductance become more important as we increase the number of cables.

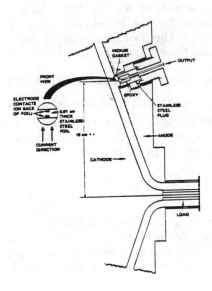

Fig. 4.6. A segmented CVR for vacuum transmission line applica-
           tions. Resistor samples a fraction of the current
           flowing in the circuit along the wall. The decay time
           may be estimated from the ratio of cavity inductance
           to resistance.

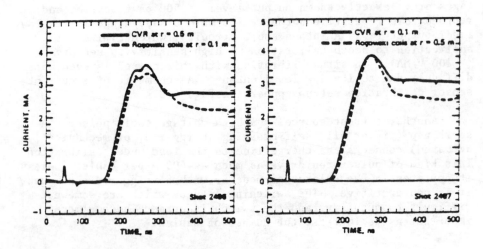

Fig. 4.7. Response of a segmented CVR on the PITHON accelerator.

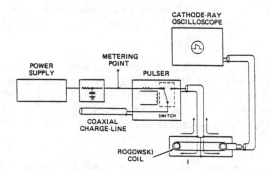

Fig. 4.8. Schematic of a current sensor calibration set-up.
Reed switch launches a wave from charge line into
transmission line with Rogowski coil at shorted end.

A third source, applicable in the microsecond timescale, is
the ringdown of a capacitor in an RLC circuit. The current is
easily obtained with some simple calculations and with measurements of voltage, period and decay rate.

A useful device for calibration is a thin foil shunt [34]
which is well characterized in the laboratory under dc and pulsed
conditions, and is used for calibrations in situ. A coaxial device of this type has been used standardly in our laboratory for
several years.

6.   CONCLUSIONS

By this time it should have become clear that there is no
single sensor that will fulfill all the instrumentation needs for
current measurements in pulse applications.

The purpose of this paper has been to survey that which is
available and objectively present the merits and limitations of
different approaches.

REFERENCES

1. Baum, C. E., Breen, E. L., Giles, J. C., O'Neill, J. and
   Sower, G. D., "Sensors for Electromagnetic Pulse Measurements Both Inside and Away From Nuclear Source Regions,"
   IEEE Transaction Electromagnetic Compatibility, EMC-20,
   1 pp. 22-34 (1978).

2. Pellinen, Donald G., Di Capua, Marco S., Sampayan, Stephen
   E., Gerbracht, Harold, and Wang, Ming, "Rogowski Coil for
   Measuring Fast, High Level Pulsed Currents," Review of
   Scientific Instruments, 51, pp. 1535-40 (1980); Rogowski
   Coil PIM-220, Physics International Company, San Leandro
   California 94577.

3. Spiegel, Ronald J., Booth, Charles, A. and Bronaugh, Edwin,
   "A Radiation Measuring System With Potential Under-Hood
   Application," IEEE Transactions on Electromagnetic Com-
   patibility," EMC-25, 2, pp. 61-69 (1983).

4. Hofer, Walter, J., "A 1 GHz Bandwidth $\dot{B}$ Probe," Lawrence
   Livermore National Laboratory Engineering News," 5, 4, p. 4
   and cover (1983).

5. Duncan, Paul, H., "Analysis of the Moebius Loop Magnetic
   Field Sensor," IEEE Transactions on Electromagnetic Compatibility, EMC-16, pp. 83-89 (1974).

6. Baum, Carl, E., Breen, Edward, L., Pitts, Felix, L., Sower,
   Gary, D. and Thomas, Mitchel, "The Measurement of Lightning
   Environmental Parameters Related to Interactions With Electronic Systems," IEEE Transactions on Electromagnetic Compatibility, EMC-24, 2, pp. 123-137 (1982).

7. MGL $\dot{B}$ Sensor (Free Field), Data Sheet 1100, EG&G WASC, Inc.
   Albuquerque, NM 87108 (1980).

8. Pellinen, Donald, G., "A Subnanosecond Risetime Fluxmeter,"
   Review of Scientific Instruments, 42, 5, pp. 667-670 (1971).

9. PIMM-199 Fluxmeters, Physics International Company, San
   Leandro, California 94577.

10. Rogowski, W., and Steinhaus, W., Archiv Elektrotechnik, 1,
    141, (1912).

11. Shannon, J., Chu, E. and Richardson, R., "Cavity Current
    Monitors," in Proc. Workshop on Measurement of Electrical

Quantities in Pulse Power Systems, National Bureau of Standards, Boulder, Colorado, NBS Publication 628, pp. 289-299 (1982).

12. Sower, Gary, "I Dot Probes for Pulsed Power Monitors," in Proceedings 3rd IEEE International Pulsed Power Conference, IEEE 81-CH1662-6, pp. 189-192 (1981).

13. Ekdahl, C. A., "Voltage and Current Sensors for a High Density Z Pitch Experiment," Review of scientific Instruments, 51, pp. 1645-1648 (1980).

14. EG&G IMM İ Sensor, Data Sheet 1123, EG&G, WASC, Inc., Albuquerque, NM 87108 (1980).

15. McDaniel, Dillon, H., SNLA, private communication (1978).

16. Anderson, John, M., "Wide Frequency Range Current Transformers," Review of Scientific Instruments, 42 pp. 915-926 (1971).

17. Anderson, John, M., "Wide Frequence Range Current Transformers and Application to Pulsed Power Systems," in Proc. Workshop on Measurement of Electrical Quantities in Pulse Power Systems, National Bureau of Standards, Boulder, Colorado, NBS Publication 628, pp. 233-243 (1982).

18. Wilmer, Michael, E. and Pearson, Paul, A., "Precise Measurement of Current in Pulsed Power Systems," in Proc. Workshop on Measurement of Electrical Quantities in Pulse Power Systems, National Bureau of Standards, Boulder, Colorado. NBS Publication 628, pp. 194-203 (1982).

19. Series CT Current Transformer, GE License, T. and M. Research Products, Inc., Albuquerque, NM 87108 (1983).

20. IPC Current Monitors, Ion Physics Company, Burlington, MA 01803.

21. Cooper, J., "On the High Frequency Response of a Rogowski Coil," Plasma Physics, 5, pp. 285-289 (1963).

22. Di Capua, Marco, S., "Rogowski Coils, Fluxmeters and Resistors for Pulsed Current Measurements," in Proc. Workshop on Measurement of Electrical Quantities in Pulse Power Systems, National Bureau of Standards, Boulder, Colorado, NBS Publication 628, pp. 175-193 (1982).

23. Current Monitor PIM-221, Physics International Company, San Leandro, California 94577.

24. Malewski, R., "New Device for Current Measurement in Exploring Wire Circuits," Review of Scientific Instruments, 39, pp. 90-94 (1968).

25. Schwab, A. J., IEEE Trans. on Power Apparatus and Systems, PAS-90, 5, pp. 225-227 (1971).

26. Malewski, R., IEEE Trans. on Power Appparatus and Systems, PAS- , pp. - (1981).

27. Malewski, R., IEEE Trans. on Power Apparatus and Systems, PAS-96, pp. 579-585 (1977).

28. Hortopan, G., and Hortopan, V., Rev. Rum. Sci. Tech. (Electrotech. et Energie), 19, pp. 41-47 (1974).

29. Park, John H. "Shunts and Inductors for Surge Current Measurements," Journal of Research of the National Bureau of Standards, 39, pp. 191-212 (1947).

30. Knoepfel, H., Pulsed Magnetic Fields, American Elsevier Publishing Company, New York, NY (1970).

31. Bailey, Vernon, et al., "REB Current Drive in the Macrotor Tokamak," Physics International Company, PIFR-1466-1, Contact Number DE-AT03-76ET-53019 (1982); also, "Intense Relativistic Electron Beam Injector System for Tokamak Current Drive," J. Appl. Phys., 54 pp. 1956-1665 (1983).

32. Di Capua, Marco, S., and Sincerny, Peter, "A Simple Current Viewing Resistor for Current Measurement in Magnetically Insulated Vacuum Feeds," 1982 IEEE International Conference Plasma Science, Ottawa, Ontario, Canada (1982).

33. Garwin, R. L., Rev. Sci. Instruments, 21, 903 (1950).

34. Donald G. Pellinen, "A High Current, Subnanosecond Response Faraday Cup," Review of Scientific Instruments, 41, pp. 1347-1348 (1970).

35. Di Capua, Marco, and Pellinen, Donald G., "Propagation of Power Pulses in Magnetically Insulated Vacuum Transmission Lines," J. Appl. Physics, 50 (5) (1979).

36. Nudelman, Aaron, Princeton Plasma Physics Laboratory, private communication (1983).

37. Wikswo, Jr., John, P., "Improved Instrumentation to Measure the Magnetic Field of Cellular Action Currents," Review of Scientific Instruments, 53, pp. 1847-1850 (1982).

NUCLEAR REACTION DIAGNOSTICS FOR INTENSE PARTICLE BEAM
MEASUREMENTS

R. J. Leeper

Sandia National Laboratories
Albuquerque, NM
USA

1.   INTRODUCTION TO NUCLEAR INTENSE ION BEAM DIAGNOSTICS

Intense pulsed ion beams, with kinetic energies ranging from
several hundred kilovolts to ten megavolts and current densities
ranging from several ten's of kiloamperes per cm^2 to almost one
mega-ampere per cm^2, have been generated by reflex triode, reflex
triode, reflex pinched, and magnetically insulated ion diodes [1,
2, 3, 4, 5].

The determination of the ion beam voltage, current, and current density for these diodes is difficult. In part, this is due
to the fact that intense ion beams are partially space-charged
and current-neutralized. These traits preclude, or make difficult, many electrical measurements involving charge collection or
the measurement of self-magnetic fields. In addition, recent
evidence indicates that a measurement of the voltage of the diode
anode cathode gap is not sufficient to determine the kinetic
energy of the accelerated ion beam [6]. It was found in these
measurements that the ion beam voltage at any given instant of
time had a spread of several hundred keV and that the beam voltage did not simply track the anode cathode gap voltage. Another
problem with the use of conventional electrical diagnostics is
that these diagnostics do not differentiate between the various
ion species or electrons that compose the beam.

In an effort to overcome many of these problems and difficulties, a great variety of nuclear reaction diagnostic techniques have been employed. These techniques fall into one or
more of three classes:

(1)   Induced nuclear activation
(2)   Prompt gamma ray production
(3)   Prompt neutron production.

Induced nuclear activation occurs when the ion beam is
allowed to strike a suitable target and a fraction of these ions
induces nuclear reactions in the target whose final product is
radioactive. Then, by measuring the amount of radioactivity
present and knowing the probability of producing the reaction, it
is possible to determine the number of ions that were incident on
the target. It should be understood that the induced activity
method gives no information about the time history of the ion
beam on target, but only gives information on the integrated beam
current.

The prompt gamma ray technique is based on the fact that,
when an ion beam strikes a suitable target, energetic gamma rays
(6- 18 MeV) are produced whose intensity and time history can be
measured with fast nuclear detectors to determine the incident
ion beam's current, or current density and sometimes voltage.
The prompt neutron technique is similar in concept to the prompt
gamma ray technique except that neutrons are produced. Both
prompt nuclear techniques suffer from the major problem of intense bremsstrahlung backgrounds from leakage electrons in the
ion diodes and, therefore, require massive detector shields. In
addition, the prompt neutron technique has the additional complication of neutron time-of-flight smear due to the neutron's non-
zero rest mass.

Though not exploited to date, it is also possible to measure
properties of an intense electron beam through the use of elec-
tron- and gamma ray-induced nuclear reactions. Details of how
this might be possible will be covered later in this chapter.

2.   REVIEW OF RELEVANT NUCLEAR PHYSICS

Nuclear reactions are changes induced in nuclei by interactions with nuclei of sufficient kinetic energy. They are thus
distinguished from radioactive decay processes, which are nuclear
disintegration events that occur spontaneously. Many different
processes may take place when two particles collide. A typical
nuclear reaction may be written

$$a + A \rightarrow b + B + Q \qquad\qquad\qquad (1)$$

If A is the symbol for the target nucleus, a that for the projectile, while B is the residual nucleus, and b is the emitted particle, this reaction would be written A(a,b)B. When specific isotopes are intended, the mass number (the total number of protons and neutrons in the nucleus) is written as a superscript to the left of the chemical symbol in this expression. Occasionally the total number of protons is written as a subscript to the left of the chemical symbol, but in most cases is omitted. Special symbols are used for "elementary" particles and the lightest nuclei; for example: e = electron, p = proton, n = neutron, d = deuteron or ^{2}H, t = triton or ^{3}H, and α = alpha particles or ^{4}He. A photon or gamma ray is signified by v. Further, either B or b, or both, may be left in excited states. Sometimes the excitation energy is given as a subscript, or a state of excitation may be simply indicated by an asterisk, B*, etc. In general, the residual nucleus B may or may not be stable with respect to radioactive decay. Of course, although we are mainly interested in nuclear reactions with 2 body final state b and B, nuclear reactions with 3 or more body final states do occur and a generalization is necessary in considering such reactions.

Conservation Laws

Nuclear reactions of interest in activation analysis obey four conservation laws:

1. Conservation of mass-energy   E

2. Conservation of momentum,   P

3. Conservation of nucleons,   A

4. Conservation of charge,   Z

5. Conservation of total angular momentum.   J

The term nucleon refers to either a proton or a neutron. Conservation laws three and four may be considered in chemical reaction analogy as the "balancing" of the reaction. The reaction is "balanced" with respect to nucleons (total number of protons plus neutrons) and charge in that the reactants and products each have the same number of nucleons and protons.

2.1 Reaction Energy

The symbol Q in Eq. (1) refers to the energy released during the reaction. Since the total energy is conserved in all reactions, $Q \neq 0$ means that kinetic energy has been converted into internal excitation energy (or rest energy) or vice versa, so that $Q = T_f - T_i$, where T_f is the total kinetic energy of the particles in the final state and T_i is the corresponding quantity in the initial state. If the Q value is positive, the reaction is said to be exothermic, while a reaction with a negative Q value is endothermic. In the latter case, a bombarding energy, above a definite threshold value, is required in order for the reaction to take place; in the center of mass system an energy T_i greater than $-Q$ is needed for $T_f > 0$.

2.2 Classes of Nuclear Reactions

There are several major classes of nuclear reactions. These are:

(i) Elastic scattering: Here b = a and B = A. The internal states are unchanged so that Q = 0 and the kinetic energy in the center-of-mass frame is the same before and after the scattering. We have $a + A \rightarrow a + A$; for example,

$$^{197}Au(p,p)^{197}Au \cdot \tag{2}$$

(ii) Inelastic Scattering: This term most often means collisions in which b = a but A has been raised into an excited state, B = A* say. Consequently $Q = -E_x$, where E_x is the excitation energy of this state. Since a is emitted with reduced energy, it is commonly written a', $a + A \rightarrow a' + A* - E_x$; for example,

$$^{197}Au(p,\ p')^{197}Au^* \cdot \tag{3}$$

(iii) Rearrangement collision or reaction: here b ≠ a and B ≠ A so that there has been come rearrangement of the constituent nucleons between the colliding pair (a transmutation). There are many possibilities:

$$a + A \rightarrow b_1 + B_1 + Q_1 \text{ or } + a + A \rightarrow b_2 + B_2 + Q_2, \text{ etc.}$$

For example,

$$^{12}C(d,n)^{13}N \cdot \tag{4}$$

This is a class of reactions important for intense beam diagnostics.

(iv) Capture reaction: This is a special case of class (3); the pair + A coalesce, forming a compound system in an excited state; this excitation energy is then lost by emitting one or more γ rays, $a + A \rightarrow \gamma + C + Q$. For example,

$$^{7}\text{Li}(p,\gamma)^{8}\text{Be} \ . \tag{5}$$

These reactions are especially important to intense beam diagnostics since almost all prompt ray reactions are capture reactions.

(v) Other reactions: We mean here reactions not included in Eq. (1) because there are more than two particles in the final state, such as $a + A \rightarrow b + c + B + Q$. For example,

$$^{40}\text{Ca}(\alpha,\alpha'p)^{39}\text{K} \ . \tag{6}$$

The types of reactions of concern to us in these discussions will mainly be the (p, γ) and (p, α, γ) capture reactions, the (p, n) and (d, n) neutron producing reactions, and (γ, n) photonuclear reactions. The neutron reactions (n, γ), (n, p), (n, α) and (n, 2n) will also be important to us since they form the basis for the operation of almost all total neutron yield activation detectors.

2.3 Nuclear Reaction Cross Section

We need a quantitative measure of the probability that a given nuclear reaction will take place. For this we introduce the concept of a cross-section which we define in the following way. Consider a typical reaction A(a,b)B. If there is a flux of I_o particles of type a, per unit area, incident on a target containing N nuclei of type A, then the number of particles b emitted, per unit time, is clearly proportional to both I_o and N. The constant of proportionality is the cross-section, σ, and has the dimensions of area. Then the cross-section for this particular reaction will be,

$$\sigma = \frac{(\text{number of particles b emitter/unit time})}{[(\text{number of particles a incident})/(\text{unit area-unit time})] \ \times \ [\text{number of target nuclei within the beam}]} \ . \tag{7}$$

A convenient unit of area for nuclear physics is the barn (symbol b: 1 barn = 10^{-24} cm^2 = 100 fm^2) and cross sections are usually given in barns or the sub-units millibarn, 1 mb = 10^{-3} b, and microbarn, 1 μb = 10^{-6}, etc.

If we ask for the number of particles b emitted, per unit
time, within an element of solid angle $d\Omega$, in the direction with
polar angles (θ, ϕ) with respect to the incident beam, clearly
this is proportional to $d\Omega$ as well as I_o and N (see Fig. 1). The
constant of proportionality in this case is the differential
cross-section, $d\sigma/d\Omega$. Since solid angles are dimensionless, this
also has the dimensions of area.

In general, the probability of emission of b, and hence, the
differential cross-section, will depend upon the angles θ and
ϕ. Only in special cases will the angular distribution be isotropic. To emphasize this, the differential cross-section $d\sigma/d\Omega$
is sometimes written as $d\sigma (\theta, \phi)/d\Omega$. However (unless the spins
of one or more of the particles are polarized), the scattering
process is quite symmetrical about the direction of the incident
beam which means that the differential cross-section cannot depend on the azimuthal angle ϕ. In that case, we may write it
simply as $d\sigma (\theta)/d\Omega$.

Clearly the two kinds of cross-sections are related by,

$$\sigma = \int_o^{4\pi} (d\sigma/d\Omega)d\Omega \tag{8}$$

or, since the solid angle $d\Omega = \sin \theta \, d\theta d\phi$,

$$\sigma = \int_o^\pi \sin \theta \, d\theta \int_o^{2\pi} d\phi(d\sigma/d\Omega) . \tag{9}$$

2.4 Radioactive Decay

It is often the case that the residual nucleus B in Eq. (1)
is radioactive. In activation diagnostics, nuclear reactions are
indeed picked so that the radioactivity of B is assured. The
disintegration of a radioactive isotope results in the release of
its excess energy in the form of nuclear radiations. The disintegration product may itself be unstable with respect to some
other product. The nuclear radiations emitted in radioactive decay are generally either α, β, or γ rays.

2.4.1. Alpha decay. Alpha decay is the spontaneous emission of
an alpha particle, the nucleus of the ^{4}He atom. Although alpha
decay is known in a few of the medium weight nuclides (e.g.,

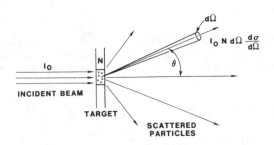

Fig. 1. Diagram for the definition of differential cross-sec-
        tion. Usually the size of the irradiated area of the
        target is small, very much smaller than the distance to
        the detector, so that the scattering angle θ is well-
        defined.

^{147}SM) it is common to the heavy nuclides with Z > 83. Alpha^{62}
decay proceeds according to the reaction,

$$^{A}_{Z}X_{N} \rightarrow \, ^{A-4}_{Z-2}Y_{N-2} \, + \, ^{4}_{2}He + Q \, . \tag{10}$$

Alpha decay is generally not important in nuclear reactions used
in intense beam diagnostics.

2.4.2. Beta decay. Beta decay is the spontaneous transformation
within a nucleus of a neutron into a proton, or of a proton into
a neutron, which results in a change of the neutron to proton
ratio (N/Z) toward a more stable configuration of the same mass
number.

When the N/Z ratio of the radioactive nucleus is greater
than that for stable nuclei of the same mass number, a neutron is
converted into a proton with the emission of a negatively charged
high-speed electron and a neutrino,

$$n \rightarrow p^{+} + e^{-} + \nu \, . \tag{11}$$

The high-speed electron is a beta particle; the neutrino
is an elusive partner (whose rest mass is about 0) required by
the laws of physics to conserve momentum and to share the decay
energy with the beta particle. Beta particles have a continuous
energy spectrum, from 0 to a maximum value, with an average energy about one-third of the maximum beta decay energy. Beta
decay for high N/Z ratio nuclides proceeds by the reaction,

$$\underset{Z}{^A}X_N \rightarrow \underset{Z+1}{^A}Y_{N-1} + \beta^- + \gamma \ . \tag{12}$$

When the N/Z ratio of the radioactive nucleus is smaller
than that for stable nuclei of the same mass number, a proton is
converted into a neutron within the nucleus by one of two processes:

1. Positron emission $p^+ \rightarrow n + \beta^+ + \nu$ (13)

2. Orbital electron capture $p^+ + e^- \rightarrow n + \nu$ (14)

In either case, the transformation results in a changed element

$$\underset{Z}{^A}X_N \rightarrow \underset{Z-1}{^A}Y_{N+1} \ . \tag{15}$$

In the first case the positive electron, when it loses its
kinetic energy, interacts with an electron in its environment and
the positive and negative electrons annihilate each other to form
two quanta of electromagnetic energy of 0.511 MeV each, equivalent to the rest mass of the electron. In the second case no
nuclear radiation, except the unmeasurable neutrino is emitted.
The radioactive decay event, however, can still be observed by
the characteristic x-rays, emitted from the resulting atom, as
atomic electron transitions occur to fill in the hole left in the
shell from which the electron was captured by the nucleus.

Positron decay is the most commonly used decay mode in intense beam diagnostics because of the positron's ease of being
counted in a highly sensitive low background coincidence counting
system. This will be discussed further in the next section.

2.4.3. Gamma decay. Gamma rays consists of electromagnetic radiation which is quantized into photons. The energy of a photon
is given by

$$E_\gamma = h\nu \ , \tag{16}$$

where h = Planck's constant, 6.625×10^{-27} erg-sec,
       ν = the frequency of the radiation.

Gamma radiation associated with radioactive decay or nuclear
reactions results from the de-excitation of product nuclei with
excess energy. In this process the energy of the nuclear transition is emitted as a discrete quantum analogous to the x-rays
emitted in orbital electron transitions. Thus, for a given transition, each gamma ray is emitted with the same energy. This
property makes gamma radiation useful for identifying specific
radionuclides, especially in the presence of many other radionuclides. The identification of several radionuclides by their

characteristic gamma-ray energies has become an important method
in radioactivation analysis. The gamma ray decay mode has sometimes been used in intense beam diagnostics.

2.4.4. Radioactive decay law. The decay of radioactive atoms is
a first order process for which the rate of change is proportional to the amount. Thus, for a radioactive source of N nuclei
of the same nuclide, the rate of decay is given by

$$- \frac{dN}{dt} = \lambda N , \tag{17}$$

where λ, the constant of proportionality called the decay constant, has a characteristic value for each radioactive nuclide.
The rate of change of N is always negative for radioactive decay
and is defined as the activity A of the source,

$$A = \frac{dN}{dt} . \tag{18}$$

Equation (17) may be solved for the initial conditions of N_o
atoms at $t = 0$ to yield,

$$N = N_o e^{-\lambda t} . \tag{19}$$

The radioactivity decay law follows from Eq. (18) such that

$$A = \lambda N = \lambda N_o e^{-\lambda t} . \tag{20}$$

Equations (17) to (21) are relationships that will continuously
be referred to throughout this entire presentation. It is also
convenient to express the characteristic decay constant of a
nuclide as a half-life, $T_{1/2}$, which is defined as the time required for any number of nuclei (or activity) to decay to half
its initial value. Thus

$$\frac{A}{A_o} = \frac{N}{N_o} = \frac{1}{2} = e^{-\lambda T_{1/2}} ; \tag{21}$$

$$T_{1/2} = \frac{\ln 2}{\lambda} = \frac{0.693}{\lambda} . \tag{22}$$

The mean life τ is defined as $\tau = \frac{1}{\lambda}$

3. DIRECT NUCLEAR ACTIVATION TECHNIQUE USING PRIMARILY THE $^{12}C(p,\gamma)^{13}N$ REACTION AS AN EXAMPLE

When a pulse of energetic ions strikes a suitable target, a small fraction of the ions induces nuclear reactions in the target. If the cross section for a given reaction is known, then the number of incident ions in the pulse can be determined from a measurement of the amount of any radioactive products formed. For ions of the light elements, there are a number of reactions which produce radioactive nuclei that decay, with half lives of a few minutes, by emitting positrons. By counting in coincidence the pairs of γ rays associated with the annihilation of the positrons, an extremely sensitive means of detecting a relatively small number of reactions is obtained. In particular, this discussion shall concentrate on the (p,γ) and (d,n) reactions from a ^{12}C target since these reactions have proved popular in diagnosing the integrated currents of intense proton and deuteron beams. Some mention will also be made of (p,γ) and (d,n) reactions from ^{10}B and ^{14}N targets. The use of any of the above reactions within a restricted range of current densities is relatively precise because the cross sections are fairly well known. Moreover, since the ions are required to have a minimum energy for the reaction to occur, this method is insensitive to low-energy plasma ions or debris that might strike the target.

There are two complications which arise when using direct nuclear activation analysis of intense ion beams [7]. First is the interference from different ions, causing reactions which produce the same radioactivity. This is the case for (p,γ) and (d,n) reactions since they both produce the same product nuclei on any target. If, for example, we have the situation of a small amount of deuterium in a proton beam (from natural abundance), then we will have interference. The amount of this interference can be important since the nuclear reactions involved can have widely different yields. The yields from (d,n) reactions are of the order of 10^4 times greater than (p,γ) yields and consequently even natural abundance contaminations of a proton beam at a level of 1.5×10^{-4} deuteron per proton can lead to major interference. To separate the contributions of the two components of the beam, either different material targets must be used or the relative abundance of protons and deuterons must be known. The second complication is caused by the loss or reaction products due to heating and evaporation of the surface of the target by the intense beam. This complication is a severe limitation of the direct activation technique and limits its use to relatively low-current density (≤ 20 kA/cm^2) beams even when attenuating metal screens and "blow-off" catchers are employed in an attempt to alleviate this problem.

3.1 Detection Apparatus for Coincidence Counting

To measure the amount of positron activity produced, standard, commercially-available, nuclear electronic instrumentation can be used to perform both pulse-height discrimination and coincidence-counting so that only 0.511 MeV γ rays, from the positron annihilation, are detected. A block diagram of this apparatus is shown in Fig. 2.

After activation, the carbon target is placed between two NaI (Tℓ) scintillation detectors. Each pulse from the detectors' photomultiplier tube is amplified by the detectors' preamp and by a double delay line (DDL) amplifier which provides a bipolar pulse, with a fast crossover, at a precise time. Pulse-height analysis is then performed by a fast-timing, single-channel analyzer (SCA) which gives an output pulse only if the input pulse amplitude is between two adjustable values (pulse height window) corresponding to the annihilation, γ ray energy of 0.511 MeV. The output pulse from each detector's SCA is then inputted into a coincidence circuit which gives an output pulse if, and only if, the pulses arrive at the inputs within a prescribed interval of time (10 nsec to 100 nsec). The output of the coincidence circuit is then scaled for a preset time. Separate outputs of the SCA's are also scaled to produce the "single's" rate for each detector.

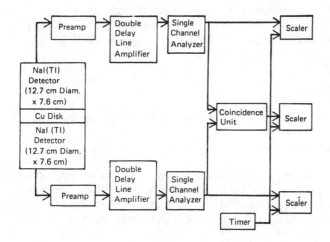

Fig. 2. Coincident γ-ray detection apparatus.

To optimize the coincidence counting, the NaI detectors should be placed as close together as possible and aligned at 180° to each other (the annihilation γ's come off at 180° relative each other). This provides the maximum solid angle for intercepting γ ray pairs by the detectors. Since detector efficiency is dependent on the radial position of the source from the cylindrical axis of the detector, the detector size should be much larger than the activated target. The thickness of the detectors should be chosen for optimium response to 0.511 MeV γ rays. Typical detector dimensions are 15.2 cm dia. by 7.6 cm thick. The detectors should be shielded from background radiation such as cosmic ray showers and other environmental radiation. To prevent drift of amplifier gains and windows, the electronics should be protected from temperature fluctuations.

For more information on nuclear detectors or nuclear electronic systems, consult any standard text on nuclear radiation detection such as Ref.[8].

The detection apparatus may be absolutely calibrated with a positron-emitting source of known activity. For example, 90% of the decay of ^{22}Na is by positron emission and 10% is by electron capture. A 1μCi source is sufficient to provide about 3×10^4 positrons/sec. A typical detection efficiency is 12%. To set the 0.511 MeV pulse-height window on the timing SCA's, a pulse-height analyzer should be employed.

3.2 Analysis

The number of radioactive nuclei dN produced, via a nuclear reaction of cross section σ(E), by N particles of energy E incident on a thin target of thickness dx, with n target nuclei/ cm^3, is given by,

$$dN = N\sigma(E) \, ndx \ . \tag{23}$$

Defining the thin target reaction yield dY as dN/N leads to

$$dY = \sigma(E) \, ndx \ . \tag{24}$$

For measurements with intense pulsed beams, thick targets are used. That is, the targets are thicker than the range of the incident particles. Consequently, the thick target yield is obtained by integrating dY over the thickness of the target to obtain

$$Y(E_i) = \int_0^{\ell} \sigma(E)ndx = n \int_0^{E_i} \frac{\sigma(E)}{dE/dx} dE \qquad (25)$$

$$= \int_0^{E_i} \frac{\sigma(E)}{\varepsilon(E)} dE . \qquad (26)$$

where E_i is the incident energy of the ions and

$$\varepsilon(E) = (1/n) \frac{dE}{dx} \qquad (27)$$

is the stopping cross section.

The stopping cross section is a slowly-varying function of energy (tabulated in many references, such as Ref. [9], so that this expression for $Y(E_i)$ can be evaluated by numerical integration if the cross section is known. At low energy, (d,n) cross sections are rapidly increasing with energy and numerical integrations are readily carried out. For resonance reactions (Briet-Wigner cross sectional behavior), such as (p,γ) proton capture reactions, the resonant behavior of the cross section simplifies the evaluation of Eq. (27).

The cross section of a Briet-Wigner resonance reaction is peaked at the resonance energy E_R. The cross section in the vicinity of a resonance of width Γ is given by

$$\sigma(E) = \frac{\Gamma^2}{4} \frac{\sigma_R}{(E-E_R)^2 + \Gamma^2/4} . \qquad (28)$$

where σ_R is the cross section at the peak of the resonance. For a resonance with a narrow width, the yield is approximated by

$$Y = \frac{1}{\varepsilon(E_R)} \int_0^{\infty} \sigma(E)dE = \frac{1}{\varepsilon} \frac{\pi}{2} \sigma_R \Gamma , \qquad (29)$$

provided the ion energy is well above the resonance energy, i.e., $E_i - E_R \gg \Gamma$. The quantity $Y_\infty = \pi\sigma_R\Gamma/2\varepsilon$ is called the "thick target step" of the resonance. If N_0 is the number of radioactive nuclei produced in the target for such a resonant reaction (say, (p,γ) reaction), then the number of incident protons is given by $N_p = N_0/Y_\infty$.

Now let us consider the case of a mixed beam or protons and deuterons incident onto a target. Then the number of radioactive nuclei induced in the target is given by

$$N_o = Y_p N_p + Y_d(E) \frac{dN_p}{dE} dE, \tag{30}$$

where Y_p is the (p,γ) thick target step, and N_p is the number of protons with energy greater than the resonance energy [7]. If the deuteron component of the beam is monoenergetic, or if the yield from the (d,n) reaction is approximately independent of energy over the energy spread of the deuteron beam, then

$$N_o = Y_p N_p + Y_d N_d . \tag{31}$$

If the deuteron fraction N_d/N_p in the beam is known (natural abundance implies $N_d/N_p = 1.5 \times 10^{-4}$), then N_p is given by

$$N_p = N_o \left[Y_p + \frac{N_d}{N_p} Y_d \right]^{-1} . \tag{32}$$

We know from the above that, given N radioactive nuclei, their decay will follow the radioactive decay law

$$\frac{dN}{dt} = -\lambda N , \tag{33}$$

where $\lambda = 1/\tau$ and τ is the mean life of the radioactivity (the half-life $T_{1/2}$ and τ are related by $\tau = T_{1/2}/0.693$). Solving this differential equation for N(t), for an initial number of nuclei N_o at t = 0, leads to

$$N(t) = N_o e^{-\lambda t} . \tag{34}$$

If the counting of the activity dN/dt is carried out over the interval between t_1 and t_2 (referenced to the end of irradiation when the number of radioactive nuclei = N_o), we have

$$C = \varepsilon \int_{t_1}^{t_2} \frac{dN}{dt} dt + B \tag{35}$$

$$C = \varepsilon \int_{t_1}^{t_2} -\lambda N dt + B \tag{36}$$

$$C = \varepsilon \int_{t_1}^{t_2} -\lambda N_o e^{-\lambda t} dt + B \tag{37}$$

$$C = \varepsilon N_o (e^{-\lambda t_1} - e^{-\lambda t_2}) + B , \tag{38}$$

where C is the total number of counts recorded, ε is the overall counting efficiency (including any self-absorption effects) and B is the number of background counts expected in $(t_2 - t_1)$. Substituting N_o from Eq. (33) and solving for the number of protons in the beam N_p we have

$$N_p = (C-B) \left\{ \left[e^{-\lambda t_1} - e^{-\lambda t_2} \right] \left[Y_p + \frac{N_d}{N_p} Y_d \right] \varepsilon \right\}^{-1} . \tag{39}$$

A similar equation results for deuterons beams containing a small fraction of protons. However, for this case, N_p/N_d must be determined by other means.

3.3 $^{12}C(p,\gamma)^{13}N$ Reaction as an Example of the Technique

As a concrete example of the direct activation technique, consider an intense proton beam incident upon a carbon target. Protons in the beam will induce the $^{12}C(p,\gamma)^{13}N$ reaction. This reaction has a resonance at 0.457 MeV with a width Γ = 37 keV and a resonance at 1.698 MeV with a width Γ = 67 keV. The thick-target step of the 0.46 MeV resonance is $(7.5 \pm 0.5) \times 10^{-10}$ $^{13}N/$ proton (measured at 1 MeV). The step of the 1.70 MeV resonance is (1.45 ± 0.003) times that of the lower resonance. The $^{12}C(p,\gamma)$ ^{13}N-thick target yield, plotted as a function of proton energy, is shown in Fig. 3 and is from Ref. [7].

As can be seen the useful energy range of this reaction to avoid the high-energy resonance is from 0.6 MeV to 1.5 MeV. This reaction has been used extensively in the diagnostics of the early pioneering work in ion intense beams and in beam transport work [4,10,11,12,13].

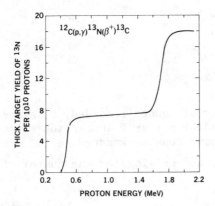

Fig. 3. Thick-target yield of the $^{12}C(p,\gamma)^{13}(\beta^+)$ reaction (from Ref. [7]).

3.4 Interference from the $^{12}C(d,n)^{13}N$ Reaction

Now let us consider the interference of the $^{12}C(d,n)^{13}N$ reaction upon the $^{12}C(p,\gamma)^{13}N$ measurement. This reaction has a threshold at 328 keV deuteron energy. Its thick target yield is presented in Fig. 4.

Following the consideration of Young [7], any proton beam extracted from a plasma produced from hydrogenic materials will always contain a small component of deuterons due to deuterium natural isotopic abundance. For an accurate determination of the numbers of protons per pulse, it is necessary to correct the activity of the target for the contributions from deuteron-induced reactions. It has been shown in Eq. (33), that the number of counts due to a given reaction is proportional to $(Y_p + (N_d/N_p) Y_d)$, where (N_d/N_p) is the ratio of the number of deuterons to protons per pulse. The deuteron correction is made assuming that, in an ion source operating in the space-charged limited regime, the current density of the various singly-ionized species is proportional to their relative abundance divided by the square root of their mass. Specifically, if a proton beam is extracted from a hydrogenic material, in which the isotopic abundance of deuterium is $\xi(D)$, the ratio of deuteron current density to that of protons is $J_d/J_p \sim \xi(D)/2$. Therefore, the deuteron correction is made by calculation of the energy dependent factor, CF, defined as

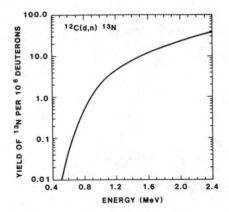

Fig. 4.   Thick-target yield for the $^{12}C(d,n)^{13}N(\beta^+)$ reaction
(from Ref. [7]

$$CF = Y_p \left\{ Y_p + \left[\xi(D)/2 \right] Y_d \right\}^{-1} . \tag{40}$$

The number of protons N_p is given by Eq. (40) and becomes

$$N_p = (C-B) \left\{ \left[e^{-\lambda t_1} - e^{-\lambda t_2} \right] \varepsilon Y_p \right\}^{-1} (CF) . \tag{41}$$

The factor, CF, is plotted in Fig. 5 for C and LiF (to be discussed in next section) targets.

3.5 <u>Target Ablation</u>

As mentioned above, the most serious problem and limitation with direct nuclear activation is loss of radioactive nuclei from the target, prior to counting, due to target "blow-off" or ablation. This happens because of the intense heating of the target surface by an intense high-current density ion beam. This problem can become important for proton beams that have current densities as low as a few 100 A/cm^2. Attenuating screens have been used in an effort to alleviate this problem, but at most can extend the range of the technique by a factor of about 25. (See Ref. [3] for an example of this technique).

Another method employed to extend the direct activation technique is the "pillbox" method illustrated in Fig. 6 [14]. In this method the target, e.g., carbon, is placed at the bottom of a metal box. The ion beam reaches the target material through

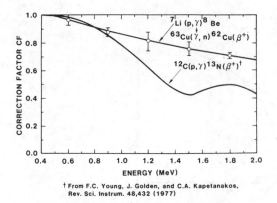

† From F.C. Young, J. Golden, and C.A. Kapetanakos,
Rev. Sci. Instrum. 48,432 (1977)

Fig. 5. Deuteron correction factors, CF, for the $^{12}C(p,\gamma)$ ^{13}N
and $^{7}Li(p,\gamma)$ ^{8}Be copper activation detectors.

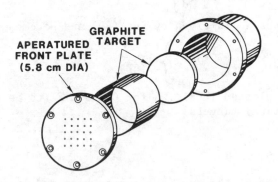

Fig. 6. Exploded view of the pillbox enclosing the graphite tar-
         get. The ion enters the pillbox through the apertured
         front plate [14].

an apertured front plate in the form of thin beamlets which expand somewhat before reaching the target. The total area of
the apertures is known accurately and constitutes a negligible
fraction of the surface of the box surround the target. Consequently, almost all of the material blown off from the traget is
redeposited on the inner walls of the box. After the shot, the
outside of the pillbox is wiped clean and counted in a standard

coincidence counting system described above. At 3 kA/cm^2 proton flux levels, the technique was shown to measure 100% of the protons in the beam. At the 25 kA/cm^2 flux level, the measured integrated proton current was down by 30% to 50%. So it is seen that above 15-25 kA/cm^2 current densities, we are in trouble with technique. In the next section, an indirect activation method employing the ^{7}Li(p,γ)^{8}Be reaction, in conjunction with a copper activation detector, will be described which completely circumvents any problems with blowoff or ablation.

3.6 Other Reactions

Finally, to end this section on direct activation, a few words should be said about other useful reactions. The ^{14}N(p,γ) ^{15}O(β+) and ^{10}B(p,γ) ^{11}C(β+) reactions have been used for proton diagnostics and the ^{10}B(d,n)^{11}C and ^{14}N(d,n)^{15}O reactions for deuteron diagnostics using BN targets. Complete details describing these reactions, plus others, may be found in Ref. [7]. Pinhole imaging the α's from the reaction ^{11}B(P,$\alpha\gamma$)2α has been used to image and measure the current density of an intense proton beam [15].

Throughout these discussions we have emphasized current measurements. Nuclear activation can also be used to determine mean ion energy. For example, simultaneous activations of carbon and aluminium, via the reactions ^{12}C(d,n)^{13}N and ^{22}Al(d,p)^{28}Al, can be used to determine mean deuteron energies. The measurement of deuteron energy is possible because the ratio of the yields of the two reactions is a sensitive function of deuteron energy. See Ref. [16] for complete details.

4.   ABLATION- PROOF LITHIUM - COPPER INDIRECT ACTIVATION TECHNIQUE TO MEASURE INTENSE PROTON CURRENT

In the last section direct nuclear activation detectors to measure intense ion currents were found to be simple, rugged, and capable of high accuracy. However, all direct nuclear activation monitors place the material to be activated directly into the ion beam where it is subject to ablation and loss of activated material. It was found that, for direct activation detectors the ablation problem becomes so severe that above beam current densities of 15-25 kA/cm^2 the technique cannot be used.

4.1 Lithium Copper Activation Technique

We present in this section a new ablation-proof indirect nuclear activation detector for the measurement of intense proton beams [17]. This detector is shown in Fig. 7.

When an intense proton beam pulse enters the 0.64 cm diam. input aperture and strikes the 2mm LiF target, all protons above 440 keV in energy may induce the nuclear reaction $^7Li(p, \gamma)^8Be$. The prompt gamma rays emitted from this reaction are produced in two energy groups of 14.7-15.2 MeV and 17.6-18.2 MeV with an average energy of 16.5 MeV. These energetic gamma rays are then allowed to strike the 1.27 diam. x 0.95 cm thick copper disk located immediately behind the LiF target. These gamma rays then induce the photonuclear reaction $^{63}Cu(\gamma,n)^{62}Cu(\beta^+)$. This reaction has an average cross section of 52.3 mb, for the 16.5 MeV average energy gamma rays, from the $^7Li(p,\gamma)^8Be$ reaction, and has a threshold of 10.9 MeV. This high energy threshold prevents the 6.1-7.1 MeV gamma rays, from the high yield prompt gamma ray reaction $^{19}F(p,\alpha\gamma)^{16}O$ (to be discussed later), from interfering with the measurement. The ^{62}Cu produced is a positron emitter with a 9.74 minute half-life and can be counted with exactly the same coincidence counting system described above. Note that the method is not sensitive to ablation, because even though the intense proton beam will heat and cause the LiF target to expand, the amount of expansion in a short pulse lasting 20ns to 50ns is small and the LiF always represents a thick target to the incoming beam. Before proceeding to describe the details and the absolute calibration of the technique, a few basic relationships governing the detector during calibration and during fast pulse activation will be presented.

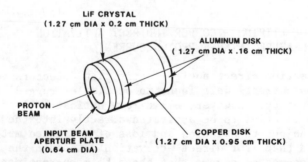

Fig. 7. Ablation-proof $^7Li(p,\gamma)^8Be$ photonuclear activation detector.

4.2 Analysis

As the first step in deriving a set of relationships that the detector will obey, consider a steady state 16.5 MeV γ ray flux ϕ, from the LiF target, to irradiate the copper sample (i.e., we are assuming a steady state proton current is striking the LiF target). Then the rate R, at which activation interactions occur per second, within the copper sample, is given by

$$R = \phi \, \varepsilon_{AB} \, N_T \, \sigma(E) \, V \, , \tag{42}$$

where
$$\varepsilon_{AB} = \text{natural abundance of } ^{63}Cu$$

$$N_T = \text{total number of Cu targets/cm}^3$$
$$\text{in the copper sample}$$

$$\sigma(E) = \, ^{63}Cu(\gamma, \, n)^{62}Cu \text{ cross section at}$$
$$\text{energy E}$$

$$V = \text{volume of copper sample .}$$

We are assuming here that the copper sample is thin enough that the gamma ray beam intensity does not decrease significantly when passing through it (this assumption will turn out to be not quite true for our detector, but any flux variation across the sample is corrected for by our calibration).

As the copper sample is irradiated, the radioactive nuclei that are formed also undergo radioactive decay. The rate of this decay is given simply by λN, where, as usual, λ is the decay constant $= 1/\tau$ and N is the total number of radioactive nuclei present. The rate of change of N is given by the difference between the rate of formation and the rate of decay given by

$$\frac{dN}{dt} = R - \lambda N \, . \tag{43}$$

Since we are assuming R is constant (steady state irradiation by the proton beam), the solution of Eq. (44) for the condition N=0 at t=0 is

$$N(t) = \frac{R}{\lambda} \left(1 - e^{-\lambda t}\right) \tag{44}$$

$$= \frac{\phi \varepsilon_{AB} N_T \sigma(E) V}{\lambda} (1 - e^{-\lambda t}) \, . \tag{45}$$

The activity A of the copper is given by λN or

$$A = \frac{dN}{dt} = \phi \varepsilon_{AB} N_T \sigma(E) V (1 - e^{-\lambda t}) \, . \tag{46}$$

We will now assume that the irradiation has proceeded for a time, at which time, the copper is removed and the total number of radioactive nuclei produced N_0 is given by

$$N_0 = N(t_0) = \frac{\phi \varepsilon_{AB} N_T \sigma(E) V}{\lambda} (1 - e^{-\lambda t_0}) . \tag{47}$$

Just as in the direct activation case, this initial number of radioactive nuclei will obey the relationship

$$N(t) = N_0 e^{-\lambda t} \tag{48}$$

during their decay. If the counting of the activity $\frac{dN}{dt} = -\lambda N$ is carried out over the interval between t_1 and t_2 (referenced to the end of irradiation), we have

$$C = \varepsilon_{Det} \int_{t_1}^{t_2} \frac{dN}{dt} \, dt + B \tag{49}$$

$$= \varepsilon_{Det} \int_{t_1}^{t_2} -\lambda N dt + B \tag{50}$$

$$= \varepsilon_{Det} \int_{t_1}^{t_2} (-\lambda) \frac{\phi \varepsilon_{AB} N_T \sigma(E) V}{\lambda} (1 - e^{-\lambda t_0}) e^{-\lambda t} dt + B \tag{51}$$

$$= \frac{\phi \varepsilon_{Det} \varepsilon_{AB} N_T \sigma(E) V}{\lambda} (1 - e^{-\lambda t_0})(e^{-\lambda t_1} - e^{-\lambda t_2}) + B , \tag{52}$$

where C is the total number of counts recorded, ε_{Det} is the overall counting efficiency (including self-absorption effects and B is the total number of background counts expected in the interval $(t_1 - t_2)$. Now, rewriting the copper V as M_{cu}/ρ, N_T as $N_0 \rho / A_T$, and rearranging, we have

$$C - B = \frac{\phi \varepsilon_{Det} \varepsilon_{AB} \sigma(E)}{\lambda} \frac{N_0}{A_T} (M_{Cu})$$

$$X (1 - e^{-\lambda t_0})(e^{-\lambda t_1} - e^{-\lambda t_2}) ; \tag{53}$$

N_o = Avogadro's number = 6.02×10^{23} atoms per mole,

A_T = copper atomic weight in grams,

M_{Cu} = mass of copper sample in grams

Now the gamma ray flux ϕ may be written as

$$\phi = I_p Y(Ep) \frac{1}{4\pi R^2} , \qquad (54)$$

where

I_p = proton current in terms of proton/sec,

$Y(E_p)$ = $^7Li(p, \delta)^8Be$ thick target yield at proton energy E_p,

R = distance to copper sample in cm.

Substituting this expression for ϕ, we have

$$C-B = \frac{I_p Y(E_p) \varepsilon_{Det} \varepsilon_{AB} \sigma(E_\gamma) N_o M_{cu}}{\lambda A_T (4\pi R^2)}$$

$$X \left(1-e^{-\lambda t_o}\right)\left(e^{-\lambda t_1} -e^{-\lambda t_2}\right) \qquad (55)$$

for the conditions of steady state irradiation (i.e., I_p = constant), for a time t_o.

Now consider the case of short irradiation time where t_o becomes very small compared to the mean life $\tau = \frac{1}{\lambda}$ of copper. In this case, the term $(1-e^{-\lambda t_o})$ simplifies to λt_o and we have

$$C - B = \frac{I_p t_o Y(E_p) \varepsilon_{Det} \varepsilon_{AB} \sigma(E_\gamma) N_o M_{cu}}{A_T (4\pi R^2)} (e^{-\lambda t_1} -e^{-\lambda t_2}). \quad (56)$$

Now define a calibration factor $F(E_p)$ to be

$$F(E_p) = \frac{Y(E_p) \varepsilon_{Det} \varepsilon_{AB} \sigma(E_\gamma) N_o}{A_T(4\pi R^2)} \qquad (57)$$

such that for steady state irradiation we have

$$C - B = I_p \frac{M_{Cu}}{\lambda} (1 - e^{-\lambda t_o})(e^{-\lambda t_1} - e^{-\lambda t_2}) F(E_p). \qquad (58)$$

It may then be seen that the calibration factor F under, steady
conditions, is directly applicable to short pulsed activation
since, under these conditions

$$C - B = I_p t_o M_{Cu} (e^{-\lambda t_1} - e^{-\lambda t_2}) F(E_p) . \qquad (59)$$

Now in the most general case of intense beam pulsed activation,
the proton beam current I_p and voltage E_p are functions of time.
Under these conditions, Eq. (60) may be generalized to

$$C - B = M_{Cu} (e^{-\lambda t_1} - e^{-\lambda t_2}) \int_{t_I}^{t_F} F(E_p(t)) I_p(t) dt , \qquad (60)$$

since the total activation can be thought of as the sum of the
activation, induced over a time dt, at energy $E_p(t)$, if beam
bunchinging is ignored (i.e., short flight paths). A generalization of Eq. (6), including beam bunching, will be presented in
the $^7Li(p,\gamma)^8Be$ prompt gamma ray section, but this complication
need not concern us in this section. Let us now see how the
above relationships may be applied to using and calibrating the
technique.

4.3 Detector Calibration

The absolute calibration of the detector was carried out on
the Sandia 2.5 MeV Van de Graaff accelerator. The detector was
irradiated by a steady state proton beam ranging in energy from
0.5 MeV to 2.0 MeV. Target currents were generally between 400nA
and 750nA. Beam runs of 30 minutes were used throughout the calibrations. After a beam run was completed the copper disk was
removed and counted for a period of 20 minutes in the coincidence
counting arrangement described above. The quantity $F(E_p)$ was
then obtained from the data and Eq. (59).

Figure 8 shows $F(E_p)$ plotted as a function of incident proton energy E_p. The absolute detection efficiency ε_{Det} for the
15.2 diam. x 7.62 cm NaI coincidence counting system is contained
in $F(E_p)$ and has not been unfolded from the data. It is clear
from Fig. 8 that $F(E_p)$ has a rather slow dependence on proton

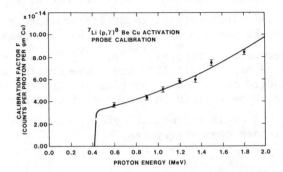

Fig. 8.   Li-Cu activation detector calibration of counts/
(proton-gm copper) versus proton energy.

energy which allows the proton current to be measured with the
detector, even if the proton voltage is only approximately known.

A complication in the use of the detector was found, during
the calibration, if it is used at proton energies above the 1.88
MeV threshold for the reaction $^7Li(p,n)^7Be$. This is because,
above threshold, the $^7Li(p,n)^7Be$ reaction produces slow neutrons
that activate the cooper via the reaction $^{63}Cu(n,\gamma)^{64}Cu(\beta^+)$.
Even though ^{64}Cu has a 12.71 hour half-life, there are so many
neutrons captured leading to ^{64}Cu that, even for a few minute
counting interval, this activity adds to the desired ^{62}Cu activity to form a substantial fraction of the total counts recorded.
This undesirable contribution to the copper activity may be corrected from the data by counting the copper sample several hours
after irradiation. The number of counts recorded at such a long
time after irradiation will, almost solely, be due to the 12.71
hour activity since the 9.74 minute activity will have almost
completely decayed. These counts may then be corrected back to
the (t_2-t_1) interval using the known 12.71 hour decay rate and
subtracted from the data. In this way the true 9.74 minute
activity, during the interval (t_2-t_1), may be obtained. Inciden-
tally, the fact that 12.71 hour activity is present means that
at least a fraction of the protons had energies above 1.88 MeV.
So this technique has a built in threshold voltage monitor.

The above complication can be eliminated entirely, if desired, by using the rare earth praseodymium Pr instead of copper
as a detector. ^{144}Pr is a 100% natural abundance isotope and has
a large photonuclear cross section of 225mb, at 16.5 MeV gamma
ray energy, for the reaction ^{141}Pr(γ,n) ^{140}Pr. ^{140}Pr is a positron emitter with a 3.39 minute half-life and consequently can be
counted using the standard coincidence counting arrangement described above. Slow neutron capture in the case of praseodymium
yields ^{142}Pr which is a β^- emitter and, therefore, easily dis-
criminated by a coincidence counting system.

4.4 Natural Abundance Deuterium Correction

Just as in the case of the ^{12}C (p,γ)^{13}N reaction, the Li-Cu
detector has some sensitivity to natural abundance deuterium
in a proton beam. This occurs because fast neutrons from the
^{7}Li(d,n)^{8}Be reaction can induce the ^{63}Cu(n,2n)^{62}Cu(β^+) reaction
in the copper sample. The extent of this background problem was
determined in a separate Van de Graaff run using a pure deuterium
beam. The activation of the pure deuterium beam was then corrected to the value that would be formed by a natural abundance
deuterium beam divided by the $\sqrt{2}$ (the square root of the mass
ratio of d to p), just as was done for ^{12}C(p,γ)^{13}N reaction
above. A deuterium correction factor (CF) was then determined as
the ratio of proton induced activity to total induced activity
from both the protons and a natural abundance proton beam. CF is
plotted as a function of incident beam energy in Fig. 5 for both
the ^{7}Li (p,γ)^{8}Be reaction and the ^{12}C(p,γ)^{13}N reaction, as noted
earlier.

4.5 Minimum Detectable Signal and Experimental Demonstration of Technique

In order to achieve a feeling for the minimum detectable
proton current onto the detector, consider a triangular proton
burst of 1.2 MeV that is 18 nsec wide at FWHM. In addition consider a typical coincidence counting system that has a background
of 22 counts per 4 minute counting interval (typical) and that we
will start counting the copper sample at 180 sec after the end of
irradiation. We will also require that C-B will be at least 3σ
above the background level. Under these conditions,

C-B = 14 counts,

$t_o = 18 \times 10^{-9}$ sec,

$M_{Cu} = 10.8$ gm

$e^{\lambda t_1} - e^{\lambda t_2} = 0.2001$, for a 4 min counting interval, starting at 180 sec

$$F(E_p) = F(1.2MeV) = 5.7 \times 10^{-14} \frac{counts}{(gm-Cu)\ Proton} ,$$

which, then substituted into Eq. (61), yields,

$$I_p = 6.3 \times 10^{21} \frac{p's}{sec} = 1.0 \text{ kA} \qquad (61)$$

for a minimum detectable current.

Finally, the detector was used to measure the proton beam current of the applied B field diode, on shot 4007, on the Sandia Proto I accelerator. The LiF front target surface was located at a radius of 3.1mm from the diode center in this experiment. In a 4 minute counting interval, starting 275.0 sec after the shot, the copper sample yielded 560 counts, with a background of 22.75 counts, for the same 4 minute counting interval. Applying Eq. (61) it was found that 7.8×10^{-4} Coulombs of protons struck the detector during the pulse. Assuming a triangular proton beam pulse shape of 18 nsec FWHM (known to be approximately correct from ion $\dot{B}$ monitors and other diagnostics), a peak current of 43.4 kA on target would be inferred, with a peak current density of 137.0 kA/cm^2, at the 3.1 mm radius. Knowing that the vertical extent of the beam is 3.5 mm (from proton induced Al Kα x-ray imaging), the total cylindrical peak beam current of the diode was calculated to be 267kA.

5. ^{7}Li(p,γ)^{8}Be PROMPT GAMMA RAY DIAGNOSTIC FOR ABSOLUTE TIME RESOLVED PROTON CURRENT MEASUREMENTS

The lack of time resolution is a major limitation of the nuclear activitation technique, i.e., the number of ions per pulse is determined rather than the current. This implies that, to measure the current with an activation detector, the temporal history of the number of ions per second striking the target must be determined with some other diagnostic (for example $\dot{B}$ monitors or a fast PIN detector into which ions are Rutherford scattered from the beam).

We present here a natural extention of the previous section which allows the direct measurement of the absolute time-resolved current, of an intense proton beam, by detecting the prompt gamma rays emitted from the $^7Li(p,\gamma)$ ^8Be nuclear reaction, induced in a lithium metal target by the beam. These prompt gamma rays are emitted on time scales much shorter than the ion pulse duration. In this section the method used to calibrate the diagnostic, as well as a demonstration of its use, will be presented.

5.1 Diagnostic Method

When an intense proton beam pulse strikes a thick lithium metal target, all protons above 440 keV in energy may induce the nuclear reaction $^7Li(p,\gamma)^8Be$. The prompt gamma rays emitted from this reaction are produced in two energy groups of 14.7 - 15.2 MeV and 17.6 - 18.2 MeV, with an average energy of 16.5 ± 0.3 MeV, independent of incident proton energy in the range 0.44 MeV to 1.8 MeV. By measuring the time history of the prompt gamma ray production with fast scintillator photomultiplier detectors, the proton current on target can be determined, provided the $^7Li(p,\gamma)$ ^8Be thick target cross-section is known. In principle, it is possible to calibrate the scintillator photomultiplier detectors directly on a Van de Graaff accelerator. In practice, however, due to the intense bremsstrahlung background associated with the ion beam production, the scintillator photomultipliers require thick shields of some 76.2 cm - 96.5 cm of concrete and 5.1 cm of lead for the preferential detection of the prompt gamma rays. The use of these massive shields would require proton beam sources beyond the capabilities of conventional accelerators. In order to achieve absolute calibration, we have used the photonuclear activation of copper via the reaction $^{63}Cu(\gamma,n)^{62}Cu$ to measure the total number of prompt gamma rays produced [4,5,19]. Then the proton current is determined by normalizing the prompt gamma ray time history signal with the total number of prompt gamma rays produced. A correction is made for dependence of the thick target yield on proton energy. The time history of the prompt gamma ray production required the use of an array of six scintillator photomultipliers to insure data of good statistical quality because of the rather low $^7Li(p,\gamma)^8Be$ production cross-section and the gamma ray attenuation in the massive shield.

5.1.1 Prompt gamma ray time history. The six detector array
used to measure the prompt gamma ray time history is shown in
Fig. 9.

The scintillator photomultiplier combinations each consist
of a 5.6 cm diameter x 6.7 thick cylinder of NE-111 plastic scintillator, viewed by an Amperex XP2020 photomultiplier tube [20].
These detectors are used in the mean level, or current mode, as
opposed to the more usual single particle counting mode.

The FWHM response of the detectors is 4.5 nsec when used to
observe a single instantaneous event such as a cosmic ray. Each
detector is capable of linear currents to 0.5 amp and is fully
saturated at 1.0 amp of current. The detectors are operated at
a sensitivity of 0.2 amps/(rad/sec) to ^{60}Co gamma rays. The
scintillators are coated with white paint except for the photomultiplier tube viewing port. The gains of each detector are
cross calibrated using the Compton edge technique. In addition,
the gain of each detector is periodically monitored for any shift
throughout an experimental run. Typically, the gain shifts observed were less than 1%.

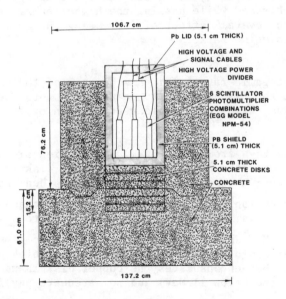

Fig. 9. Schematic of prompt gamma ray time history detector.

The six detector array is required in present experiments
to insure data of good statistical quality. It is found that on
an individual shot (at an average signal level of 6 volts) there
are variations of $\pm$ 80% between individual detectors. By simply
averaging the six detectors together we produce a signal that is
statistically better by a factor of the 6, resulting in a net
signal good to $\pm$ 33%. It should be noted that a secondary benefit of a multiple detector array is that it allows the statistical precision of the data to be measured experimentally, whereas
with a single detector, this is more difficult to determine.

The massive concrete shield shown in Fig. 9 is necessary
to attenuate the large x-ray bremsstrahlung background that is
generally produced by electron loss in an ion diode. Whereas an
ion diode may be 50%-80% efficient in producing ions, there is an
accompanying electron current which can amount to several 100 kA
(20%-50% of the diode total current at voltages exceeding 1.5 MV).
Experimentally, it is found that 76.2 cm - 96.5 cm of concrete
and 5.1 cm Pb are enough shielding to reduce this bremsstrahlung
background to less than 35% of the prompt gamma ray signal for
the 1.8 MV,1TW Proto I accelerator. The reason concrete is
chosen as the primary shielding material is because its gamma ray
attenuation coefficient if a factor 2.1 smaller at 15 MeV -
18 MeV than it is at 2.0 MeV [21]. Consequently, the concrete
shielding will preferentially attenuate the lower energy x-ray
bremsstrahlung background while passing the higher energy prompt
gamma rays, with the overall effect of increasing the prompt
gamma ray signal to noise ratio.

5.1.2 Total prompt gamma ray yield. A schematic of the experimental arrangement that is employed to measure the total number
of prompt gamma rays, using photonuclear activation of copper, is
shown in Fig. 10.

The energetic gamma rays from the $^7Li(p,\gamma)^8Be$ reaction are allowed
to strike a 7.6 cm diameter by 1.0 cm thick copper disk located
19.1 cm from the lithium target. These gamma rays then induce
the photonuclear reaction $^{63}Cu(\gamma,n)$ $^{62}Cu(\beta^+)$. As discussed
above, this reaction has, on the average, a 52.3 mb cross section
for the 14.7 - 15.2 MeV and 17.6 - 18.2 MeV $Li(p,\gamma)^8Be$ gamma
rays, and has a threshold of 10.9 MeV. The 10.2 cm thick polyethylene moderator that encapsulates the copper disk is used to
partially moderate (to below 11 MeV) and to attenuate any fast
neutrons that might be produced in the reaction $^7Li(d,n)^8Be$ by
the natural abundance of deuterium in the proton beam. This is

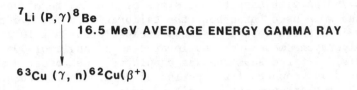

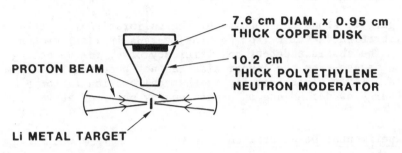

Fig. 10. Experimental arrangement for measuring the total number
          of prompt gamma rays using the photonuclear activation
          of copper.

important since neutrons above 11.0 MeV may produce ^{62}Cu via the
reaction ^{63}(n,2n)^{62}(β+) just as in the case of the LiF - Copper
detector. This thickness of moderator will attenuate these fast
neutrons by a factor of approximately 18.5 and make their contribution to the activation negligible [22]. Note that photonuclear
neutron production in the moderator, due to the bremsstrahlung
background, does not occur since the bremsstrahlung end point
energy is below any photonuclear reaction threshold. After activation, the copper disk is removed from the accelerator vacuum
chamber and placed between two 12.7 cm diameter by 7.6 cm thick
NaI (Tℓ) scintillator photomultiplier detectors. The two 0.511
MeV annihilation gamma rays from the ^{62}Cu (9.8 minute half-life)
decay are then energy selected and counted in a standard coincidence counting arrangement. The background count rates are generally between 4 and 5 counts per minute.

5.2 Diagnostic Calibration

The absolute calibration of the copper disk was carried out
on the Sandia 2.5 MeV Van de Graaff accelerator using the same
method as employed for the LiF - Copper detector. A thick
lithium metal target was irradiated by protons ranging in energy

from 0.5 MeV to 1.5 MeV. Target currents were generally 500 nA
to 750 nA with target spot sizes of 9 mm^2. Beam runs of 30 minutes were used throughout the calibration. The 7.6 cm diameter
by 1 cm thick copper disk was encapsulated in a 10.2 cm thick
polyethyene moderator and located at an angle of 90° to the beam
axis. After a beam run was completed, the copper disk was removed and counted for a period of 20 minutes in the coincidence
counting arrangement described above.

Figure 11 shows the results of the calibration. Plotted is
the calibration factor F from Eq. (59) versus incident proton
energy. The absolute detection efficiency of 6% for detecting
positron decays from the copper disk has not been unfolded from
the data. It is clear from Fig. 3 that calibration factor F has
the rather slow dependence on proton energy observed above.

5.3 Experimental Demonstration

The current of a proton beam onto a 19 mm diameter by 12.7
mm tall thick lithium metal target has been measured as an experimental demonstration of this diagnostic method. The proton
beam is generated by a radial, magnetically insulated, ion diode
powered by the Sandia Proto I generator. A schematic of this
diode is shown in Fig. 12. The copper activation sample is
located 19.1 cm from diode center and along the vertical axis of
the diode. The six detector array is positioned along the vertical axis at a distance of 1.8 m from diode center. A description
of the operation of the diode is given in Reference [3].

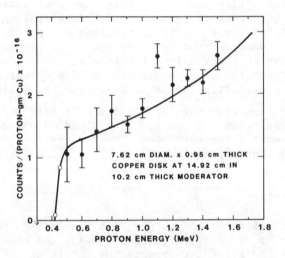

Fig. 11. Copper calibration factor Y_{CU} versus proton energy.

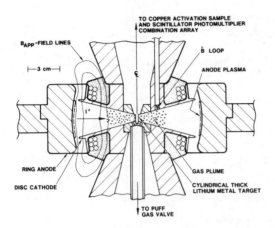

Fig. 12.   Schematic of the Proto I radial magnetically insulated
            ion diode.

The larger of the two waveforms shown in Fig. 13 is the
signal obtained upon averaging the signal from each detector,
of the six detector array, for the case of the 19 mm diameter
lithium target shot. The copper disk activity on this shot was
128 counts/min, measured starting 115 sec after the shot. The
smaller waveform in Fig. 5 shows the average signal from an identical shot, for the case of a 17 mm diam. aluminum target. This
signal is due to bremsstrahlung production from electron loss in
the anode cathode gap of the diode. The lower production cross
section and lower energy gamma rays from proton induced prompt
gamma emission, from Al, combine to make their contribution to
the signal neglible. The copper disk activity for this shot was
8 counts/min, at 130 sec, after the shot. As can be seen, the
peak of the signal from the aluminum target comes earlier than
the lithium target signal and has an amplitude equal to 35% of
the lithium target signal. The delay between the signals is due
to time-of-flight of the protons transversing the 6 cm distance
from the anode to the lithium target. Consequently the smaller
aluminum target signal, the delay between the aluminum and lithium signals, and the absence of any counts above background on
the copper disk for the case of the aluminum target, when taken
together, serve as positive evidence that we are indeed ob-
serving $^{7}Li(p,\gamma)^{8}Be$ prompt gamma rays.

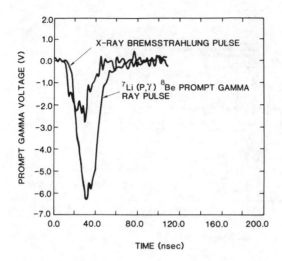

Fig. 13.   Average $^7\text{Li}(p,\gamma)^8\text{Be}$ prompt gamma ray signal and x-ray
bremsstrahlung background signal.

To determine the proton current from the normalization of
the measured prompt gamma time history and copper activation
count, a correction must be made for the flight times of different energy (voltage) protons from the diode to the lithium target. Then the time variation of the prompt gamma ray detector
voltage $V_{Det}(t)$ is related to the variation of the proton current
$I_p(t)$ in the following way

$$V_{Det}(t) = K \int_0^\infty Y\left(V(t')\right) I_p(t') \, \delta\left(t - t' - \frac{\ell}{v(t')}\right) dt', \quad (62)$$

where: K is a constant specified by the detector geometry, sensitivity, and target-to-detector distance; $V(t')$ is the induc-
tively corrected diode voltage at time t'; $Y(V(t'))$ is a quantity
pro-portional to the thick lithium target yield for protons of
energy V (number of prompt gamma rays per proton); $I_p(t')$ is the
proton current at time t' (number of protons per second); ℓ is
the proton flight path from the anode to the target; and v is the
velocity of protons of energy $V = 1/2 \, M_p v^2$, emitted at time t',
from the anode [23]. The calculated copper activation count
above background $(C-B)_{Cal}$ is obtained by adapting to our case the -
standard activation formula of Eq. (61). This relationship is

$$(C-B)_{Cal} = M_{Cu} \ (e^{-\lambda t_i} - e^{-\lambda t_2})$$

$$X \int_{t_i}^{t_f} \int_0^\infty Y_{Cu} \ (V(t')) \ Ip(t') \ \delta \left(t-t'-\frac{\ell}{\overline{v(t')}}\right) dt'dt \ . \tag{63}$$

Here M_{Cu} is the mass of the copper sample, λ is the copper decay constant, t_1 is the time interval between the end of irradiation to the start of counting, t_2 is the time interval from the end of irradiation to the end of counting, t_i is the time the voltage pulse starts, t_F is the time at the end of the voltage pulse, $Y_{Cu}(V(t'))$ is the absolute copper activation yield curve given in Fig. 11, and all other variables are the same as those defined for Eq. (63).

The first step in the data reduction routine we have adopted is to determine the temporal shape of the proton current $I_p(t)$ from the prompt gamma ray time history signal $V_{Det}(t)$. In practice, this is determined by iterating on Eq. (63) until the prompt gamma ray response calculated, with an assumed proton current, agrees in temporal shape with the measured prompt gamma ray response. We use the absolute copper activation yield of Fig. 11 for $Y(V(t'))$ since it is a quantity directly proportional to the lithium thick target yield. The initial proton current assumed is the total diode current corrected for electrons lost. Often this initial guess is sufficient to give good agreement between the measured and calculated prompt gamma responses. Once the temporal shape of $I_p(t)$ has been determined, it is substituted into Eq. (64) which is then integrated for the duration of the voltage pulse. The net result of the calculation of Eq. (64) is the total number of copper activation counts above background, $(C-B)_{Cal'}$, for the counting interval t_1 to t_2, for the initial assumed proton current $I_p(t)$. The calculated copper activity $(C-B)_{Cal}$ is then normalized to the experimentally observed copper activity, $(C-B)_{Exp'}$ to yield a final proton current given by

$$I_p^{Final}(t) = \frac{(C-B)_{Exp}}{(C-B)_{Cal}} I_p(t) \ . \tag{64}$$

Applying the above data reduction procedure to the present
data yields the results presented in Fig. 14. Shown in Fig. 14
are the final proton current and energy (diode voltage) used in
Eq. (63) to calculate the prompt gamma ray signal presented in
Fig. 14b. The experimental signal in Fig. 14b is obtained by
subtracting the aluminum target signal from the lithium target
signal in Fig. 13. In this way, the bremsstrahlung contribution
to the lithium signal is subtracted away. The small prepulse
in front of the larger prompt gamma ray signal in Fig. 14b is
due to this subtraction not being perfect. This prepulse being
neglected, it is observed that there is excellent agreement between the calculated and experimental prompt gamma ray pulses and
that there is 353 $\pm$ 54 kA peak current on the 19 mm diameter
lithium target. The error quoted is the addition in quadrature
of a $\pm$ 5.4% statistical error from the copper activity used to
obtain $(C-B)_{Exp}$ and an estimated error of $\pm$ 14.2% in the value of
the quantity $(C-B)_{Cal}$. This latter error is estimated from a
calculation of $(C-B)_{Cal}$ in which values of $Y_{Cu}(V(t))$ in Eq. (64)
were allowed to deviate one standard deviation from the mean. We
note, however, that this rather modest error in $(C-B)_{Cal}$ is due
in part to the unique behavior of the voltage and current in this
experiment that just happens to emphasize the higher precision
data points of Fig. 11. In an experiment that has high currents
at substained low voltage this error will be somewhat larger. In
yet another parameter sensitivity calculation, it is found that
varying the voltage by $\pm$ 10% resulted in a variance of $\pm$ 7.6% in
the quantity $(C-B)_{Cal}$. No error in the quantity $(C-B)_{cal}$, due to
voltage variations, was included in the present error analysis.
Finally, it should be pointed out that the prompt gamma ray measurements only sample the proton current above 440 keV. Therefore, the measurement is insensitive to the current profile for
proton energies below this threshold.

Figure 15 shows data obtained on shot 1047 on the 3.5 TW
Proto II accelerator with an applied B field diode.

The target in this experiment was 2 cm diameter by 2cm tall.
As can be seen, 1.2 MA peak current was incident on the target.

5.4 Thick Target Yield Dependence On Target Temperature

Before leaving the subject of ^{7}Li$(p,\gamma)^8$Be prompt gamma ray
diagnostics a word should be said about the effect of target
temperature on the amount of dE/dx loss, of the beam, in the target. There have been theoretical studies that suggest that dE/dx
will at first increase, over its classical value, as the target

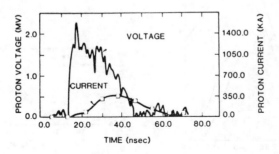

(a)

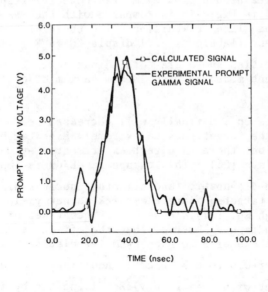

(b)

Fig. 14. The measured prompt gamma ray detector response (solid
          line) in Fig. 14b is compared with the response (open
          square points) calculated using the proton voltage and
          current displayed in Fig. 14a.

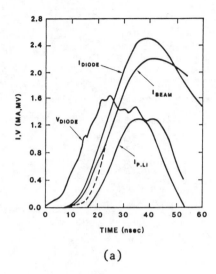

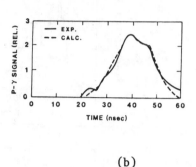

(a)   (b)

Fig. 15.   The measured prompt gamma ray detector response (solid
           line) in Fig. 15b is compared with the response
           (dashed line) calculated using the proton voltage and
           current (labeled $I_{p,Li}$) displayed in Fig. 15a.   Other
           quantities shown are total diode current and ion beam
           current measured with 13.5cm radium dB/dt loops.

temperature goes up and finally will decrease, below its classical value, as the target goes to extremely high temperatures [24,
25].   In addition, there is direct experimental evidence that
this effect exists [26].   This change in dE/dx is important to
the $^7Li(p,\gamma)^8Be$ diagnostic (and all other nuclear diagnostics
that use thick targets) since the thick target yield is dependent
on dE/dx through Eq. (27).   However, a recent set of dE/dx measurements at Sandia has observed that the effect is small and only
occurs near the end of a 20 nsec FWHM irradiation, at a peak
power density of 0.6 to 0.7 TW/cm^2.   Additional measurements have
been carried out with the $^7Li(p,\gamma)^8Be$ diagnostic by first measuring the current of a highly focused proton beam, on a large
lithium target, where the power density was low and the target
was cold, and then measuring the current on a small target where
the power density was 0.6 TW/^{cm2} and the target was hot.   The
same current was measured in both cases.   So our conclusion is
that the dependence of dE/dx on target temperature is unimportant

at the 0.6 TW/cm^2 level on lithium targets. But in future high power density experiments this effect could become of increased importance and should be carefully evaluated.

6. ^{19}F(p,$\alpha\gamma$)^{16}O PROMPT GAMMA RAY DIAGNOSTICS FOR TIME RESOLVED PROTON CURRENT MEASUREMENTS

Only a brief account of the ^{19}F(p,$\alpha\gamma$)^{16}O prompt gamma ray diagnostic technique will be given here since conceptually it is identical to the ^{7}Li(p,γ)^{8}Be prompt gamma technique. Details concerning the technique may be found in references [18] and [27].

6.1 Diagnostic Method

The ^{19}F9p,$\alpha\gamma$)^{16}O reaction is resonant with low energy pro-tons ($E_p \geq 0.34$ MeV) and produces rays with energies of 6 to 7 MeV. The most intense gamma ray is 6.13 MeV, but 6.92 and 7.12 MeV gamma rays are also produced [18,27]. The thick target yeild for this reaction is given in Fig. 16.

Prompt gamma rays, from this reaction, in a Teflon (CF$_2$) target can be measured with a scintillator-photomultiplier detector of good time resolution (≤ 5 nsec FWHM). If the detector output is calibrated absolutely, a measurement of the prompt gamma ray re-sponse can be used to determine the total number of protons and the time variation of the proton current.

6.2 Absolute Calibration

Young, et al., in Ref. [27] have directly calibrated two scintillator-photomultiplier combinations, mounted in two differ-ent size lead shields, that had end plates 3.2 cm in diameter and 5.1 cm thick. Calibration of both prompt gamma ray detectors were carried out on a Van de Graaff accelerator by integrating directly the photomultiplier anode output current and the target beam current, on a teflon target, with current integrators. The calibration factor was then simply the ratio of the integrated detector output current to the integrated beam current on target. These detectors have been used to measure 40 nsec FWHM proton beam pulses, whose peaked voltage and current were 550 kV and 1.1 kA. For proton pulses of higher voltage and current, which would require more massive shields, because of the increase in x-ray bremsstrahlung associated with the beam, this method of direct calibration would be increasingly difficult to carry out

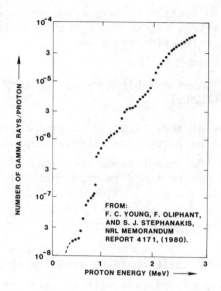

Fig. 16.   Measured thick-target yield Y(E) for the $^{19}F(p,\alpha\gamma)^{16}O$
reaction on a Teflon target (from Ref. [27].

because of the increased attenuation of the detector shield.
Also, it is clear from the thick target yield in Fig. 16, this
diagnostic requires an accurate determination of beam voltage for
reasonably accurate current measurements. The major advantage
of the diagnostic over the $^{7}Li(p,\gamma)^{8}Be$ reaction is a factor of
approximately 30 to 50 increase in thick target yield.

6.3 Relative Measurements

Because of the problems associated with the absolute calibration of massively shielded detectors and the requirement of
an accurate determination of beam voltage, this diagnostic has
been used at Sandia for only relative measurements. In particular, it has been used to study proton beam focusing by comparing
detector signals on large and small targets, where beam voltages
were known to be essentially identical.

7.   PROMPT NEUTRON MEASUREMENTS

In addition to prompt gamma ray measurements presented
above, prompt neutron measurements may be used to determine ion
beam current and voltage. The $d(d,n)^{3}He$ reaction has been used

to determine those quantities for a deuteron beam [28] and the ^{7}Li(p,n)^{7}Be reaction (threshold = 1.88 MeV) for a proton beam [29]. Another interesting and important use of prompt neutron measurements is the measurement of dE/dx in a high temperature target [26]. Ion beam voltage is measured by first measuring the prompt neutron energy by time-of-flight methods and then relating the measured neutron energy to the ion energy beam energy by employing reaction kinematics. The neutron response in a time resolved detector provides a direct measure of ion beam current, if the width of the response results primarily from the duration of neutron emission.

7.1 Neutron Time-of-Flight Method to Measure Mean Ion Energy

Consider first the neutron time-of-flight to determine ion beam voltage. The pulsed nature of these ion beams makes the time-of-flight (TOF) method ideal for measuring neutron energies. The neutron energy is determined from a measurement of the neutron flight time for a known flight path. The ion energy is then determined from the neutron-producing reaction kinematics. Using fast scintillator photomultiplier combinations, a detector time resolution of 4-5 ns FWHM may be obtained. Several meter flight paths are used to separate the neutron pulse from the x-ray bremsstrahlung due to election loss in the ion diode. The neutron energy resolution is given by $\frac{\Delta E}{E} = \frac{2\Delta t}{t}$, where t is the flight time of the neutrons and Δt is the time spread of the neutron response in the detection system. Since the flight path ℓ is related to neutron velocity v by $\ell = vt$, we have $\frac{\Delta E}{E} = \frac{2v\Delta t}{\ell}$. Consequently, as the flight path is increased, the neutron energy resolution scales inversely with the flight path, for a fixed time resolution, but the number of neutrons incident on a given detector decreases quadratically with distance. A compromise must be made between signal intensity and energy resolution [30].

The time spread of the neutron response in a TOF detector is composed of three contributions which add in quadrature: (1) the detector response; (2) the energy dispersion of the neutrons (which translates into a time dispersion because of the finite mass of the neutron, as compared to the massless gamma ray in the prompt gamma techniques); and (3) the duration of emission of the neutrons. Care must be taken in the contribution from the detector response, since the total response of the detector includes any dispersion from the shielding surrounding the scintillator photomultipliers used. Since the required shielding in most experiments is massive, this contribution to the time response of the detector can be sizeable. The shielding

dispersion effect has been measured for 2.45 MeV and 14.1 MeV
neutrons by Leeper and Chang, for lead shielding thicknesses of
up to 20.3 cm [31], with a special 3 ns FWHM neutron source developed for this purpose [32].   Ion energy measurements require
a long neutron flight path so that the time spread from energy
dispersion dominates the duration of neutron emission.   For ion
current measurements, a short flight path is required so that the
time spread from energy dispersion is minimized relative to the
duration of neutron emission.

The neutron time-of-flight method to obtain the energy and
emission time of the most energetic deutrons from an ion diode
can be illustrated with some measurements by Young, et al., using
the $d(d,n)^3He$ reaction [28].   Kinematics for this reaction are
shown in Fig. 17.

As can be seen, the neturon energy depends most sensitively
on the deutron energy for neutrons emitted in the forware direction.   To take advantage of this fact the (two neutron) TOF detectors used in this measurement were deployed at 15°.   The
deuteron beam was directed onto a thick CD_2 target.   The response
of the two neutron detectors is shown in Fig. 18.

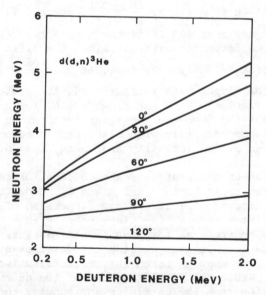

Fig. 17.   Dependence of the outgoing neutron energy, at different angles on the incident deuteron energy, for the
$d(d,n)^3He$ reaction (from Ref. [28].

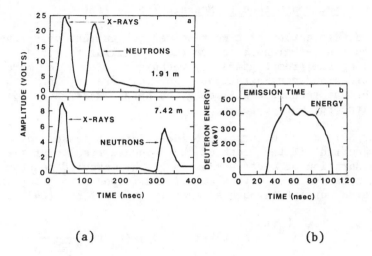

(a)   (b)

Fig. 18.   (a) Neutron TOF measurement for two detectors at 15° to
the deuteron beam direction and at flight paths of
1.91 m and 7.42 m respectively. The maximum deuteron
energy corresponds to the most energetic neutrons which
are detected at the leading edge of the TOF traces.
(b) The emission time of these deuterons is compared
with the deuteron energy trace (from Ref. [28].

Both detectors record an initial pulse, resulting from diode
bremsstrahlung, before the neutron responses are recorded. The
neutron signal on the 7.42 m detector is delayed relative to
1.91 m detector, as expected. The most energetic neutrons have a
flight time corresponding to the half-height of the leading edge
of the neutron signal. The neutron time-of-flight for each detector is measured relative to the peak of the x-ray signal and
corrected for the x-ray flight time. The difference of the neutron flight times implies a neutron energy of 3.41 ± 0.15 MeV.
The corresponding deuteron energy is 0.45± 0.10 MeV, based on the
d(d,n)^{3}He kinematics at 15°. The uncertainties in these values
are the result of uncertainties in the time and distance measurements. The neutron flight time versus distance is extrapolated
to zero distance to determine the emission time for the most
energetic neutron. Also included in Fig. 15 is the deuteron energy, as determined from a inductance corrected voltage monitor.
As can be seen, the maximum deuteron energy for this trace is
0.46 MeV, in good agreement with the energy determination by
TOF. More details on these measurements may be found in Reference [28]. In addition, similar measurements can be carried out
with the ^{7}Li(p,n)^{7}Be [29].

7.2 Prompt Neutron Method to Measure Ion Current

The determination of time resolved ion current using the
prompt neutron technique is conceptually similar to its determination using the prompt gamma ray technique, with a few more
complications and considerations. For this reason, only a brief
account of the technique will be given here. Reference [28] may
be consulted for more details.

The neutron in a TOF detector provides a direct measure of the
deuteron current if the width of the response results primarily
from the duration of neutron emission. The measured neutron response can be compared with a calculated response to determine
the deuteron current through a generalization of Eq. (64) to,

$$R_n(t_n,\theta_n) = \int_o^{E_d(t)} I_d(t) Y(E_d(t),\theta_n) \delta\left(t_n - t - \frac{\ell_n}{v_n} - \frac{\ell_d}{v_d}\right) dt, \quad (65)$$

R_n = neutron TOF detector response

t_n = neutron time of arrival at the TOF detector

θ_n = angle of neutron detector relative to the
direction of deuteron beam

$E_d(t)$ = incident neutron energy at time t

$I_d(t)$ = deuteron current at time t

$Y_d(E_d(t),\theta_n)$ = d(d,n)^{3}He thick target yield for deutron of
energy E_d at an angle = θ_n

ℓ_n = neutron flight path

v_n = neutron velocity

ℓ_d = deuteron flight path

v_d = deuteron velocity

Details of the deviations of Eq. (66) may be found in Ref. [28].
The additional complexities in Eq. (66), as compared to the
prompt gamma case, arise from a velocity dispersion and, hence, a
time dispersion due to the slowing down of the deutrons in a
thick target and the angular dispersion of deuterons on target
(the latter leads to a time dispersion because outgoing neutron
velocity is related to incident deuteron angle by reaction kinematics). These complications do not arise for prompt gamma rays
because gamma rays travel at the speed of light irregardless of

their energy. A straight forward inversion of Eq. (66) for deuteron current I_p is not possible. Instead, the right side of Eq. (66) is calculated and $I_p(t)$ iterated until suitable agreement with the measured response is obtained. In all of this, it must be remembered that the time response of the detectors, including any shielding dispersion, must be deconvoluted from the data before any attempt to obtain $I_p(t)$. In addition, the technique may be used in a similar fashion to obtain time resolved proton current via the $^7Li(p,n)^7Be$ reaction.

Before leaving the subject of prompt neutron reactions, it should be pointed out that is possible to measure an x-ray bremsstrahlung spectrum (of short duration) by measuring the neutron spectrum from (γ,n) photonuclear reactions. This has not been used in intense beam work. The scheme is to measure the energy of the neutron emitted with the time-of-flight technique. The energy of the neutron that is produced from a given energy gamma ray is given by the relation [33]

$$E_n = \frac{A-1}{A} \left[E_\gamma - E_T - \frac{E_\gamma^2}{1862(A-1)} \right] + \delta . \tag{66}$$

This relation is deduced from the conservation of energy and momentum in the interaction of the proton with the nucleus. The quantity A is the mass of the target nucleus, E_γ the energy of the gamma ray in MeV, and E_T the threshold energy in MeV for the (γ,n) reaction in the target nucleus. The correction term δ is given approximately by

$$\delta = E_\gamma \cos\theta \left[\frac{2(A-1)(E_\gamma - E_T)}{931A^3} \right]^{\frac{1}{2}} , \tag{67}$$

where θ is the angle between the path of the gamma ray and the direction of the emitted neutron. Note the very small dependence on θ.

The photodisintegrations of deuterium and beryllium, via the reaction $d(\gamma,n)p$ and $^9Be(\gamma,n)^8Be$ which have thresholds E_T of 2.2 MeV and 1.63 MeV, respectively, are possible examples of reactions that could be used. It should be relatively easy, for example, to measure the approximate end point energy of the spectrum to obtain the electron energy in the beam. In any case, it

should be possible to exploit this technique to measure some
properties of an intense x-ray or electron beam.

8.   INVERSE REACTIONS TO MEASURE HIGH Z BEAMS

Recent theoretical results [34] indicate that power density
on target for light ion diodes should increase approximately as
V^2. In order to take advantage of this scaling, it has been
suggested that intense lithium beams at 30 MeV or intense carbon
beams at 60 MeV could be used to increase power density on target
in particle beam inertial confinement fusion (ICF) experiments.
The range of proton beams at these high voltages would be much
too long to efficiently couple to a 50μm-100μm shell thickness of
an ICF target.

The simplest way to employ nuclear reactions to diagnose Li
as C beams is to simply reverse the roles of projectile and target in the nuclear reactions we have already discussed. For
example, instead of considering the reaction $^7Li(p,\gamma)^8Be$, we consider the reaction $p(^7Li, \gamma)^8Be$, where p is now the target and
^7Li the beam. The term inverse reactions is applied to reactions
in which the traditional roles of target and projectile are reversed. In order to investigate these reactions further, we will
now consider some simple kinematic realtionships for inverse reactions.

8.1 Inverse Reactions Kinematics

Part of the energy of the incident particle, carried into
a nuclear reaction, must be used to conserve momentum of the
colliding system. If we consider a target nucleus of mass M_A,
standing at rest, and an incident particle of mass M_a approaching
it with velocity V_a, we can picture the motion of the system before and after collision as follows:

Before Collision   After Collision

$$M_a \xrightarrow{V_a} M_A \qquad\qquad M_a + M_A \rightarrow V_{a+A}. \tag{68}$$

a   A   a+A

Conservation of momentum in the reaction system requires

$$M_a V_a = (M_a + M_A) V_{a+A} \ . \tag{69}$$

After the collision, we are considering the motion of the total mass of the system $(M_a + M_A)$. Squaring both sides of Eq. (70), multiplying by 1/2, and using the definitions of kinetic energy K.E. $= 1/2MV^2$, we have,

$$M_a E_a = (M_a + M_A) E_{a+A} \ . \tag{70}$$

Thus, the recoil energy is

$$E_{a+A} = \frac{M_a}{(M_a + M_A)} E_a \ , \tag{71}$$

and the kinetic energy of particle a, available to a nuclear reaction is,

$$E_{Available} = E_{cms} = \frac{M_A}{(M_a + M_A)} E_a \ . \tag{72}$$

$E_{Available}$ is also the total kinetic energy available in the center-of-mass frame (see, for example, any standard test on nuclear physics such as Ref. [35]. Since the cross sectional energy dependence of a nuclear reaction is only dependent on the total energy made available to the reaction in the center-of-mass frame, we can set the energy available for the reaction A(a,b)B to the energy available for the reaction a(A,b)B to obtain,

$$\left(\frac{M_A}{M_a + M_A} \right) E_a = \left(\frac{E_a}{M_a + M_A} \right) E_A \ , \tag{73}$$

where E_a and E_A are understood to be the incident kinetic energy of particle a or A. These quantities are related by

$$E_A = \left(\frac{M_A}{M_a} \right) E_a \ . \tag{74}$$

Now let us apply these considerations for the reactions $^{7}Li(p,\gamma)^{8}Be$ and $P(^{7}Li, \gamma)^{8}Be$. In the normal direction, for protons incident on ^{7}Li, we have a threshold of 0.440 MeV, as discussed above. If a Li beam is incident on a target of protons we have,

$$E_{Li} = \left(\frac{M_{Li}}{M_p}\right) E_p = (\tfrac{7}{1}) (0.440) \tag{75}$$

$$= 3.08 \text{ MeV} .$$

The threshold for the inverse reaction is, therefore, seven times its threshold in the "forward" direction.

8.2 $d(^{7}Li,n)^{8}Be$ Diagnostic for Intense Li Beams

Even though for eventual ICF applications we will need Li beams in the 30 MeV range, initial experiments will be carried out at much lower voltage of 1.5 MeV to 2 MeV. The $d(^{7}Li,n)^{8}Be$ reaction could be useful in this range. This reaction has a positive Q value of 15.02 MeV and consequently has no kinematic threshold. But for deuterons incident on lithium there is an effective coulomb barrier threshold of approximately 0.15 MeV. This translates into $(7/2)(.15) = 0.53$ MeV for the inverse reaction. So this reaction should be useful above this energy.

The $d(^{7}Li,n)^{8}Be$ thick target calibration factor F, versus incident Li energy, is plotted in Fig. 19. This was measured using the University of Iowa's Van De Graaff. As can be seen, the thick target yield has a strong energy dependence at low lithium voltages, but above approximately 4.5 MeV, the energy dependence starts to become flat. This weak dependence on lithium voltage makes this diagnostic ideal for lithium current measurements at lithium energies above 4.5 MeV. Below 4.5 MeV lithium energy, the voltage of the lithium beam must be known well before any accurate determination of lithium current can be made using the $d(^{7}Li, n)^{8}Be$ reaction.

8.2 $p(^{7}Li, \gamma)$ Diagnostics for Intense Li Beams

Just as the $^{7}Li(p,\gamma)^{8}Be$ reaction has proven to be extremely useful for proton diagnostics, the $p(^{7}Li,\gamma)^{8}Be$ reaction should also be useful for Li beam measurements. As pointed out above,

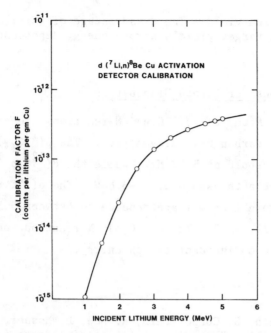

Fig. 19. Thick-target yield of the d(^{7}Li,n)^{8}Be reaction.

the p(^{7}Liγ)^{8}Be reaction has a threshold of 3.08 MeV. The ^{63}Cu(γ,n)^{62}Cu activation detector for the 16.5 MeV, emitted from the p(^{7}Li,γ)^{8}Be, reaction has a threshold of 11 MeV. So above 11 MeV the copper activation detector would suffer from direct acti- vation from the x-ray bremsstrahlung associated with beam produc- tion. However, for Li beam between 3.1 MeV and 11 MeV, this diagnostic should perform well and have the same relatively weak voltage dependence (proton energy scale converted to Li energy scale by Eq. (75) as that described above for the ^{7}Li(p,γ)^{8}Be re- actions. Prompt gamma ray time resolved diagnostics would become increasingly difficult as the beam voltage is increased since the x-ray bremsstrahlung background will become harder and closer in energy to the 16.5 MeV gamma rays emitted in the reaction.

8.3 p(^{7}Li,n)^{7}Be Diagnostic for Intense Li Beams

The p(^{7}Li,n)^{7}Be reaction could be useful for Li beam diagnos- tics above its threshold of 13.16 MeV (threshold = 1.88 MeV in forward direction). The prompt neutron emissions could be ob- served with a time resolved detector to measure Li current and voltage. In addition, neutron activation could be employed to measure total neutron yield. The disadvantage of this reaction would be its thick target yield's strong energy dependence on Li beam energy.

8.4 Other Reactions for Carbon Diagnostics

The $p(^{12}C,\gamma)^{13}N$ and the $d(^{12}C,n)^{13}N$ reactions are of possible interest for carbon beam diagnostics. The $p(^{12}C,\gamma)^{13}N$ reaction has a threshold of 5.52 MeV, while the $d(^{12}C,n)^{13}N$ reaction would have a threshold of 1.93 MeV. The $p(^{12}C,\gamma)^{13}N$ reaction would have a flat voltage dependence between 5.52 MeV and approximately 19.2 MeV. The $d(^{12}C,n)^{13}N$ reaction would have a strong dependence on incident carbon energy.

REFERENCES

1. S. A. Goldstein, G. Cooperstein, R. Lee, D. Mosher, and S. J. Stephanakis, "Focusing of Intense Ion Beams from Pinched-Beam Diodes," Phys. Rev. Lett. 40 1504 (1978).

2. J. Golden, C. A. Kapetanakis, S. J. Marsh, and S. J. Stephanakis, "Generation of 0.2-TW Proton Pulses," Phys. Rev. Lett. 38, 130 (1977).

3. D. J. Johnson, G. W. Kuswa, A. V. Farnsworth, Jr., J. SP. Quintenz, R. J. Leeper, E. J. T. Burns, and S. Humpries, Jr., "Production of 0.5-TW Proton Pulses with a Spherical Focusing, Magnetically Insulated Diode," Phys. Rev. Lett. 42, 610 (1979).

4. D. J. Johnson, E. J. T. Burns, A. V. Farnsworth, Jr., R. J. Leeper, J. P. Quintenz, K. W. Bieg, P. L. Dreike, D. L. Fehl, J. R. Freeman, and F. C. Perry, "A Radial Ion Diode for Generating Intense Focused Proton Beams," J. Appl. Phys. 53, 4579 (1982).

5. D. J. Johnson, P. L. Dreike, S. A. Slutz, R. J. Leeper, E. J. T. Burns, J. R. Freeman, T. A. Mehlhorn, and J. P. Quintenz, "Applied-B Field Ion Diode Studies at 3.5 TW," J. Appl. Phys. 54, 2230 (1983).

6. R. J. Leeper, J. R. Lee, D. J. Johnson, W. A. Stygar,
    D. E. Hebron, and L. D. Roose, "Direct Measurement of the
    Energy Spectrum of an Intense Proton Beam," Proc. 5th Int.
    Top. Conf. on High Power Particle Beams, San Francisco,
    California, p. 514 (1983).

7. F. C. Young, J. Golden, and C. A. Kapetanakis, "Diagnostics
    for Intense Pulsed Ion Beams," Rev. Sci. Instrum. 48, 432
    (1977).

8. Glenn F. Knoll, Radiation Detection and Measurement,
    John Wiley and Sons, New York (1979).

9. H. H. Anderson and J. F. Ziegler, The Stopping and Ranges
    of Ions in Matter, Pergamon, New York, 1977, Vol. 3.

10. D. J. Johnson, A. V. Farnsworth, Jr., D. L. Fehl,
    R. J. Leeper, and G. W. Kuswa, "A Dual-current-feed Mag-
    netically Insulated Light-ion Diode," J. Appl. Phys. 50,
    4524 (1979).

11. J. P. VanDevender, J. P. Quintenz, R. J. Leeper,
    D. J. Johnson, and J. T. Crow, "Self-magnetically Insulated
    Ion Diode," J. Appl. Phys. 52, 4 (1981).

12. J. N. Olsen, D. J. Johnson, and R. J. Leeper, "Propagation
    of Light Ions in a Plasma Channel," Appl. Phys. Lett. 36,
    808 (1980).

13. J. N. Olsen and R. J. Leeper, "Ion Beam Transport in Laser-
    initiated Discharge Channels," J. Appl. Phys. 53, 3397
    (1982).

14. A. E. Blaugrund and S. J. Stephanakis, "Improved Technique
    for Measuring Intense Pulsed Ion Beams by the Nuclear Acti-
    vation Method," Rev. Sci. Instrum. 49, 866 (1978).

15. S. Miyamoto, T. Ozaki, K. Imasaki, S. Higaki, S. Nakai, and
    C. Yamanaka, "A Diagnostic for Intense Focused Proton Beams
    Using the ^{11}B(p, $\alpha\gamma$)2α Reaction," J. Appl. Phys. 53, 5440
    (1982).

16. F. C. Young and M. Friedman, "A New Approach to ION Accel-
    eration with Electron Beams," J. Appl. Phys. 46, 2001
    (1975).

17. R. J. Leeper, J. Chang, and D. E. Hebron, "A New Ablation
    Proof Nuclear Activation Detector for the Measurement of In-
    tense Proton Currents," Bull. Am. Phys. Soc. 27 984 (1882).

18. J. Golden, R. A. Mahaffey, J. A. Pasour, F. C. Young, and
     C. A. Kapetanakos, "Intense Proton Beam Current Measurement
     Via Prompt Rays from Nuclear Reactions," Rev. Sci. Instrum.
     49, 1384 (1978).

19. R. J. Leeper, E. J. T. Burns, D. J. Johnson, and
     W. M. McMurtry, "Proton Current Measurements Using the
     Prompt Gamma Ray Diagnsotic Technique," Proceedings of the
     Workshop on Measurement of Electrical Quantities in Pulsed
     Power Systems, National Bureau of Standards Special Publica-
     tion 628, p. 267 (1982).

20. Available from EG&G, Inc., Las Vegas, Nevada, as Model
     NPM-54.

21. R. E. Lapp and H. L. Andrews, Nuclear Radiation Physics,
     Prentice Hall, Inc., Englewood Cliffs, NJ, p. 424 (1972).

22. S. S. Nargolwalla and E. P. Przybylowica, Activation
     Analysis with Neutron Generators, John Wiley and Sons,
     New York, p. 30 (1973).

23. F. C. Young, F. Oliphant, and S. J. Stephanakis, "Absolute
     Calibration of a Prompt Gamma Ray Detector for Intense
     Bursts of Protons," NRL Memorandum Report, 4171 (1980).

24. T. A. Mehlhorn, "A Finite Material Temperature Model for Ion
     Energy Deposition in Ion-driven Inertial Confinement Fusion
     Targets," J. Appl. Phys. 52, 6522 (1981).

25. E. Nardi, E. Peleg, and Z. Zinamon, "Plasma Effects in the
     Interaction of Intense Light Ion Beams with Light Targets,"
     Appl. Phys. Lett. 39, 46 (1981).

26. F. C. Young, D. Mosher, S. J. Stephanakis, S. A. Goldstein,
     and T. A. Mehlhorn, "Measurements of Enhanced Stopping of
     1-MeV Deuterons in Target-ablation Plasmas," Phys. Rev.
     Lett. 49, 549 (1982).

27. F. C. Young, W. F. Oliphant, S. J. Stephanakis, and
     A. R. Knudson, "Absolute Calibration of a Prompt Gamma-ray
     Detector for Intense Bursts of Protons," IEEE Trans. Plasma
     Sci. PS-9, 24 (1981).

28. F. C. Young, D. Mosher, S. J. Stephanakis, S. Goldstein,
     and D. Hinshelwood, "Temporal Deuteron Current Determina-
     tions Using Neutron Time-of-Flight," NRL Memorandum Report
     3823 (Washington, D. C., August, 1978).

29. R. A. Meger, F. C. Young, A. T. Drobot, G. Cooperstein, S. A. Goldstein, and D. Mosher, "High-impedance Ion-diode Experiment on the Aurora Pulser," J. Appl. Phys. 52, 6084 (1981).

30. F. C. Young, "Neutron Diagnostics for Pulsed Plasma Sources," IEEE Trans. Nucl. Sci. NS-22, 718 (1975).

31. R. J. Leeper and J. Chang, "The Response of Heavily Shielded Plastic Scintillator-Photomultiplier Combinations to Nano-second Neutron Pulses," IEEE Trans. Nucl. Sci. NS-29, 798 (1982).

32. J. Chang and R. J. Leeper, "Intense 3 NS Neutron Source," Rev. Sci. Instrum. 51, 715 (1980).

33. L. F. Curtiss, Introduction to Neutron Physics, D. Van Nostrand Company, Inc., New York, p. 95 (1959).

34. G. W. Kuswa, J. P. Quintenz, D. B. Seidel, D. J. Johnson, C. W. Mendel, Jr., E. J. T. Burns, D. Fehl, R. J. Leeper, F. C. Perry, P. A. Miller, M. M. Widner, and A. V. Farnsworth, Jr., "Scalability of Light Ion Beams to Reach Fusion Conditions," Proc. 4th Int. Top. Conf. High Power Electron and Ion Beam Research and Tech., Palaiseau, France, 1981.

35. H. Enge, Introduction to Nuclear Physics, Addison-Wesley, Reading, Massachusetts (1966).

PARTICLE ANALYZER DIAGNOSTICS FOR INTENSE PARTICLE BEAM
MEASUREMENTS

R. J. Leeper, J. R. Lee, L. Wissel, D. J. Johnson,
and W. A. Stygar

Sandia National Laboratories
Albuquerque, NM
USA

1. INTRODUCTION TO THE FUNDAMENTALS OF CHARGED PARTICLE
    ANALYZERS FOR INTENSE BEAM MEASUREMENTS

One of the most important measurements that can be performed
on an intense high-current density ion beam is a direct measurement of the beam's time-resolved energy spectrum. This measurement allows a direct determination of the current density, power
density, current, and voltage of the beam as a function of time.
These quantities are all of obvious importance in characterizing
any intense beam. In addition, a comparison of a direct measurement of an ion energy spectrum, with the ion energy distribution,
inferred from a line voltage monitor allows a study of A-K gap
energy loss mechanisms and the effects of any A-K gap neutral gas
on diode performance.

A straightforward and precise method for making energy and
momentum measurements on charged particles involves observation
of their deflection in applied electric and magnetic fields.
These fields produce a force ez $(\vec{E} + \vec{v} \times \vec{B}/c)$, where e is electronic charge and z is the atomic number, which acts to alter
the trajectory of the particle of charge ez in proportion to its
energy or momentum. An instrument using an appropriate arrangement of fields and apertures for the selection of the particle
energy or momentum is called an analyzer or spectrometer.

In the process of making an energy or momentum analysis of
charge particles, it is necessary to select a well-collimated
beam to input to the analyzer field. This has been the major
source of difficulty in the use of particle analyzers in the

measurement of intense beams. These difficulties have occurred
because an intense high-current density beam, upon striking the
input apertures of the analyzer, can turn the edges of these
apertures into a plasma. This plasma will start to move during
the pulse in such a way as to close off the apertures and to
cause large dE/dX losses of the input beam. Another problem in
using particle analyzers in intense beam measurement is that the
beamlet, selected by the aperture system, to be analyzed by the
analyzer field, may be of such high-current density that it will
"blow up" in crossing the analyzer field. The beamlet "blow up"
is due to space charge forces.

    In an effort to overcome these difficulties, we present here
the idea of using Rutherford scattering from a thin target to reduce the particle flux of an intense beam to a level suitable for
charge particle analysis [1]. Since Rutherford scattering is an
elastic process, it follows that Rutherford scattered ions will
suffer little energy loss in scattering from thin, high z scattering foils. This method of reducing beam flux intensity is
generally applicable to any particle analyzer system.

1.1 Electrostatic Deflection

    Electric fields provide a force $ze\vec{E}$ for charged-particle
energy analysis. An electric field oriented arbitrarily to the
velocity $\vec{v}$ would lead to a relatively complicated equation of
motion including a change in energy T of the particle. However,
a system designed such that $\vec{E} \cdot \vec{v} = 0$ maintains T = Constant [2].
Under the condition of small deflection, a simple form of electrostatic analysis, where these conditions are approximately
meant is shown in Fig. 1.

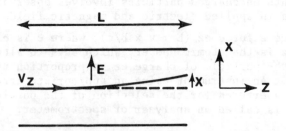

Fig. 1. Schematic of a simple electrostatic analyzer system.

Here, a collimated beam of the charged particles to be analyzed is passed between charged parallel plates, of length L, which maintain a constant electric field $\vec{E}$. The initial velocity vector $\vec{v}$ of the charged particle beam is directed along the z-axis and the electric field vector E is directed along the x-axis.

Under these conditions, the equation of motion of the charge particles of mass M in our analyses is given by,

$$M \frac{d^2x}{dt^2} = ezE .$$ (1)

Integrating Eq. (1) for the initial conditions of x=0 at t=0 we have for the deflection x,

$$x = \left(\frac{ez}{M}\right) \left(E\right) \left(\frac{t^2}{2}\right) .$$ (2)

The time t that a charged particle spends in the electric field is simply $t = \frac{L}{v}$. Substituting and using the relation $T = \frac{1}{2}Mv^2$ we have

$$x = zeE \left(\frac{L^2}{4T}\right) .$$ (3)

We see that the amount of deflection is inversely proportional to the kinetic energy of the charged particle beam. The details of other more complicated electrostatic analyzers, are discussed in Ref. [2]. These include 90° deflection analyzers as well 127° 17' focusing spectrometers.

1.2 Magnetic Deflection

Magnetic analyzers provide a means of passive momentum analysis of charge particles. A particle of net charge ze, moving with a velocity v, in a uniform transverse magnetic field B, obeys the equation of motion $Mv^2/R = ezv\,B/c$ (in CGS units). This means that the product of BR (where R is the radius of curvature of the trajectory) is proportional to the momentum per unit charge. In the common flat type of spectrometer or spectrograph the input beam is collimated by an aperture system, before passing through a region of well defined length L with a transverse magnetic field B, where the beam is deflected onto an array of detectors.

The general case of magnetic deflection, in a transverse magnetic field, for any incident angle can be understood from Fig. 2.

In this figure a uniform transverse magnetic field B is located at all points, between the lines labeled I and II, for a distance L along the z axis as shown. Now consider a particle of charge ez and velocity v whose trajectory enters the field with angle β_i relative to the z axis. Upon entering the field the particle will follow a circular path of radius R and exit the field with an angle of β_f and a deflection of y. The radius of the trajectory is given by

$$R = \frac{mvc}{ezB} = \frac{pc}{ezB} ,$$ (4)

where the momentum p=mv and c is the speed of light. From inspection of Fig. 2 it is found that,

$$\sin \beta_f = \frac{L + R\sin\beta_i}{R} .$$ (5)

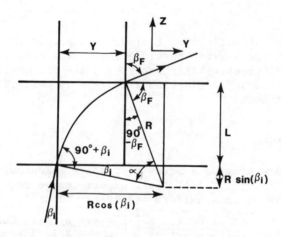

Fig. 2.   Schematic of the orbit of a charged particle in a transverse magnetic field.

Substituting R from Eq. (4), we have,

$$\sin \beta_f = \frac{ezBL}{pc} + \sin\beta_i .$$ (6)

For small deflection angles β_f and $\beta_i=0$, Eq. (6) becomes,

$$\beta_f = \sin \beta_f = \frac{ez\ BL}{pc} .$$ (7)

Now $\beta_F \simeq \frac{Y}{L}$, so for small deflections, we have,

$$y = \frac{ezBL^2}{pc} + \frac{ezBL^2}{\sqrt{2mT}\ c} ,$$ (8)

where the relation $\frac{p^2}{2m}=T$ has been used.

Another useful relationship from Fig. 2 is that,

$$y = R\ (\cos\beta_i - \cos\beta_f)\ \text{or,}$$ (9)

$$Y = \frac{pc}{ezB}(\cos\beta_i - \cos\beta_f) .$$ (10)

Finally, before leaving the subject of magnetic deflection, it is useful to express Eq. (4) in a system of practical units. Equation (4) becomes

$$R(cm) = \frac{3.3355\ P(^{MeV}/c)}{z\ B(kg)} ,$$ (11)

where z is the integer charge of the particle being deflected.

1.3 Deflection in a Uniform Parallel Magnetic and Electric Field

An important case of combined E and B field analysis is the case of where E and B fields are applied in a well-defined region, parallel to each other, and perpendicular to the charged particle beam under analysis. This field configuration was first introduced by J. J. Thomson in 1911 and has become known as the Thomson parabola field for reasons that shortly will become apparent.

As a first step in analyzing the deflection of a charge particle in such a field pattern, consider Fig. 3. Here the charge particles enter the field of length L, along the z axis. The parallel E and B fields point along the x-axis. Now let

us consider the small angle deflection of the charge particle
beam onto a detector (for example, either a photographic plate
or a nuclear track recording plate), located in the x, y plane,
at the output side of the analyzing field. The deflection due to
E field is along the x-axis and is given by Eq. (3) as,

$$x = zeE \left(\frac{L^2}{4T} \right) \; . \tag{12}$$

The deflection due to the B field is along the y-axis and is
given by Eq. (8) as,

$$y = \frac{ezBL^2}{(2mT)c} \; . \tag{13}$$

Combining Eqs. (12) and (13) and eliminating the kinetic energy
T, we obtain the well-known equation of a parabola as

$$y^2 = (2L^2) \left(\frac{ez}{mc^2} \right) \left(\frac{B^2}{E} \right) x \; . \tag{14}$$

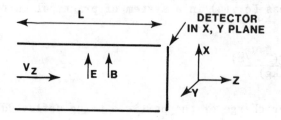

Fig. 3. Schematic of a simple Thomson parabola analyzer.

Thus the detector plate in the x, y plane will show a separate
parabolic trace for particles of each z/m ratio present, with an
extent from the z axis dependent upon the geometry, the fields,
and the energy range of that z/m particle. An interesting variation of the usual Thomson parabola analyzer has been built by
Dreike, et. al, [3]. In this analyzer, the static E field has
been replaced by a time changing E field in an effort to obtain
time resolved information with the analyzer.

1.4 Beam Intensity Reduction

As discussed above, an intense particle beam must be reduced in intensity before it can be inputted into a particle analyzer. It also was pointed out that Rutherford scattering was an ideal method to achieve such a beam intensity reduction. In this section we will discuss the details of the Rutherford scattering cross-section and kinematics.

1.4.1 Rutherford Scattering Cross-Section. The Rutherford scattering cross-section is given by [4]

$$\frac{d\sigma}{d\Omega} = \frac{0.8139\ Z^2 z^2}{2\pi E^2 \sin^4(\theta/2)} \times 10^{-26} \ \frac{cm^2}{nucleus} \ , \tag{15}$$

where   Z = Atomic number of target
             z = Atomic number of projectile
             E = Projectile energy in MeV
             θ = Laboratory scattering angle in degrees.

The number of Rutherford scattered particles, into solid angle element $d\Omega$, is given by

$$\frac{dN}{d\Omega} = \frac{d\sigma}{d\Omega} \left(\frac{I}{A}\right) \ nA\ \Delta x \ , \tag{16}$$

where   I = # beam particles/sec
             A = Area of beam in cm^2
             n = # target atoms/cm^3
             Δx = target thickness in cm

$\frac{d\sigma}{d\Omega}$ = Differential Rutherford cross-section in cm^2 .

The quantity $dN/d\Omega$ may also be written as,

$$\frac{dN}{d\Omega} = \left(\frac{d\sigma}{d\Omega}\right) \left(\frac{I}{A}\right) \left(\frac{N_A \rho}{M}\right) A\ \Delta x \ , \tag{17}$$

where   M = gram-atomic weight of target material
             N_A = avogadro's number
             ρ = target density in g/cm^3

The normal application of Eq. (17) is that we have a final aperture of area dA, at a distance of R from the target that defines the solid angle acceptance of a collimator system. Then $d\Omega = \dfrac{dA}{R^2}$ and we can apply Eq. (17) to directly calculate the number of particles that are scattered into an analyzer system.

1.4.2 Kinematics. Consider the elastic scattering of two particles whose masses are m_1 and m_2. Further let T_i represent the initial kinetic energy of m_1 and T_F its final kinetic energy. Let θ_c be the center-of-mass scattering angle. Then it can be shown [5] that the ratio of the final kinetic to the initial kinetic is given by

$$\frac{T_F}{T_I} = \frac{m_1^2 + m_2^2 + 2m_1m_2\,\cos\theta_c}{(m_1 + m_2)^2} . \tag{18}$$

When $M_2 \gg m_1$, the laboratory scattering angle θ_{lab} is essentially equal to θ_c. So for the case of protons scattering from gold at 90° we have,

$$\frac{T_F}{T_I} = \frac{(1)^2 + (197)^2 + 2(1)(197)(\cos 90°)}{(1 + 197)^2} \tag{19}$$

$$= 0.99 .$$

2. INTENSE PROTON BEAM ENERGY SPECTRUM MEASUREMENT WITH A MAGNETIC SPECTROGRAPH EMPLOYING A RUTHERFORD SCATTERED INPUT

A magnetic spectrograph employing a Rutherford scattered input has recently been used to make a time resolved energy spectrum measurement of an intense proton beam (0.6 TW/cm^2 level). The proton beam was generated with the applied B field magnetically insulated ion diode coupled to Sandia's Proto I accelerator. This proton energy spectrum measurement is the first ever of an intense ion beam. In what follows, a detailed description of the magnetic spectrograph employed, the method of data analysis, and the rather surprising results that were obtained will be presented.

2.1 Apparatus

A schematic of the experimental arrangement used to make the proton energy spectrum measurement is shown in Fig. 4. As can be seen, the magnetic spectrograph is mounted vertically into the applied B field diode and is operated in the diode vacuum chamber, at diode vacuum. The operation of this diode is described in detail in Ref. [6]. Sufficient for our purposes is that the diode generates an intense proton beam that is directed radially inward from the cylindrical anode surface. This beam then enters a gas cell, through a 2 μm thick mylar window, located just behind the cathode rings and is ballistically focused to the center of the diode. The gas cell is nominally filled with 6 Torr of Argon gas.

Located at the center of the diode is a gold Rutherford scattering foil mounted at an angle of 45° to the vertical axis. The nominal thickness of the foil is 2500 $\overset{o}{A}$ (the thickness of the gold varied from 2300 $\overset{o}{A}$ to 2600 $\overset{o}{A}$ during the course of the experiment). The foil was mounted on a 2μm Mylar backing for strength. The target holder in which the foil was placed was designed to accept a sector of beam from the cylindrical anode surface that was 90° azimuthally and ±15° vertically (referenced to the midplane of the diode). Those protons, which Rutherford scattered through an average angle of 90°, then passed through a second 2μm thick Mylar gas cell output window and into the spectrograph's collimator system. This collimator system consisted of two 1 mm diameter apertures separated by 12.3 cm. The apertures were fabricated from brass and had edges that were 0.5 mm thick to minimize collimator edge scattering.

The beam, after passing through the collimator system, then enters the magnetic field region of the spectrograph. The magnetic field is supplied by a samarium-cobalt permanent magnet which has a field value of 6.65 kg, uniform to within ±1% over the entire volume of the magnet, excluding a fringe field region which is located within the first 3 mm distance from the magnet's edge. The magnet had circular pole pieces that were 12.7 cm in diameter. The pole pieces were separated by 3.2 mm. Typical proton orbits through the magnetic field are shown in Fig. 4.

Upon exiting the magnetic field, the deflected protons are recorded on either the nuclear track recording media CR-39 or on eight, 1.13 mm diameter by 35 m thick, PIN diode nuclear detectors. The PIN diode detectors are located behind the CR-39 strip and view the protons through 3 mm diameter holes drilled through the CR-39. CR-39 records protons through a polymer radiation damage mechanism, induced by the ionization energy loss of the protons striking it. The CR-39 may then be etched in a solution

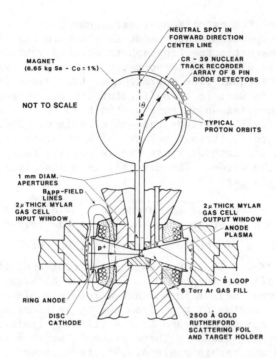

Fig. 4.   Schematic of experimental arrangement used to make
          intense proton beam spectra measurements.

of sodium hydroxide which preferentially etches along the proton
damage region to make the proton tracks visible under a microscope.   The number of proton tracks per unit area may then be
counted and the CR-39 record can be used to generate high quality
time integrated proton energy spectrum.

The PIN diode detectors are standard solid state nuclear detectors [7].   One electron-hole pair is formed in the detector
for every 3.62 ev of energy deposited.   The PINS used in this experiment had rise times of 450 ps and FWHM widths of 1.1 ns, as
measured by single 5.6 MeV [241]Am alpha particles.   However, because of the long cables (and biasing network used), the system
response time of these detectors was 3.17 nsec FWHM.   The PINS
has an entrance dead layer thickness of 0.5μm.

Finally, a small number of the protons, passing through the
spectrograph, charge exchange and become neutral.   These neutrals

pass through the spectrograph in the forward direction, unde-
flected, and are recorded in the CR-39. This mark of forward
direction on the CR-39 is then used as an origin for all measurements of proton deflection in the experiment.

2.2 Data Analysis

Essential to the analysis was a Monte Carlo computer simulation of the experiment. The variables and geometry used in the
Monte Carlo code are shown in Fig. 5.

The simulation generated protons uniformly on a 90° sector
of the anode, which had a ±15° vertical extent. Each proton was
tracked through the 2μm thick Mylar gas cell input window and
then through the gas cell filled with 6 Torr Argon gas, to an
interaction point located within a 3.0 mm diameter sensitive area
on the gold foil. It was found that 6% of the interactions occur
in the 2 μm Mylar backing of the gold foil. After Rutherford
scattering, the proton was tracked through the gas cell to a second 2 μm thick Mylar gas cell output window at the front of the
magnetic spectrograph. The proton was then tracked through the
spectrograph to a detecting surface at the position of the CR-39
strip and the PIN diode array. All positions along this detecting surface were defined by the angle $\theta = X/R$ whose vertex is
located at the center of the circular pole pieces as shown in
Fig. 5. The origin of the coordinate X is at the neutral density spot, in the forward direction, as discussed above. The Y
coordinate system is located along the direction perpendicular to
the direction of magnetic deflection and is, therefore, normal to
the X-coordinate system.

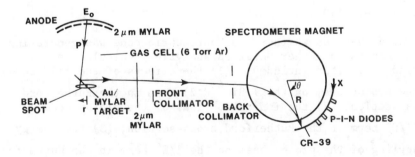

Fig. 5. Schematic of geometry used in Monte Carlo calculations.

For each segment of the proton trajectory, multiple scattering, ionization energy losses, and energy straggling are taken into account. All ionization energy losses were obtained from the table of Janni in Ref. [8]. The Bohr theory of straggling was used given by

$$\frac{dn}{d(\Delta T)} \propto e^{- (\Delta T)^2/2\Omega^2} ,$$ (20)

where

$$\Omega = 8.85 \cdot \sqrt{2Z/A} \cdot \sqrt{\rho \Delta x} \quad keV .$$

This theory describes the probability of energy straggling, an amount ΔT, while passing through a foil of density ρ, thickness Δx, an atomic number of Z, and an atomic weight A. Details of this theory of straggling is given in Ref. [9]. The multiple scattering relation used was [10],

$$\frac{dn}{d\theta_{Proj}} \propto e^{- (\theta_{Proj}^2/2 \, \theta_{ms}^2)} ,$$ (21)

where

$$\theta_{MS} = \frac{15MeV}{p\beta} \sqrt{L/L_{Rad}} \left[1 + 1/9 \, \log_{10} \, (L/L_{Rad})\right] .$$ (22)

Here $dN/d\theta_{Proj}$ describes the probability of multiple scattering through an angle θ_{Proj}. The variables p and β are the proton's momentum and velocity given by $\beta = v/c$. L is the foil thickness and L_{Rad} is the foil materials radiation length, given for example in Ref. [10].

As a check on whether the Monte Carlo code was generating valid results, a rather detailed analytical calculation was carried out. This included: (1) the effects of the ±15° spread in the proton beam on the $1/\mathrm{Sin}^4(\theta/2)$ term in the Rutherford cross section; (2) the ionization loss of the proton beam on the $1/E^2$ term in the Rutherford cross section; (3) the energy straggling of the proton beam on the $1/E^2$ term in the Rutherford cross section; and (4) the effect of target thickness spread due to the finite angular extent of the proton beam of ±45° azimuthally and ±15° vertically. The results of the analytical

model were within 5% of those predicted by the code. Consequently, it was verified that the code is certainly calculating
reasonable results.

2.3   Results of Monte Carlo Calculations

A few results calculated with the Monte Carlo code will now
be presented. Consider, for illustrative purposes, a monoenergetic proton beam of 1.5 MeV. Figure 6 shows the plot of events/
proton versus the angle θ.

For purposes of these calculations, an event is any proton
that makes it to the detecting surface located at the position
of the CR-39 strip and the PIN array as described above. As can
be seen, there is a strong, sharp peak located at a mean angle of
29.04° and a small, broad peak in the region of 32°. This small
peak is due to events that are scattered from the 2 μm Mylar
backing of the gold scattering foil. These events are located at
a larger angle (i.e., lower energy) than the mean since they are
due to elastic scattering from the lower mass target atoms of
carbon and oxygen instead of gold. Also tabulated in the plot is
the number TOTAL = 2.48 x 10^{-10}. This number is the total number
of events/proton, summed over the histogram, and represents the
total number of events/proton at 1.5 MeV that enter the magnetic
field of the spectrograph and strike the detector plane somewhere. Plotted in Fig. 7 are the same events plotted as a function of diode voltage. These events can also be plotted against
the X coordinate as shown in Fig. 8.

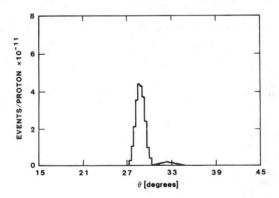

Fig. 6.   Monte Carlo calculation of events/proton, versus the
          angle θ, at 1.5 MeV proton energy.

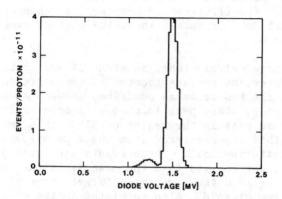

Fig. 7.   Monte Carlo calculation of events/proton, versus the
          diode voltage, at 1.5 MeV proton energy.

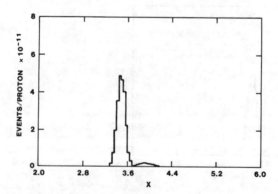

Fig. 8.   Monte Carlo calculation of events/proton, versus the
          coordinate X, at 1.5 MeV proton energy.

Figure 9 shows the Y distribution of these same 1.5 MeV events.
The r^2 distribution at the beam focus (r defined in Fig. 5) of
the 1.5 MeV events is shown in Fig. 10.

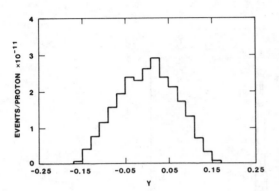

Fig. 9. Monte Carlo calculation of events/proton, versus the
coordinate Y, at 1.5 MeV proton energy.

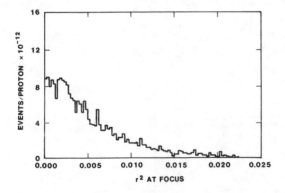

Fig. 10. Monte Carlo calculation of events/proton, versus r^2,
at 1.5 MeV proton energy.

The time-of-flight of the 1.5 MeV protons from the anode surface is shown in Fig. 11.

By combining the results of many Monte Carlo runs for different values of E_o we can obtain many important relationships. Figure 12 shows a plot of θ versus energy E_o. Also indicated on the graph are the locations of the eight PIN diode detectors.

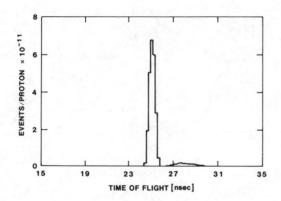

Fig. 11.   Monte Carlo calculation of the 1.5 MeV proton time-of-flight from the anode to the detector plane.

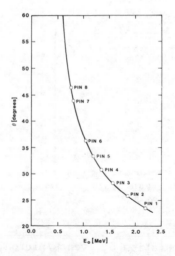

Fig. 12.   Monte Carlo calculated plot of θ versus energy E_o.

Figure 13 shows a plot of the proton energy loss in going through the spectrograph versus E_o. These data take on a particularly simple form when plotted against θ as shown in Fig. 14.

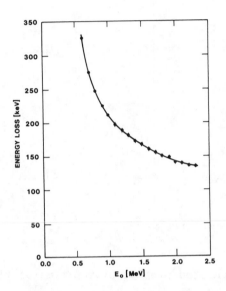

Fig. 13.   Monte Carlo calculated plot of proton energy loss, in going through the spectrograph, versus energy E_o.

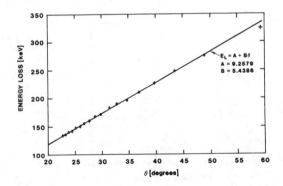

Fig. 14.   Monte Carlo calculated plot of proton energy loss, in going through the spectrograph, versus θ.

Figure 15 shows a plot of proton time-of-flight versus E_o.   The
time-of-flight is seen to be a linear function of θ from Fig. 16.

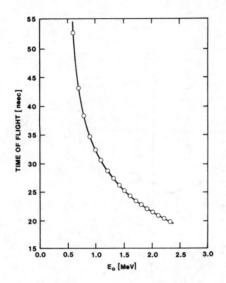

Fig. 15.   Monte Carlo calculated plot of proton time-of-flight
           versus E_o.

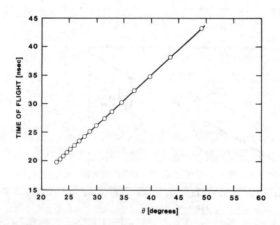

Fig. 16.   Monte Carlo calculated plot of proton time-of-flight
           versus θ.

The sensitivity versus E_o is plotted in Fig. 17 and it is plotted versus θ in Fig. 18.

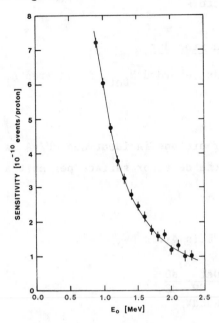

Fig. 17.   Monte Carlo calculated plot of sensitivity S versus E_o.

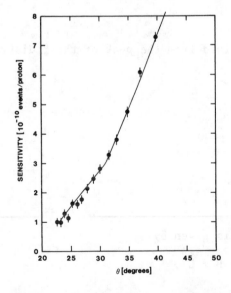

Fig. 18.   Monte Carlo calculated plot of sensitivity S versus θ.

Armed with the power of the Monte Carlo code, let us now consider how it may be applied to the data reduction for the CR-39 and PIN diode detectors.

2.4 CR-39 Data Reduction Procedure

The number of protons recorded N_{Det}, in an area A, on the CR-39, is given by

$$N_{Det} = N_p \frac{dP}{dA} A, \tag{23}$$

where N_p is the number of protons incident and dP/dA is the probability of reaching the detector surface per unit area. Now we have

$$dA = R \, d\theta \, dY . \tag{24}$$

For a small area A, dN_p/dE is given by,

$$\frac{dN_p}{dE} = \frac{dN_p}{d\theta} \frac{d\theta}{dE} = \frac{N_{Det}}{d\theta A \frac{dP}{Rd\theta dY}} \frac{d\theta}{dE} . \tag{25}$$

$$\frac{dN_p}{dE} = \frac{R \left(N_{Det}/A \right)}{\frac{dE}{d\theta} \frac{dP}{dY}} . \tag{26}$$

Now (dP/dY) can be obtained from the peak of the Y distribution and is found to be

$$\frac{dP}{dY} = (5.30 \text{ cm}^{-1}) \text{ S} . \tag{27}$$

Then we have,

$$\frac{dN_p}{dE} = \frac{R N_{Det} /5.30A}{S \frac{dE}{d\theta}} . \tag{28}$$

For our spectrograph E is given by

$$E = A \cot^2 (\theta/2) + B + C , \tag{29}$$

where

$$A = 8.796 \times 10^{-2}$$
$$B = 9.2579 \times 10^{-3}$$
$$C = 3.116 \times 10^{-1} \quad .$$

The absolute value of $dE/d\theta$ is

$$\frac{dE}{d\theta} = \frac{A}{\tan(\theta/2)\sin^2(\theta/2)} - C \quad . \tag{30}$$

Now define the weight function $W(\theta)$ as

$$W(\theta) = S \left| dE/d\theta \right| \quad . \tag{31}$$

Equation (28) can now be written as,

$$\frac{dN}{dE} = \frac{R \, N_{Det}/5.30A}{W(\theta)} \quad . \tag{32}$$

This is the desired relationship that connects the number of protons measured N_{Det}, in an area A, to dNp/dE. $W(\theta)$ is plotted as a function of θ in Fig. 19. Note the suppressed zero.

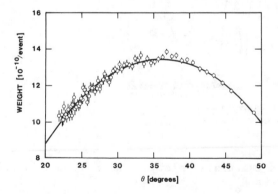

Fig. 19.   Monte Carlo calculated plot of the weight function $W(\theta)$ versus θ.

2.5 PIN Data Reduction Method

The time resolved PIN data reduction was carried out using the Sandia developed unfold code, UFO (Unfold Operation) in conjunction with Monte Carlo calculated response functions. UFO unfolded the eight relations (one for each PIN)

$$V_i(t) = \sum_{i=1,8} R_i(E)\phi_p(E,t)dE , \tag{33}$$

for the best estimate of the proton energy spectrum $\phi_p(E,t)$, at time t, where,

$$V_i = \text{ith PIN detector voltage at time} = t$$

$$R_i(E) = \text{ith PIN detector response function}$$

$$E = \text{proton voltage .}$$

The response functions $R_i(E)$ were calculated with many runs of the Monte Carlo code and are displayed in Fig. 20. The response functions, integrated over energy, are displayed in Fig. 21.

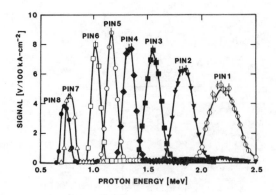

Fig. 20.   Monte Carlo calculated plot of the response functions $R_i(E)$ versus E_o.

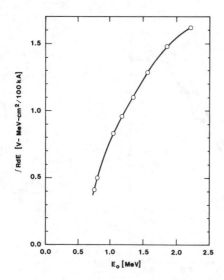

Fig. 21. Monte Carlo calculated plot of the integrated response
          functions versus E_o.

2.6 Energy Resolution of Spectrograph

The energy resolution of the spectrograph for the PIN and
CR-39 detectors was calculated with the Monte Carlo code and is
shown in Fig. 22. Also shown is σ_{KE} which represents the spread
in the kinetic energy of the protons as they enter the magnetic
field to be analyzed. σ_{KE} is due to straggling in the Mylar
foils, gold foil, and Ar gas. The resolution for the CR-39 is
calculated assuming that we can locate protons on the CR-39
arbitrarily well. As can be seen, the energy resolution is dominated by the geometry of the experiment at high energies, but
energy straggling becomes significant at about 1.5 MeV and dominates the resolution below 1.0 MeV. Plotted in Fig. 23 is σ_E/E
for the PINS and CR-39 as function of E_o.

2.7 Results

Applying the above data reduction procedures to the CR-39
and PIN data obtained on the Proto I accelerator, for each shot,
several quantities of interest were calculated. These include
time integrated proton energy spectra, time resolved proton
energy spectra mean, proton energy, proton current density, and

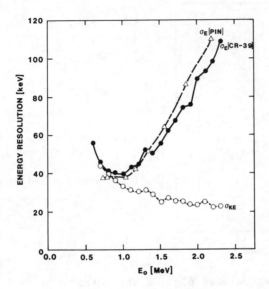

Fig. 22.   Monte Carlo calculated plot of energy resolution
           versus E_o.

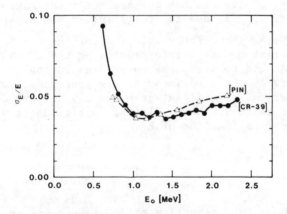

Fig. 23.   Monte Carlo calculated plot of the quantity σ/E_o
           versus E_o.

proton power density. But before giving some examples of these
results, let us consider the following simple Y distribution consistency check on the Monte Carlo code.

2.7.1 Y Distribution Consistency Check. As a consistency check,
the Y distribution of the protons on the CR-39 was calculated
with the Monte Carlo code and compared to the measured Y distribution (solid circle data) on Proto I shot 3957. The results are
shown in Fig. 24. The good agreement between the Monte Carlo
calculated results and the data is significant. First, it shows
that the Monte Carlo technique is calculating reasonable results
and that the geometry in the code is correct. Second, since our
apertures are circular, when this geometric spot size in the Y
dimension is rotated to the X dimension, it is directly relates
to the geometric contribution of our resolutions. At high energy, where geometry dominates our resolution, it has in essence
verified experimentally our resolution.

2.7.2 Time Integrated Proton Energy Spectrum. The time integrated proton energy spectrum for Proto I shot 3957 is shown in
Fig. 25 [11].

The solid circle data were obtained from the CR-39 by counting
the number of protons per unit area. The solid square data were
obtained from the time integrated PIN signals. The agreement between the two sets of data is good, as can be seen.

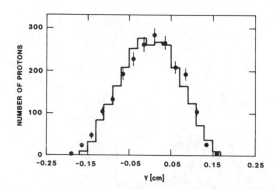

Fig. 24. Comparison of Monte Carlo calculated Y distribution
          with the experimentally obtained Y distribution for
          Proto I shot 3957.

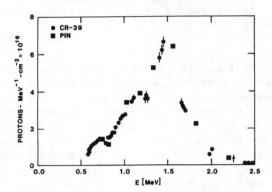

Fig. 25.   Time integrated proton energy spectrum from CR-39 data
           and time integrated PIN data for Proto I, shot 3957.

2.7.3 Time Resolved Proton Energy Spectrum.   The signals from
the eight PIN diode detectors from Proto I shot 3957 are multi-
plotted in Fig. 26.

These signals have been time-of-flight corrected, back to the
anode. The significant overlap of the signals and their rather
broad widths are indicative of a proton spectrum at each instant
in time.

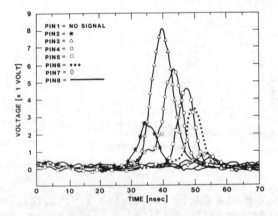

Fig. 26.   Multi-plot of the signals from the eight PIN diode de-
           tectors, from Proto I, shot 3957.

By measuring the signal level of each PIN detector at a
given time and then using the UFO based PIN data reduction procedure described above, the time resolved energy spectrum, shown
in Fig. 27 was obtained for shot 3957. As can be seen, there is
indeed a proton energy spectrum at each instant in time with a
width of some 150 keV to 250 keV. This is a most unexpected result, since previously it was expected that the protons in an ion
diode would simple "track" the diode voltage. The spectrum was
also observed on every Proto I shot.

Other quantities of interest may be caluclated from the data
of Fig. 27. The first of these is the mean proton energy, as a
function of time, shown in Fig. 28. The current density at any
instant in time may be calculated from the proton energy spectrum
by integrating it over energy. The results are shown in Fig. 29.
Finally it is possible to calculate the power density, shown in
Fig. 30 as a function of time.

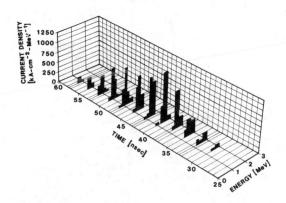

Fig. 27.   Time resolved proton energy spectrum from Proto I
           shot 3957.

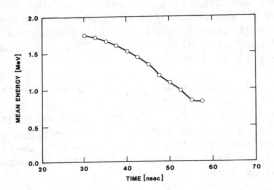

Fig. 28.   Plot of mean proton energy, versus time, from Proto I
shot 3957.

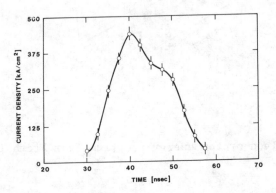

Fig. 29.   Plot of proton current density, versus time, from
Proto I shot 3957.

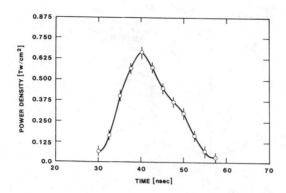

Fig. 30. Plot of proton power density, versus time, from Proto I
shot 3957.

3. MAGNETIC SPECTROGRAPH ARRANGEMENT USED TO MEASURE PULSED
POWER ACCELERATOR LINE VOLTAGE USING PROTON TEST PARTICLES

In addition to the direct beam measurements, employing the
gold Rutherford scattering foil on Proto I, experiments to measure the voltage applied to the OBI diode have been carried out on
PBFA. In this case, the magnetic spectrograph was mounted on the
triplate feed line to the OBI diode and energy analyzed the
emitted protons and carbon ions which were accelerated by the
full line potential. The experiment arrangement is shown in
Fig. 31. The protons and carbon ions were emitted from a CH
screen located on the center (+) conductor of the triplate. Because of an expected 8° bending of protons, traversing through
the insulating magnetic field of the triplate line at peak voltage, a 1.35 mg/cm^2 thick gold foil was placed in front of the
input aperture of the spectrograph. The purpose of this foil was
to elastically in-scatter a large fraction of those protons bent
at an angle of 8° into the acceptance of the spectrograph. It
should be noted that a 1.35 mg/cm^2 gold foil will only lower the
energy of a proton traversing it by 74.4 keV.

The magnetically deflected proton and carbon ions were once
again detected by the nuclear track recording medium CR-39 and
five 125 μm thick PIN diode detectors. The time integrated CR-39
data are shown in Fig. 32. A peak voltage of 1.45 ± 0.10 MeV
was obtained on OBI shot #606. The criterion employed to determine the output peak voltage was the half-height of the cutoff

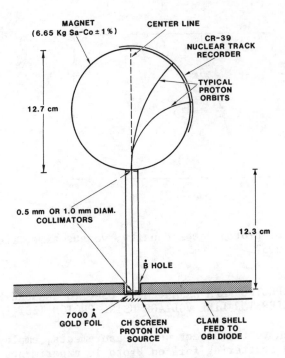

Fig. 31. Experimental arrangement used to measure pulse power
          accelerator line voltage using proton test particles.

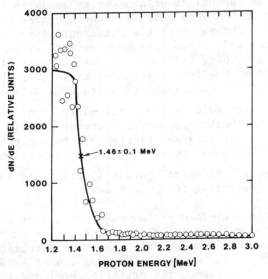

Fig. 32. Time integrated CR-39 data obtained in accelerator line
          voltage measurement.

edge, of the plot of the number of protons or carbons per unit
energy bin, versus energy. The CR-39 was read out at Los Alamos,
on a computer aided track counting system. The cutoff edges from
C^{+1}, C^{+2}, C^{+3}, and the protons all gave a consistent peak voltage.

X-ray bremsstrahlung induced 4 to 8 volt signals on the
125 μm PINS and made data reduction of the time resolved data
analysis more complex. It was performed by subtracting, from the
data, the background from the x-ray signal, obtained on a PIN detector not struck by protons. Future plans call for using the
35 μm PINS used on Proto I which should reduce the x-ray background to the 1 to 2 volt level.

4.   FUTURE OF PARTICLE ANALYZER INTENSE BEAM DIAGNOSTICS

We believe the future of particle analyzer intense beam
diagnostics is very bright indeed. We have already seen the
power of a simple deflection type magnetic spectrograph in making
rather precise direct measurements of time integrated and time
resolved proton energy spectra and the associated quantities of
mean energy, current density, and power density. Although not
emphasized above, particle analyzers based on thin target Rutherford scattering are free of range shorting effects and target
temperature induced changes in dE/dx that may cause some problems
for standard nuclear diagnostics at high power density.

One area which needs to be investigated is the use of the
Rutherford scattering techniques for high Z beams. Upon first
examining this problem, the use of a Rutherford scattering diagnostic looks complicated and intractable, since the original
charge of a high Z beam can be altered upon striking the target,
due to charge exchange. For example, an Li^{+1} beam could have its
original charge state altered in the target to a fraction of Li^{+2}
and/or Li^{+3}. Furthermore, in general, this process would be dependent on the high Z beam's voltage, which would make magnetic
analysis and data unfolding very difficult. However, for any
high Z beam of sufficiently high kinetic energy, almost 100% of
the Rutherford scattered high Z beam will come off the target
foil fully stripped. (See Ref. [12].) A lithium beam, for example, will become fully stripped at approximately 7 MeV. This
full stripping results in a considerable simplification since all
of the high Z beam, regardless of its initial charge state, will
be magnetically deflected in its fully stripped charge state.
This means that each charge state composing the accelerated beam
will be deflected in separate regions in the magnetic field because each charge state will have a kinetic energy equal to its

charge state times the diode potential. This offers a means of measuring the fraction of the original beam that is in each state. So, in summary, in the high voltage limit, the Rutherford scattering diagnostic can be applied to high Z beams.

Another application of the diagnostic could be for intense electron beam measurements. Here the complications revolve around the large self-magnetic fields associated with more intense electron beams and how the beam can be "extracted". In addition, multiple scattering in the target foil can be a large problem for electrons. Nevertheless, it is probable that the technique can be applied to at least high voltage electron beams.

REFERENCES

1. R. J. Leeper, J. R. Lee, D. J. Johnson, W. A. Stygar, D. E. Hebron, and L. D. Roose, "Direct Measurement of the Energy Spectrum of an Intense Proton Beam," Proc. 5th Int. Top. Conf. on High Power Particle Beams, San Francisco, California, p. 514 (1983).

2. J. E. Osher, "Particle Measurements," in Chapter 12 of Plasma Diagnostic Techniques (R. H. Huddlestone and S. L. Leonard, eds.), Academic Press, New York, (1965).

3. P. L. Dreike and L. P. Mix, "PBFA-I Diode Voltage Measurements Using a Fast Proton Energy Analyzer," Proc. 4th IEEE Pulsed Power Conference, Albuquerque, New Mexico, June 6-8, 1983.

4. E. Segre, Nuclei and Particles, Benjamin/Cummings Publishing Company, Inc., Reading, Massachusetts (1977), p. 26.

5. W. M. Gibson, The Physics of Nuclear Reactions, Pergamon Press, New York (1980).

6. D. J. Johnson, E. J. T. Burns, A. V. Farnsworth, Jr., R. J. Leeper, J. P. Quintenz, K. W. Bieg, P. L. Dreike, D. L. Fehl, J. R. Freeman, and F. C. Perry, "A Radial Ion Diode for Generating Intense Focused Proton Beams" J. Appl. Phys. $\underline{53}$, 4579 (1982).

7. Glenn F. Knoll, Radiation Detection and Measurement, John Wiley and Sons, New York (1979).

8. J. F. Janni, "Proton Range - Energy Tables," Atomic and Nuclear Data Tables $\underline{27}$ (1982).

9. H. Enge, Introduction to Nuclear Physics, Addison-Wesley, Reading, Massachusetts (1966).

10. Particle Data Group, "Review of Particle Properties", Rev. Mod. Phys. 52 No. 2, S44 (1980).

11. It must be understood that when discussing absolute energy spectra, current density, or power density, a surface for the beam to pass through must be defined. The original Monte Carlo code calculations used the disk surface shown in Fig. 5. However, because of the cylindrical geometry of the diode, it was desired to calculate the current density on a cylindrical surface onto which the full beam was incident. From Monte Carlo calculations and analytical calculations, the conversion between the two geometries is related by multiplying the disk geometry by a factor of 2 to obtain the cylindrical geometry. All results are reported in terms of cylindrical geometry.

DATA ACQUISITION

SOFTWARE CORRECTION OF MEASURED PULSE DATA

N. S. Nahman

Electromagnetic Fields Division
National Bureau of Standards
Boulder, CO
USA

1.   INTRODUCTION

1.1 Overview

Generally, the fundamental concern in the software correction of measured pulse waveform data is the solution of an ill-posed deconvolution problem which arises when one (or both) of the known waveforms is (are) corrupted by errors due to interference, noise, instrumentation drift, etc. The convolution equation for a measurement system is

$$y(t) = \int_{o}^{t} x(\lambda)h(t-\lambda)d\lambda, \tag{1}$$

where the casual waveforms $h(t)$, $x(t)$, and $y(t)$ are the measurement system impulse response, input, and output waveforms, respectively [1], Fig. 1. When one of the integrand functions is unknown, while the other two functions are known, the equation becomes the deconvolution integral equation for the unknown waveform. Solution of an ill-posed deconvolution problem is obtained by signal processing or filtering and at most yields an estimate for the unknown waveform. The filtering is necessary to yield a stable and physically consistent result.

Our objective in this paper is to bring out the ideas of ill-posedness and to give examples of applications to pulse measurement problems which require deconvolution, i.e., the removal (correction) of pulse source effects and/or measurement

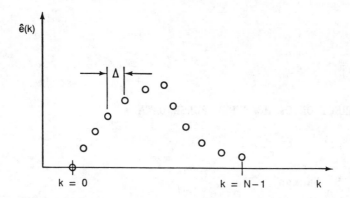

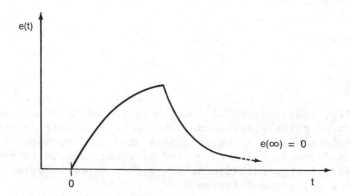

Fig. 1.   A discrete representation (upper figure) $\hat{e}(k)$ having a
          sequence of N values with spacing Δ for the continuous
          function $e(t)$ (lower figure).

system effects as encountered in signal pulse waveform measurements and system impulse response measurements.

    The following discussion is divided into eight sections, in
addition to the remainder of the present introduction. The sections are as follows:

1.   Introduction

2.   Deconvolution

3.   The Ill-Posed Problem

4.   Solution of the Ill-Posed Problem

5. Measurement of System Impulse Response

6. Impulse Response of the Regularization Filter

7. The Effective Bandwidth of a Measurement System

8. Measurement of a Pulse Waveform

9. Summary and Conclusions

A. Appendix

Sections 2 through 4 develop the concept of the ill-posed problem and what is meant by a solution to an ill-posed problem.

In Section 5 an optimal DFT (Discrete Fourier Transform) domain deconvolution method is used to obtain the insertion impulse response of a low-pass filter having a 3-dB (voltage) cut-off frequency of 3.5 GHz. The optimum solution of a set of optimal solutions is defined here as the one which minimizes the standard deviation of the fluctuation of the imaginary part of the impulse response.

Because the impulse response must be real, the imaginary part should be zero. It is postulated that the remaining flucuations are due to the computation round-off errors which contain a component dependent upon the smoothness of the solution (lack of irregular variation in the resultant solution's waveform). This idea is examined in Section 6.

In Section 7, optimum deconvolution is used to examine the capability of a time domain measurement system to repeat a given measurement. Because the measured or acquired data differs from trial to trial, the net result is a limitation on the resolution of the system which corresponds to limiting the system bandwidth.

Next, in Section 8, the optimum deconvolution method is applied to the measurement of the pulse waveform from a reference waveform generator. In this application, the distorting effect of the measurement system is removed from the measured data (waveform) of a picosecond domain step-like waveform.

Finally, a summary of the discussion and the conclusions drawn are given in Section 9.

Also, in the remainder of this introduction, two topics are briefly discussed. First, we discuss the difference between a continuous casual function and any numerical sequence which can only approximately represent the continuous function; this is done for general background information and to explain why only sequences are used in all of the computations herein. Second, we briefly discuss the convolution summation which is the sequence or numerical form of the convolution integral.

1.2   Continuous Functions/Numerical Sequences

In practice, a deconvolution problem involving the physical-continuous excitation and response of a physical system is solved using numerical-discrete or sampled data. The sequence of numerical values representing a given continuous waveform is an approximation to the waveform. For example, in Fig. 1, the continuous function e(t) is casual (zero, for t less than zero) and therefore not bandlimited, and decays to zero in the limit as t approaches infinity. On the other hand, the sequence is not continuous and is truncated to zero at all values of k greater than N-1.

When viewed in terms of a continuous function, a sequence is an approximation. Commonly, the Nyquist sampling-rate [2] is invoked to establish if a sequence adequately represents a continuous function. The criterion is that the waveform's spectrum be bandlimited and the sampling rate (frequency) be at least twice that of the highest frequency contained in the waveform's bandlimited spectrum. A causal signal or waveform can not be bandlimited, nor can a duration-limited waveform because in either case the Paley-Wiener criterion would be violated [3]. That is, if $\alpha(\omega)$ is the attenuation (-20 log of the magnitude) of the spectrum of a causal signal, then the Paley-Wiener criterion requires that

$$\int_{-\infty}^{\infty} \frac{\alpha(\omega)}{1 + \omega^2} \, d\omega < \infty \, . \tag{1}$$

For the integral to converge to a finite value, $\alpha(\omega)$ can increase no faster than ω^m, $0 \leq m \leq 1$, for increasing ω. Consequently, $\alpha(\omega)$ cannot be infinite over a band of values (two or more successive values of ω).

Thus, a causal sequence and a limited or finite number of sequence values can only approximate continuous causal and duration-limited waveforms, respectively. When the Discrete Fourier Transformation (DFT) is applied to sequences representing such continuous waveforms, aliasing is always present; this simply means that the DFT discrete spectrum is always different than the continuous Fourier Transformation (CFT) at corresponding discrete spectral frequencies. The error is easily calculated

when an exact mathematical equation is known for the continuous
function [4,5].

Aliasing error only exists in the context of a continuous
function and a numerical sequence approximating the continuous
functions. When we work only with sequences we are working with
discrete systems, i.e., systems whose operations are only described by a sequence of values (with nothing being defined between any two values of the sequence). In all of the discussion
which follows we will be talking about sequences; thus questions
of aliasing have no influence on the origin of ill-posedness in
our discrete systems or analyses.

1.3 Discrete (Sequence) Convolution

For discrete systems, the convolution operation corresponding to the convolution integral (1) is the convolution sum of two
causal sequences $\hat{h}(k)$ and $\hat{x}(k)$ which is given by

$$\hat{y}(k) = \sum_{i=0}^{k-1} \hat{h}(i)\hat{x}(k-i), \qquad k = 0,1,\ldots, M+P-1. \qquad (2)$$

The carat over y and x indicates they are sequences. $\hat{y}(o)$ and
$\hat{y}(1)$ are zero because for $k \leq 0$, $\hat{h}(k)$ and $\hat{x}(k)$ are zero. For
brevity we may also write

$$\hat{y}(k) = \hat{h}(k) * \hat{x}(k), \qquad k = 0,1,\ldots,M+P-1. \qquad (3)$$

The terms in the sum (2) are identical to the coefficients of the
(M+N)-order polynomial generated by the product of M- and N-order
polynomials, one having the set of "$\hat{h}(k)$" values as coefficients,
while the other set the "$\hat{x}(k)$" values. Notice that the arguments
in (2), i and k-i, can be interchanged without altering the result, $\hat{y}(k)$. When two sequences, each having first (k=0) and last
values of zero and a sequence duration of M and N, respectively,
their convolution sum yields N + M values with the first and last
values being zero, and the first value appearing at k=1, Fig. 2.

2. DECONVOLUTION

Three sequences which are exactly related to each other
through the convolution sum are defined here to be consistent sequences. In this part we will present examples of time domain
deconvolution using consistent sequences with the direct and
iterative methods, respectively. Also, we will briefly discuss
the options for implementing deconvolution in the time domain and
in transformation domains.

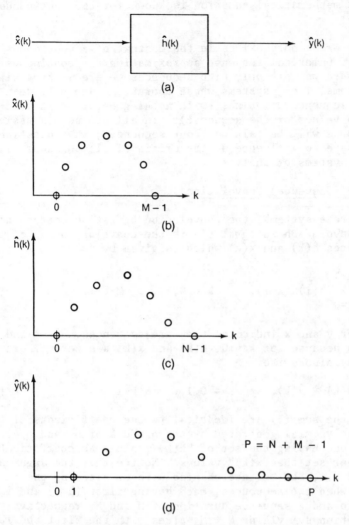

Fig. 2.   The response ŷ(k) of a linear discrete system of impulse
response ĥ(k), to an excitation sequence x̂(k), ŷ(k), is
the convolution sum x̂(k) * ĥ(k).

2.1   Consistent Sequences

Three sequences x̂(k), ĥ(k), and ŷ(k), which are exactly related through the convolution sum (2), are defined here to be
consistent. By forming the convolution sum using the sequences

$$\hat{x}(k) = e^{-0.05k}(1-e^{-0.05k}), \qquad k = 0, 1, \ldots, 1023, \qquad (4)$$

and

$$\hat{h}(k) = e^{-0.15k}(1-e^{-0.05k}), \quad k = 0, 1, \ldots, 1023, \tag{5}$$

we can obtain a $\hat{y}(k)$ consistent with the given sequences $\hat{x}(k)$ and $\hat{h}(k)$, i.e.,

$$\hat{y}(k) = \hat{x}(k) * \hat{h}(k). \tag{6}$$

Though not being germane here, we now take the opportunity to point out that the result $\hat{y}(k)$ would not be equal to the values of the continuous function $y(t)$, which could be analytically obtained with the convolution integral, and the continuous functions

$$x(t) = e^{-0.05t}(1-e^{-0.05t}) \, u(t), \tag{7}$$

and

$$h(t) = e^{-0.15t}(1-e^{-0.05t}) \, u(t), \tag{8}$$

where $u(t)$ is the unit step function. The reason that the two results would not be equal at discrete points is that $x(t)$ and $h(t)$ are not bandlimited functions and thus no finite number of samples could represent $y(t)$ (refer to Section 1.2).

2.2 Direct Time Domain Deconvolution

By expanding the convolution sum (2) a set of equations for $\hat{y}(k)$ is obtained; then solving for $\hat{h}(k)$ we obtain the causal sequence $\hat{h}(k)$ whose first non-zero term is $\hat{h}(1)$.

$$h(k) = \frac{1}{\hat{x}(1)} \left[\hat{y}(k) - \sum_{1=1}^{k-1} \hat{h}(i)\hat{x}(k+1-i) \right]; \quad k = 0,1,\ldots,N-1. \tag{9}$$

Similarly, by interchanging the known and unknown functions under the sum (2) we obtain

$$\hat{x}(k) = \frac{1}{\hat{h}(1)} \left[\hat{y}(k) - \sum_{i=1}^{k-1} \hat{x}(i)\hat{h}(k+1-i) \right]; \quad k = 0,1,\ldots, N-1 \tag{10}$$

In terms of our brief notation, we write

$$\hat{h}(k) = \hat{y}(k)(1/*)\hat{x}(k) \tag{11}$$

$$\hat{x}(k) = \hat{y}(k)(1/*)\hat{h}(k) \tag{12}$$

Notice that the symbol (1/*) corresponds to the deconvolution operation.

We now give an example of direct deconvolution using (9) to obtain the impulse response of our discrete system defined by (4) through (6). Taking only 50 points (0 ≤ k ≤ 49), uniformly spaced over the time window defined by $\bar{\tilde{y}}(k)$ (cf, Fig. 3), we obtain the numerical result for $\hat{h}(k)$, Fig. 4.

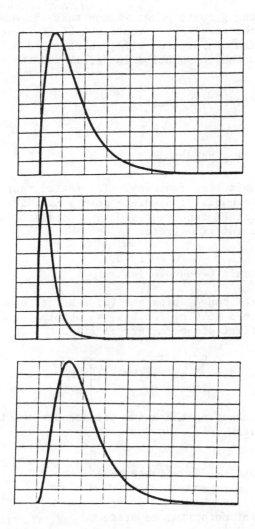

Fig. 3.   A consistent set of sequences, $\hat{y}(k)$, being formed by the convolution sum $\hat{x}(k) * \hat{h}(k)$. Maximum values: $x_m = 0.25$, $h_m = 0.105$, and $y_m = 0.35$.

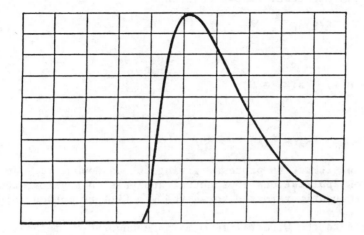

Fig. 4. The impulse response $\hat{h}(k)$ obtained by the direct discrete deconvolution method from the exact signals $\hat{x}(k)$
and $\hat{y}(k)$. No. of points 50; $\hat{h}(k)$ min. and max., 0.000
and 0.10518; vertical scale, 0.0105181/div.; horiz.
scale, 1 ns/div.

2.3   Iterative Time Domain Deconvolution

This method is based upon successive approximations which
converge to the unknown, $\hat{h}_1(k)$. As a first approximation let

$$\hat{h}_1(k) = \hat{y}(k), \tag{13}$$

then

$$y_1(k) = \hat{h}_1(k) * \hat{x}(k). \tag{14}$$

For a second approximation, we take

$$\hat{h}_2(k) = \hat{h}_1(k) + \hat{y}(k) - y_1(k), \tag{15}$$

which is

$$\hat{h}_2(k) = \hat{h}_1(k) + \hat{y}(k) - \hat{h}_1(k) * \hat{x}(k). \tag{16}$$

For the i-th approximation, we have

$$\hat{h}_i(k) = \hat{h}_{i-1}(k) + \hat{y}(k) - \hat{h}_{i-1}(k) * \hat{x}(k). \tag{17}$$

Inspection of this result shows that when

$$\hat{h}_{i-1}(k) = \hat{h}(k), \tag{18}$$

then

$$y(k) - h(k) * x(k) = 0,$$

and necessarily,

$$\hat{h}_i(k) = \hat{h}_{i-1}(k) = \hat{h}(k) \tag{19}$$

which ends the iterative process, giving the unknown $\hat{h}(k)$. In deconvolution applications to spectral analysis in physics, this method is frequently referred to as the Van Cittert method [6].

Compared to the Direct Time Domain Deconvolution method, this iterative method requires a very large amount of computation time due to the many numerical convolutions. When this method is applied to the same 50-term sequences as used in the direct deconvolution example in Section 2.2, it requires 1000 iterations (convolutions) to yield essentially the same result, Fig. 5. Graphically, the results (Figs. 4 and 5) are identical.

2.4 Transform Domain Deconvolution

The consistent sequences $\hat{x}(k)$ and $\hat{y}(k)$ define the linear discrete system $\hat{h}(k)$. Because the system is linear, $\hat{x}(k)$, $\hat{y}(k)$, and $\hat{h}(k)$ can be transformed by linear transformations into domains other than the time sequence domain. Two very useful transformations are the Discrete Fourier Transformation (DFT) and the Z-transformation.

The DFT is conveniently implemented by the Fast Fourier Transformation (sometimes called the Finite Fourier Transformation) and transforms a real sequence into a complex sequence [7]. On the other hand, the Z-Transformation transforms a real sequence into a complex function. In relation to the Laplace Transformation which transforms a real function into the complex-frequency plane ($s = \sigma + j\omega$), the Z-transformation maps the left-hand-plane of the s-plane into the interior of the unit-circle in the complex z-plane [8].

Both transformations possess the properties that in their transform-domains, convolution corresponds to multiplication and deconvolution to division. Thus, transformation to either domain provides simple means to perform these operations, but with the

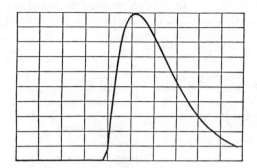

Fig. 5.   The impulse response ĥ(k) obtained by the discrete
          iterative deconvolution method from the exact signals
          x̂(k) and ŷ(k).   1000 iterations; No. of points,
          50; ĥ(k) min. and max., 0.000 and 0.10518; vert. scale,
          0.010518/div.; horiz. scale, 1ns/div..

penalties of having to perform the transformations and subsequent
inverse transformation.   Table 1 provides a summary of convolu-
tion/deconvolution in these different domains.

3.   THE ILL-POSED PROBLEM

When three sequences are not exactly related to each other
through the convolution sum, they are defined here to be inconsistent sequences.   The consequences of inconsistency will not be
demonstrated.

3.1   Exact or Well-Posed Problems

In Section 2.1 we constructed a consistent set of sequences
(4 through 6) using two given sequences, x̂(k) and ĥ(k), and the
convolution sum to specify the third sequence, ŷ(k).   Next, using
two different deconvolution methods we calculated ĥ(k) from x̂(k)
and ŷ(k) and obtained the same result in each case.   Having constructed a consistent set of sequences we knew the proper result
for ĥ(k); therefore, the deconvolution problem was solved using
exact data for the x̂(k) and ŷ(k) sequences.

Similarly, we could have constructed a consistent set of
equations by starting with given ĥ(k) and ŷ(k) sequences and then

Table 1. Convolution and elementary-deconvolution methods in different domains. Elementary denotes that the data is acted upon only by the convolution/deconvolution operations and that stablizing procedures are not used.

	DISCRETE TIME DOMAIN	TRANSFORM DOMAIN
	$\hat{f}(k) \equiv$ Sequence of N real numbers $k = 0, \ldots, N-1$ "^" carat denotes a sequence.	1. Z transform: z−domain, $\bar{F}(z)$ is a continuous function of the complex variable z. 2. DFT, Discrete Fourier Transform, $\hat{F}(n)$ is a sequence of N complex numbers. $n = 0, \ldots, N-1$.
CONVOLUTION	$\hat{y}(k) = \hat{x}(k) \cdot \hat{h}(k)$ $= \sum_{i=0}^{k-1} \hat{h}(i)\hat{x}(k-i)$ $k = 0, \ldots, N-1$. x(k), h(k), & y(k) are the linear system input, system impulse response, & output sequences, respectively.	1. $\bar{Y}(z) = \bar{X}(z)\bar{H}(z)$, a function. 2. $\hat{Y}(n) = \hat{X}(n)\hat{H}(n)$, a sequence. NOTE: the symbols $\bar{X}, \bar{H}, \bar{Y}$ are not the same as $\hat{X}, \hat{H}, \hat{Y}$, thus they denote different mathematical quantities.
DECONVOLUTION Shown for $\hat{h}(k)$ solution. For $\hat{x}(k)$ solution interchange x with h and X with H.	$\hat{h}(k) = \hat{y}(k)(1/^*)\hat{x}(k)$ A. DIRECT $h(k) = \dfrac{1}{\hat{x}(1)}\left[\hat{y}(k) - \sum_{i=1}^{k-1} \hat{h}(i)\hat{x}(k+1-i)\right]$ $\hat{x}(1) \neq 0, k=0, \ldots, N-1.$ B. ITERATIVE − j denotes the j−th iteration. $\hat{h}_{j+1}(k) = \hat{y}(k) - [h_j(k)^*\hat{x}(k)] + h_j(k)$ when $\hat{h}_j(k)^*\hat{x}(k) = \hat{y}(k),$ $\hat{h}_j(k) = \hat{h}_{j+1}(k) = \hat{h}(k).$	1. $\bar{H}(z) = \bar{Y}(z)/\bar{X}(z)$ 2. $\hat{H}(n) = \hat{Y}(n)/\hat{X}(n)$
DECONVOLUTION IMPLEMENTATION	A & B: All operations performed in the time domain. B requires a very large amount of time to compute the final result.	1. Two Z transformations, complex division in z−domain, and one Z^{-1} transformation. 2. Two DFT's, complex division in n-sequence domain, and one DFT^{-1}

determining $\hat{x}(k)$ by the deconvolution sum. Again, using exact
data we would obtain the proper $\hat{x}(k)$ specified by the exact numerical sequences $\hat{h}(k)$ and $\hat{y}(k)$.

3.2   Inexact or Ill-Posed Problems

From our knowledge of and experience with causal physical
systems, if given two sequences, $\hat{x}(k)$ and $\hat{y}(k)$, whose values were
obtained by measurements on a linear causal physical system, we
expect that the impulse response $\hat{h}(k)$ would commence at zero, de-
part from zero, and vary in some manner, and then return to zero.
Our expectations are based upon fundamental physical constraints
which require $\hat{h}(k)$ to have the general properties of beginning
at zero and finally ending at zero. However, in practice, using
measured or inexact data for $\hat{x}(k)$ and $\hat{y}(k)$, we may find that the
computed sequence for $\hat{h}(k)$ does not meet our expectations nor is
it apparently constrained as in causal linear physical system
theory.

The reason for such irregular results lies in the fact that
small changes to one or both of exact sequences for $\hat{x}(k)$ and $\hat{y}(k)$
can produce large changes in $\hat{h}(k)$, thus giving a result which no
longer agrees with physical theory and our expectations. The
problem posed to compute $\hat{h}(k)$ from inexact data for $\hat{x}(k)$ and $\hat{y}(k)$
is said to be an ill-posed problem, in contrast to a well-posed
problem, which employs exact data.

Because in practical applications the data is measured, or
inexact, data, the two deconvolution problems

$$\hat{h}(k) = \hat{y}(k)(1/*)\hat{x}(k) \tag{20}$$

and

$$\hat{x}(k) = \hat{y}(k)(1/*)\hat{h}(k) \tag{21}$$

are in general ill-posed problems.

To illustrate the ill-posed problem, we now destroy the consistency between the three sequences (4 through 6) by adding
small random sequences to both $\hat{x}(k)$ and $\hat{y}(k)$ which gives

$$\hat{x}_1(k) = \hat{x}(k) + \hat{n}_x(k) \tag{22}$$

and

$$\hat{y}_1(k) = \hat{y}(k) + \hat{n}_y(k). \tag{23}$$

Numerically, both $\hat{n}_x(k)$ and $\hat{n}_y(k)$ have a mean value of zero, i.e.,

$$\bar{\hat{n}}_i = \frac{1}{50} \sum_{k=0}^{49} \hat{n}_i(k) = 0. \tag{24}$$

Their respective standard deviations are such that

$$20 \log \left[\frac{[\hat{x}]max}{\sigma_x} \right] = 40 \text{ dB}, \tag{25}$$

and

$$20 \log \left[\frac{[\hat{y}]max}{\sigma_y} \right] = 40 \text{ dB}, \tag{26}$$

where

$$\sigma_i = \left[\frac{1}{50} \sum_{k=0}^{49} [\hat{n}_i(k)]^2 \right]^{1/2}. \tag{27}$$

The σ_i is the rms value of the noise (voltage) sequence; thus the rms value of the departure from the exact value is one percent of the maximum value in each case (25) and (26).

Upon using $\hat{x}_1(k)$ and $\hat{y}_1(k)$ in the direct deconvolution sum (9), we obtain an irregular result $\hat{h}_1(k)$ which does not resemble what we would expect for $\hat{h}(k)$ (compare Fig. 6 with Fig. 4). Repeating the deconvolution using the iterative method obtains a very similar result, Fig. 7.

Recognizing that the precise numerical values of $\hat{x}_1(k)$ and $\hat{y}_1(k)$ are different from those of $\hat{x}(k)$ and $\hat{y}(k)$, respectively, it is clear that what has happened is that we have computed a numerical sequence $\hat{h}_1(k)$ which is consistent with the inexact data $\hat{x}_1(k)$ and $\hat{y}_1(k)$. We have done so by using a deconvolution algorithm which forces $\hat{h}_1(k)$ to be consistent with $\hat{x}_1(k)$ and $\hat{y}_1(k)$, but not with $\hat{x}(k)$ and $\hat{y}(k)$.

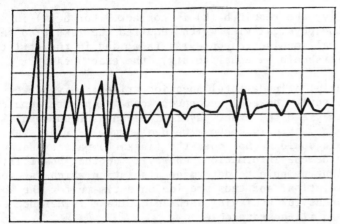

Fig. 6.   The impulse response $\hat{h}_1(k)$ obtained by the direct discrete deconvolution method from the corrupted input and
output signals $\hat{x}_1(k)$ and $\hat{y}_1(k)$; 40 dB signal (max.
voltage)-to-noise (RMS volts) ratio for both signals.
No. of points 50; $\hat{h}_1(k)$ min. and max., -1.5909 and
1.8156; vert. scale, 0.340651/div.; horiz. scale, 1 ns/
div..

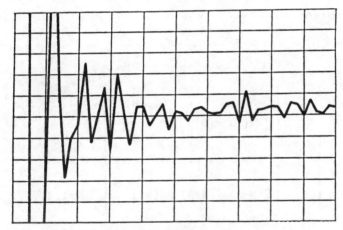

Fig. 7.   The impulse response $\hat{h}_1(k)$ obtained by the discrete
iterative deconvolution method from the corrupted input
and output signals $\hat{x}_1(k)$ and $\hat{y}_1(k)$.   40 dB signal (max.
voltage)-to-noise (RMS volts) ratio for both signals.
750 iterations; No. of points 50; $\hat{h}_1(k)$ min. and max.,
-1.6000 and 1.8000; vert. scale, 0.3400/div.; horiz.
scale, 1 ns.div..

Numerically, the result $\hat{h}_1(k)$ is correct. But $\hat{h}_1(k)$ is incorrect in the physical view which required the result to resemble a physical impulse response waveform; and in the strictest requirement, it should be equal to $\hat{h}(k)$, the exact result.

Consequently, a fundamental question arises: Given two measured (inherently inexact) sequences, can we operate upon or process them to yield the exact physical deconvolution result? The answer is "no", simply because we don't know a-priori what the exact result would be nor does the (inexact) pair of data sequences provide the information. However, another possibility exists - we might be able to determine a solution which is close to the exact solution, and thus provide an estimate for the solution. The process for achieving such an estimate is defined as a solution to an ill-posed problem.

Before going on to the solution of ill-posed problems, as a matter of perspective we wish to point out the similarity between the irregular result $\hat{h}_1(k)$, and ill-conditioned calculations in the solution of sets of linear equations. The convolution sum (2) can be written as a set of linear equations; in terms of matrices we have for

$$\hat{y}(k) = \hat{x}(k) * \hat{h}(k) \tag{28}$$

the matrix equation

$$[\hat{y}] = [\hat{x}] \cdot [\hat{h}], \tag{29}$$

given by

$$
\begin{bmatrix}
\hat{y}(1) \\
\hat{y}(1) \\
\cdot \\
\cdot \\
\cdot \\
\hat{y}(N-1)
\end{bmatrix}
=
\begin{bmatrix}
\hat{x}(0) & 0 & \ldots & 0 \\
\hat{x}(1) & \hat{x}(0) & \ldots & 0 \\
\cdot & \cdot & & \cdot \\
\cdot & \cdot & \ldots & \cdot \\
\cdot & \cdot & \ldots & \cdot \\
\hat{x}(N-2) & \hat{x}(N-3) & & \hat{x}(0)
\end{bmatrix}
\cdot
\begin{bmatrix}
\hat{h}(1) \\
\hat{h}(2) \\
\cdot \\
\cdot \\
\cdot \\
\hat{h}(N-1)
\end{bmatrix}
$$

where $[\hat{x}]$ is an (N-1) by (N-1) square matrix, and $[\hat{y}]$ and $[\hat{h}]$ are (N-1) column matrices. Alternately, we could have expressed $[\hat{y}]$ in terms of an $\hat{x}$ column matrix and an $\hat{h}$ square (N-1) matrix, if we wish to solve a signal deconvolution problem for $\hat{x}$ (meaning $\hat{y}$ and $\hat{h}$ are the known quantities).

The solution to (29) yields the deconvolution result,

$$[\hat{h}] = [\hat{x}]^{-1}[\hat{y}] \tag{31}$$

In the solution of linear equations, small changes in the values of the elements in $[\hat{x}]$ can cause very large changes in the solution for $[\hat{h}]$. When this occurs, the matrix $[x]$ is said to be ill-conditioned (or the set of linear equations are said to be ill-conditioned).

4.   THE SOLUTION OF ILL-POSED PROBLEMS

The presence of errors in the data-pair creates an ill-posed problem. For our example of an ill-posed problem we added random variations to the data pair to establish errors in the data-pair. In this section we attempt to remove the errors by simple smoothing, and demonstrate that such filtering is a small step in the right direction, but that we can never get back to the original error-free deconvolution result, $\hat{h}(k)$. We then go on to describe an optimal filtering process which in some defined sense yields an optimum estimate for $\hat{h}(k)$.

4.1 Simple Linear Filtering

A linear filter is a filter whose characteristics do not depend upon the signal which is being filtered. When we speak of the linear system

$$\hat{y}(k) = \hat{x}(k) * \hat{h}(k), \tag{32}$$

we are actually saying that the linear filter $\hat{h}(k)$ filters the signal $\hat{x}(k)$ to yield the filter output signal $\hat{y}(k)$. Note that $\hat{h}(k)$ is independent of $\hat{x}(k)$.

We now filter or smooth each one of the corrupted sequences (22) and (23) by the convolution of each sequence with the simple four-point rectangular sequence,

$$\hat{h}_2(k) = \left\{ \begin{array}{ll} 1/N, & k = 0, 1, 2, 3 \\ 0, & 4 \leq k \leq N-1 \end{array} \right\} \tag{33}$$

Note that this filter has four sequential values equal to the constant 1/N, and all other values are zero. This class of filter is commonly called a running average filter and in the present case has a duration (length) of four. This smoothing filter and some more sophisticated filters are discussed in [9].

Applying the filter (33) to the corrupted sequences we obtain

$$\hat{x}_2(k) = \hat{x}_1(k) * \hat{h}_2(k) \tag{34}$$

and

$$\hat{y}_2(k) = \hat{y}_1(k) * \hat{h}_2(k) \; ; \tag{35}$$

thus we filter, or smooth, $\hat{x}_1(k)$ and $\hat{y}_1(k)$ with the running average filter $\hat{h}_2(k)$.

4.2 Deconvolution Using the Smoothed Sequences

We now apply the direct and iterative deconvolution methods, respectively, using the smoothed sequences $\hat{x}_2(k)$ and $\hat{y}_2(k)$. With each method we perform the deconvolution and obtain the two results,

$$\hat{d}_D(k) = \hat{y}_2(k)(1/*)_D \hat{x}_2(k) \tag{36}$$

and

$$\hat{d}_I(k) = \hat{y}_2(k)(1/*)_I \hat{x}_2(k), \tag{37}$$

where the subscripts D and I denote the direct and iterative deconvolution methods, respectively (Figs. 8 and 9). Notice that after 500 iterations, $\hat{d}_I(k)$ is very close (identical graphically) to $\hat{d}_D(k)$. These two results are estimates of the true impulse response $\hat{h}(k)$, i.e., the impulse response consistent with the uncorrupted sequences $\hat{x}(k)$ and $\hat{y}(k)$. We see that the filtering has stabilized or regularized the solution in that (a) the solutions no longer oscillate widely with increasing oscillation amplitude and (b) a causal impulse response sequence superimposed upon a noise-sequence has emerged from the data, $\hat{d}_D(k)$ and $\hat{d}_I(k)$.

Summarizing, we have seen that two different deconvolution methods achieve the same result and that the filtering or smoothing has partially reversed the corruption of the original seauences. In our earlier experiments of Sections 2.2 and 2.3 with consistent sequences, we saw that the results were the same using two different deconvolution methods (direct and iterative, respectively). Similarly, we saw that the results for the inconsistent (corrupted) sequences were essentially the same, Section 3.2. This independence of the result with respect to the deconvolution method emphasizes that the irregular results are due to inconsistent sequences, or in other words, an ill-posed problem rather than the deconvolution method.

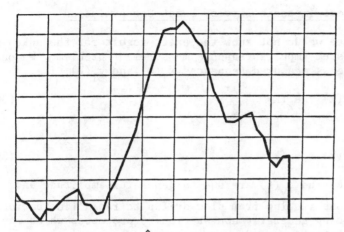

Fig. 8.   The estimate $\hat{d}_D(k)$ of the impulse response $\hat{h}(k)$ obtained by the direct deconvolution method using the smoothed sequences $\hat{x}_2(k)$ and $\hat{y}_2(k)$. Smoothing filter sequence, 0.25(1,1,1,1); $\hat{d}_D(k)$ min. and max., -0.001 and 0.10501; vert. scale, 0.011502/div; horiz. scale, 1.0 ns/div..

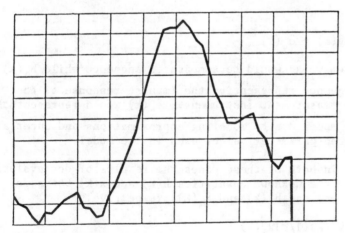

Fig. 9.   The estimate $\hat{d}_I(k)$ of the impulse response $\hat{h}(k)$ obtained by the iterative deconvolution method using the smoothed sequences $\hat{x}_2(k)$ and $\hat{y}_2(k)$. Smoothing filter sequence, 0.25(1,1,1,1); 500 iterations; $\hat{d}_I(k)$ min. and max., -0.01000 and 0.10501; vert. scale, 0.011502/div.; horiz. scale, 1.0 ns/div..

4.3 The Inherent Impossibility of Exact Deconvolution

In practice we do not know the exact result for the unknown $\hat{h}(k)$ in the system impulse response measurement problem. We only know the erroneous measured sequences $\hat{x}_1(k)$ and $\hat{y}_1(k)$, i.e.,

$$\hat{x}_1(k) = \hat{x}(k) + \hat{e}_x(k) \tag{38'}$$

and

$$\hat{y}_1(k) = \hat{y}(k) + \hat{e}_y(k). \tag{39}$$

The errors $\hat{e}_x(k)$ and $\hat{e}_y(k)$ are due to generator amplitude and timing variations ranging from slow drifts to rapid fluctuations, all of which being generally non-stationary processes which may contain some stationary random components.

The same situation holds true for the signal measurement problem in which we do not know the exact result for the unknown signal $\hat{x}(k)$, but only know the erroneous sequences $\hat{h}_1(k)$ and $\hat{y}_1(k)$,

$$\hat{h}_1(k) = \hat{h}(k) + \hat{e}_h(k) \tag{40}$$

and

$$\hat{y}_1(k) = \hat{y}(k) + \hat{e}_y(k). \tag{41}$$

In this problem $\hat{y}_1(k)$ would be a measured sequence while $\hat{h}_1(k)$ would be a sequence representing the impulse response of the model for the measurement instrument. $\hat{e}_h(k)$ would represent the errors in the model due to modeling approximations and uncertainties in measured parameter values used in the model.

Clearly, in both of these cases the results of deconvolution will not yield the unknown quantities $\hat{h}(k)$ and $\hat{x}(k)$ but some other quantities, say $\hat{h}_3(k)$ and $\hat{x}_3(k)$, respectively:

$$\hat{h}_3(k) = \hat{y}_1(k)(1/*)\hat{x}_1(k) \tag{42}$$

$$\hat{x}_3(k) = \hat{y}_1(k)(1/*)\hat{h}_1(k). \tag{43}$$

Taking the DFT (cf. Table 1) of (38), (39), and (42), and writing the transformed dependent variables in the corresponding upper case letters gives:

$$\hat{H}_3(n) = \frac{\hat{Y}(n) + \hat{E}_y(n)}{\hat{X}(n) + \hat{E}_x(n)} \, , \tag{44}$$

While for (40), (41), and (43) we obtain

$$\hat{X}_3(n) = \frac{\hat{Y}(n) + \hat{E}_y(n)}{\hat{H}(n) + \hat{E}_h(n)} \, . \tag{45}$$

We can write

$$\hat{H}_3(n) = \hat{H}(n) \, \frac{1 + \hat{E}_y(n)/\hat{Y}(n)}{1 + \hat{E}_x(n)/\hat{X}(n)} \tag{46}$$

and

$$\hat{X}_3(n) = \hat{X}(n) \, \frac{1 + \hat{E}_y(n)/\hat{Y}(n)}{1 + \hat{E}_x(n)/\hat{H}(n)} \, , \tag{47}$$

because

$$\hat{H}(n) = \hat{Y}(n)/\hat{X}(n) \tag{48}$$

and

$$\hat{X}(n) = \hat{Y}(n)/\hat{H}(n) \, . \tag{49}$$

Thus, we see that both $\hat{H}_3(n)$ and $\hat{X}_3(n)$ are proportional to the exact values of the unknowns $\hat{H}(n)$ and $\hat{X}(n)$, respectively, i.e.,

$$\hat{H}_3(n) = \hat{H}(n) \, \hat{M}(n) \tag{50}$$

and

$$\hat{X}_3(n) = \hat{X}(n) \, \hat{P}(n) \, . \tag{51}$$

Typically, we then have in either case the situation depicted in Fig. 10, where we see that the DFT domain spectra of the exact result $\hat{H}(n)$ overlaps that of the inherent error term $\hat{M}(n)$. Because of this overlap Δ, it is impossible to separate $\hat{H}(n)$ and $\hat{M}(n)$, since in general we do not know $\hat{H}(n)$ and $\hat{M}(n)$; we only know their product, $\hat{H}_3(n)$.

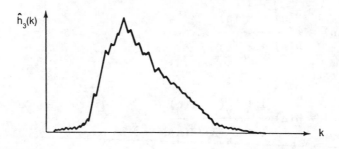

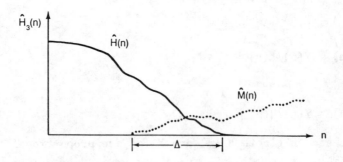

Fig. 10. The deconvolution result for the impulse response both
          in the time domain (upper fig.) and in the frequency
          domain (lower fig.). Δ is the spectral overlap between
          the exact result and error components.

Summarizing, the result $\hat{H}_3(n)$ is the result computed from the acquired waveforms $\hat{x}_1(k)$ and $\hat{y}_1(k)$, (38) and (39). When $\hat{x}_1(k)$ and $\hat{y}_1(k)$ are repetitive signals, the acquisition process usually includes ensemble averging of 100 waveforms or so to remove stationary random components which increases the signal-to-noise ratio. Nonetheless, the errors wich remain are usually large enough to create an ill-posed problem. In the non-repetitive case ensemble averging cannot be implemented, but analog filtering may be used in the signal acquisition channel. In both cases, filtering, averging, etc. cannot remove all the errors; thus the ill-posed problem cannot be solved to yield the desired exact result. Furthermore, filtering acts on the error components as well as the desired signal both in the course of signal acquisition and in the regularizing or stabilizing of a deconvolution operation. Consequently, exact deconvolution is inherently impossible.

4.4 Redefinition of the Concept of a Solution: Optimal Deconvolution.

Since we cannot obtain an exact solution, we redefine what we shall mean by a solution to an ill-posed problem. Rigorous mathematical developments of methods for solving ill-posed problems have been created by the mathematician A. V. Tikhonov. References to his work and the work of others are given in the English translation of Tikhonov, and Arsenin's book which was originally in Russian [10]. Other references of interest are found in [11, 12].

Most importantly, Tikhonov makes the point that in general the solution to an ill-posed problem requires a subjective judgement as to when a solution is satisfactory. An ill-posed problem is defined as a problem in which apparently well-behaved data in one domain, upon transformation to a second domain, yields a widely varying result which does not represent the true or physical situation. The solution of an integral equation of the first kind of the convolution type is such a transformation problem; deconvolution is the inverse transformation of the convolution integral transformation. Tikhonov's point regarding the requirement for a subjective judgement is based on the fact that the most an ill-posed problem solution can yield is a result that in some measure is close to the correct result. Furthermore, he points out that a problem may appear to be representative of a physical one but is not in fact. For example, take the deconvolution problem for a signal waveform measurement in which the known quantities are an instrument function (impulse response), an observed signal, and an observed signal-to-noise ratio.

Tikhonov points out that the signal to noise ratio is generally not observable; what is observable is only signal plus noise (see page 153 in [10]).

Returning to how we shall define when our deconvolution problem is solved, our choice will be arbitrary as there are an infinite number of definitions available to us.

To develop a reasonable basis for our choice initially we consider the performance measures

$$E_h = \sum_{k=0}^{N-1} [\hat{y}_1(k) - \hat{d}_h(k) * \hat{x}_1(k)]^2 \tag{52}$$

and

$$E_x = \sum_{k=0}^{N-1} [\hat{y}_1(k) - \hat{d}_x(k) * \hat{h}_1(k)]^2 , \tag{53}$$

where $\hat{y}_1(k)$, $\hat{x}_1(k)$, and $\hat{h}_1(k)$ are erroneous data sets. $\hat{d}_h(d)$ and $\hat{d}_x(k)$ are our solutions to these ill-posed problems; they are estimates (approximations) to the desired exact results, $\hat{h}(k)$ and $\hat{x}(k)$.

Generally, we view the $\hat{d}(k)$'s as filtered versions of the results we obtain directly from the erroneous data sets, i.e., we filter (50) and (51). Writing (50) and (51) to explicitly show the corresponding time domain expressions we have

$$\hat{h}_3(k) = [DFT]^{-1} \{\hat{H}_3(n)\} = [DFT]^{-1} \{\hat{H}(n)\hat{M}(n)\} \tag{54}$$

and

$$\hat{x}_3(k) = [DFT]^{-1} \{\hat{X}_3(n)\} = [DFT]^{-1} \{\hat{X}(n) \hat{D}(n)\}. \tag{55}$$

Upon filtering we have, respectively,

$$\hat{d}_h(k) = \hat{h}_3(k) * \hat{r}_h(k) = [DFT]^{-1} \{\hat{H}(n)\hat{M}(n)\hat{R}_h(n)\} \tag{56}$$

and

$$\hat{d}_x(k) = \hat{x}_3(k) * \hat{r}_x(k) = [DFT]^{-1} \{\hat{X}(n)\hat{P}(n)\hat{R}_x(n)\} , \tag{57}$$

where $\hat{r}_n(k)$, $\hat{R}_h(n)$ and $\hat{r}_x(k)$, $\hat{R}_x(n)$ are the corresponding time and DFT domain filters, respectively, for each case. Thus, we see that the filtering process attempts to suppress the effects of the factors $\hat{M}(n)$ and $\hat{P}(n)$ which cause the irregular or noisy results. Fig. 6 is a typical example of a strongly irregular result, while Fig. 8 is typical of a mildly irregular (noisy) result.

Returning to the performance expressions (52) and (53), we see that E_h and E_x are zero when we do not filter, i.e., when $\hat{R}_h(n)$ and $\hat{R}_x(n)$ are equal to unity for all n. Consequently, these performance measures express the fact that the irregular results $\hat{h}_3(k)$ and $\hat{x}_3(k)$ are the exact results which fit the erroneous data sets $\hat{x}_1(k)$, $\hat{y}_1(k)$, and $\hat{h}_1(k)$, as the case may be. However, such results are not what we wish to achieve; what we want are smooth estimates to the exact results $\hat{h}(k)$ and $\hat{x}(k)$. Consequently, we must use a different form of performance measure.

Following the reasoning of Twomey [11] we choose performance criteria which include a constraint for smoothness; we then have

$$P_h = E_h + \gamma_h S_h \tag{58}$$

and

$$P_x = E_x + \gamma_x S_x, \tag{59}$$

where the γ_i are real positive numbers which weight the degree of smoothness. We may use the subscript "i" to represent either h or x in the subsequent text. S_h and S_x are performance components due to the smoothing process and are given in the time and DFT domains by:

$$S_h = \sum_{k=0}^{N-1} [C(k) * \hat{d}_h(k)]^2 \tag{60}$$

$$= \frac{1}{N} \sum_{n=0}^{N-1} [\hat{C}(n) \hat{D}_h(n)]^2 \tag{61}$$

and

$$S_x = \sum_{k=0}^{N-1} [\hat{c}(k) * \hat{d}_x(k)]^2 \tag{62}$$

$$= \frac{1}{N} \sum_{n=0}^{N-1} [\ \hat{C}(n)\ \hat{D}_x(n)\ ]^2 \tag{63}$$

The relation between the two domains is given by Parseval's Theorem for a sequence and its DFT [13]. The sequence $\hat{c}(k)$ is the second backward-difference operator sequence

$$\hat{c}(k) \equiv \nabla^2 = 1,\ -2,\ 1,\ 0,\ 0,\ \ldots,\ 0,\ N-1, \tag{64}$$

in which the first three terms are non-zero while the rest of the (N-3) terms are zero. The square of the magnitude of the DFT of $\hat{c}(k)$ is

$$\left|\hat{C}(n)\right|^2 = 16\ \text{Sin}^4(\pi n/N),\ 0,\ 1,\ 2\ldots N-1. \tag{65}$$

The last two equations are derived in the Appendix of the present paper; also included there are the fundamental equations for the DFT, its inverse, and Parseval's Theorem. In terms of a continuous function analogy, we have included in our performance criterion a term proportional to the second derivative of our deconvolution estimate; the inclusion weights through the arbitrary constant γ_i.

If we make γ_i large so that the constraint term $(\gamma_i S_i)$ dominates the performance measure P_i, then the result would be very smooth at the expense of any rapidly changing pulse features in the deconvolution result. On the other hand, if se set γ_i equal to zero and do not filter, i.e., $\hat{R}_i(n) = 1$, then P_i will be zero, but the deconvolution result will be the ill-posed result. Thus we see that we can not accept P_i being zero because we must filter, i.e., $0 \le \hat{R}_i(n) < 1$, to obtain a useful result.

Functionally, the performance measures appear to depend upon the $\hat{R}_i$,

$$P_h = E_h(\hat{D}_h) + \gamma_h\ S_h(\hat{C}\hat{D}_h)$$

$$= E_h(\hat{R}_h \hat{H}_3) + \gamma_h S(\hat{C}\hat{R}_h \hat{H}_3)$$

$$= E_h(\hat{R}_h) + \gamma_h S_h(\hat{R}_h)$$

$$P_x = E_x(\hat{D}_x) + \gamma_x S_x(\hat{C}\hat{D}_x)$$

$$= E_x(\hat{R}_x \hat{X}_3) + \gamma_x S_x(C\hat{R}_x \hat{X}_3)$$

$$= E_x(\hat{R}_x) + \gamma_x S_x(\hat{R}_x)$$

However, we shall see that the performance measures are actually functions of the γ_i, i.e.,

$$P_h = E_h(\gamma_h) + \gamma_h \, S_h(\gamma_h) \tag{66}$$

$$P_x = E_x(\gamma_x) + \gamma_x \, S_x(\gamma_x) \tag{67}$$

Because we must filter, we choose our solution to be an optimal one, that is, we select the filter that minimizes the performance measures under consideration [14]. We minimize a function of the form

$$P_i = E_i + \gamma_i S_i$$

in the DFT domain by differentiating with respect to R. In the DFT domain by Parseval's Theorem we have for (58)

$$P_h = E_h + \gamma_h S_h = \frac{1}{N} \sum_{n=0}^{N-1} [\hat{E}(n) \, \hat{E}^{\,*}(n)$$

$$+ \gamma_h \, \hat{S}(n)\hat{S}^{\,*}(n)], \tag{68}$$

where

$$\hat{E}(n) = [\hat{Y}_1(n) - \hat{X}_1(n) \, \hat{D}_h(n)] \tag{69}$$

and

$$\hat{S}(n) = \hat{C}(n) \, \hat{D}_h(n). \tag{70}$$

In (68) the asterisk superscript denotes the complex conjugate, and should not be confused with the in-line asterisk or reciprocal-asterisk (1/*) which denote convolution or deconvolution, respectively.

Putting

$$\hat{D}_h(n) = \hat{R}_h(n) \, \hat{H}_3(n) \tag{71}$$

into (69) and (70), and the result into (68) we obtain

$$P_h = \frac{1}{N} \sum_{n=0}^{N-1} \left| \hat{Y}_1(n) \right|^2 \{ 1 + \left| \hat{R}_h(n) \right|^2$$

$$+ \frac{\gamma_h \left| \hat{C}(n) \right|^2 \left| \hat{R}_h(n) \right|^2}{\left| \hat{X}_1(n) \right|^2} - 2 \left| R_h(n) \right| \}. \tag{72}$$

To obtain the $\left| \hat{R}_h(n) \right|$, which will minimize or maximize (72), take the partial derivative of (72) with respect to $\left| R_h(n) \right|$, set it equal to zero, and then solve for $\left| \hat{R}_h(n) \right|$. We obtain

$$\hat{R}_h(n, \gamma_h) = \frac{\left| \hat{X}_1(n) \right|^2}{\left| \hat{H}_1(n) \right|^2 + \gamma_h \left| \hat{C}(n) \right|^2}. \tag{73}$$

Because the second partial derivative of P_h with respect to $\hat{R}_h(n)$ is equal to or greater than zero, $\hat{R}(n, \gamma_h)$ minimizes (72). Similarly, we can obtain

$$\hat{R}_x(n, \gamma_x) = \frac{\hat{H}_1(n)^2}{\hat{H}_1(n)^2 + \gamma_x \hat{C}(n)^2}. \tag{74}$$

Notice that the γ_i are arbitrary, i.e., given an γ_i, the filter $R_i(n, \gamma_i)$ minimizes P_i. Also, the filters are real, and the γ_i set the roll-off characteristics of the filters.

The solutions obtained using the regularization filter $\hat{R}_h(n)$ or $\hat{R}_x(n)$ are called optimal solutions. The selection of a specific value of γ_i in general requires a subjective judgement. In summary, we have defined that a deconvolution problem is optimally solved when a deconvolution result is regularized by the filter $\hat{R}_h(n)$ or $\hat{R}_x(n)$, as the case may be. Furthermore, our solution is conveniently implemented in the DFT domain using the filters $\hat{R}_h(n)$ and $\hat{R}_x(n)$, (73) and (74). Thus, we have for the optimal solutions, in the two cases of deconvolution, the following formulas:

$$\hat{d}_h(k) = [DFT]^{-1} \left\{ \frac{\hat{Y}_1(n)}{\hat{X}_1(n)} \hat{R}_h(n) \right\} \tag{75}$$

for the impulse response measurement problem, and

$$\hat{d}_x(k) = [DFT]^{-1} \left\{ \frac{\hat{Y}_1(n)}{\hat{H}_1(n)} \hat{R}_x(n) \right\} \tag{76}$$

for the signal-waveform measurement problem.

One add-tional comment. $\hat{C}(n)$ in our solution is the DFT of the second-difference operator. Tikhonov [15] developed a more general continuous function filter in which $C(\omega)$ included the Fourier transform of any even-order derivative operator. This corresponds to the DFT of any even-order difference operator; consequently, $\hat{C}(n)$ could be any even-order difference operator, not just the second.

4.5 The Optimum Solution

When a specific value of γ_{io} of the arbitrary constant γ_i is selected, we define the resulting solution as the optimum solution $\hat{d}_{io}(k)$. In general, the selection of γ_{io} requires a subjective decision as to what constitutes an acceptable solution.

As indicated in (75) and (76) the regularization filtering is conveniently implemented in the DFT domain, and the result is then transformed into the time domain by the inverse DFT. In studies applied to data from laboratory measurement experiments, M. E. Guillaume and N.S. Nahman noticed that the imaginary parts of the inverse DFT's were very small, but not zero, and their statistical parameters varied with the filtering. In particular, the standard deviation passed through a minimum with increasing γ_i, beginning from γ_i equal to zero. Thus, after the inverse DFT is performed, we actually have for the time domain impulse response the complex function

$$\hat{d}_{hc}(k,\gamma) = \hat{d}_h(k,\gamma_h) + j \hat{d}_{hI}(k,\gamma_h), \tag{77}$$

where $\hat{d}_{hI}(k,\gamma_h)$ is very small compared to $\hat{d}_h(k,\gamma_h)$, except when $\hat{d}_h(k,\gamma_h)$ is down in the noise level. The standard deviation of $\hat{d}_{hI}(k,\gamma_h)$ is given by

$$\sigma_{hI}(\gamma_h) = \sqrt{\frac{1}{N} \sum_{k=0}^{N-1} [d_{hI}(k,\gamma_h) - \overline{d_{hI}(k,\gamma_h)}]^2} \quad , \tag{78}$$

where $\overline{d_{hI}(k,\gamma_h)}$ is the mean value of $\hat{d}_{hI}(k,\gamma_h)$. Similarly, in the signal deconvolution case we have

$$\hat{d}_{xc}(k,\gamma) = \hat{d}_x(k,\gamma_x) + j\, d_{xI}(k,\gamma_x) \tag{79}$$

and

$$\sigma_{xI}(\gamma_x) = \sqrt{\frac{1}{N} \sum_{k=0}^{N-1} [d_{xI}(k,\gamma_x) - \overline{d_{xI}(k,\gamma_x)}]^2} \tag{80}$$

We define the optimum solution as that solution obtained for the value of γ_i which minimizes the standard deviation of the imaginary part of the inverse DFT transform of the deconvolution result. Consequently, we write for our optimum solutions

$$\hat{d}_h(k,\gamma_{ho}) = [DFT]^{-1} \left\{ \frac{\hat{Y}_1(n)}{\hat{X}_1(n)} \hat{R}_h(n,\gamma_{ho}) \right\} \tag{81}$$

and

$$\hat{d}_x(k,\gamma_{xo}) = [DFT]^{-1} \left\{ \frac{\hat{Y}_1(n)}{\hat{H}_1(n)} \hat{R}_x(n,\gamma_{xo}) \right\} \tag{82}$$

This particular definition for the optimal solution was first reported by Guillaume and Bizeul [16] and later discussed in greater detail by Nahman and Guillaume [17].

The choice of this method for defining the optimum solution was based upon a series of experiments in which different scientists and engineers performed the deconvolution for an impulse response, each using the same set of data and their own subjective decision as to the best solution. When their selections for γ_{ho} were compared to the standard-deviation-vs-γ curve, it was found that their choices were clustered about the minimum of the curve. Upon interviewing the participants separately, it was determined that their individual subjective judgements were based upon two physical facts which controlled the degree of filtering they applied:

a. They sought to achieve the highest peak value for $\hat{d}_h(k)$ without any visible noise on the peak.

b. Simultaneously, they observed the 50% duration of $\hat{d}_h(k)$ and sought to achieve the smallest duration while eliminating the noise on the peak of the impulse response.

Their rationale for using such criteria was that the regularization filtering was then primarily operating on the irregularities and not upon the true impulse response $\hat{h}(k)$.

Before discussing the reasons for a minimum to appear in the standard deviation of the imaginary part of the time domain deconvolution result, we shall go on to applications of this deconvolution method and then resume our discussion in the light of experimental data.

5.   MEASUREMENT OF SYSTEM IMPULSE RESPONSE

5.1 Measurement Method

The quantity being determined in this experiment is the insertion impulse response of a network referenced to a 50-ohm characteristic impedance coaxial transmission system. The insertion impulse response is the inverse Fourier transform of the scattering parameter, $S_{21}(j\omega)$. The measurement system is shown in Fig. 11. Two waveforms are acquired: (a) the reference waveform $\hat{e}_1(k)$ being obtained when the pulse source is directly connected to the measurement system, and (b) the response waveform $\hat{e}_2(k)$ being obtained when the network is inserted between the pulse source and the measurement system. The optimum estimated insertion impulse response $\hat{d}_h(k,\gamma_o)$ is obtained by the regularized deconvolution.

$$\hat{E}_1(n) = [DFT]\ \hat{e}_1(k) \tag{83}$$

$$\hat{E}_2(n) = [DFT]\ \hat{e}_2(k) \tag{84}$$

$$\hat{D}_h(n,\gamma_h) = \frac{\hat{E}_2(n)}{\hat{E}_1(n)}\ \hat{R}_h(n,\gamma_h) \tag{85}$$

$$\hat{d}_h(k,\gamma_h) = [DFT]^{-1}\ \hat{D}_h(n,\gamma_h) \tag{86}$$

$$\hat{d}_h(k,\gamma_{ho}) = [DFT]^{-1}\ \hat{D}_h(n,\gamma_{ho}) \tag{87}$$

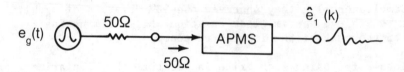

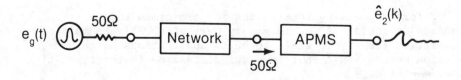

Fig. 11.   The insertion measurement method.

5.2   The Measurement System

The NBS Automatic Pulse Measurement System (APMS) [18] was used to acquire the insertion waveforms. This system is a computer-controlled sampling oscilloscope system having a step-response-to-(10-90%)-transition duration of about 20 ps. For the measurements discussed here and in Sections 6 and 7, the total duration of each acquired or computed sequence was 512. Each acquired waveform sequence was obtained by averaging an ensemble of 100 waveforms.

5.3   The Experiment

A low-pass filter was used for the network; the filter had a cut-off frequency of 3.5 GHz. The acquired reference and response $\hat{e}_1(k)$ and $\hat{e}_2(k)$ are shown in Fig. 12. The DFT's of $\hat{e}_1(k)$ and $\hat{e}_2(k)$ were taken to give $E_1(n)$ and $E_2(n)$, respectively; then $D_h(n, \gamma_h)$ (85), and $d_h(k, \gamma_h)$ (86), were computed. Finally, the optimum solution $d(k, \gamma_{ho})$ (87) was determined by adjusting γ_h in (85), and then calculating (86) and (78) until the minimum in $\sigma_{hI}(\gamma_h)$ was found.

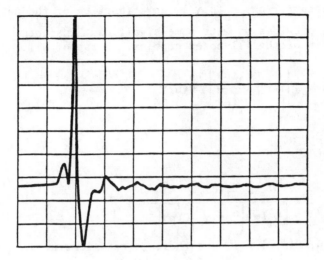

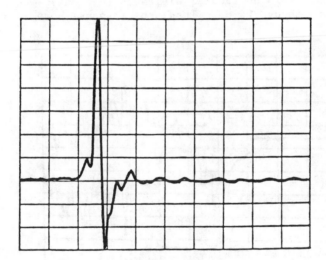

Fig. 12.   Insertion Waveforms, signal-(max. voltage)-to-noise
(RMS volts) ratios, about 50 dB; no. of points, 512.
(Upper fig.) Reference waveform $\hat{e}_1(k)$:   min. and max.,
-0.685×10^4 mV and 0.189×10^5 mV; vert. scale,
0.258×10^4 mV/div.; horiz. scale, 0.999 ns/div. (Lower
fig.) Response waveform $\hat{e}_2(k)$:   min. and max., $-0.724 \times$
10^4 and 0.164×10^5 mV; vert. scale, 0.237×10^4 mV/
div.; horiz. scale, 0.999 ns/div..

The behavior of $\left| \hat{D}_h(n, \gamma_h) \right|$ as a function of the parametric variable γ provides us with some significant insights into the basic origin of the ill-posed problem. First of all

$$\left| \hat{D}_h(n, \gamma_h) \right| = \left| \frac{\hat{E}_2(n)}{\hat{E}_1(n)} \right| \hat{R}(n, \gamma_h) \tag{88}$$

where

$$R(n, \gamma_h) = \frac{\left| E_1(n) \right|^2}{\left| E_1(n) \right|^2 + \gamma_h \left| C(n) \right|^2} . \tag{89}$$

When $\gamma_h = 0$, $R(n,0) = 1$, and there is no filtering. In Fig. 13, $\left| \hat{D}(n,0) \right|$ is the uppermost curve. Starting from the left side of the figure follow the uppermost curve. The response is smooth out to about 3-2/3 horizontal divisions (4.6 GHz), where it has decreased to about -25 dB. Then, the response becomes irregular

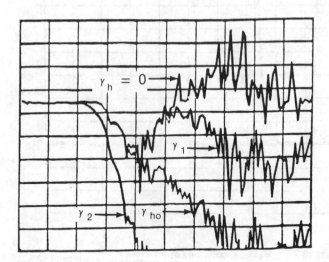

Fig. 13. The magnitude of the impulse response estimate $\hat{D}(n, \gamma_h)$ in the DFT domain as a function of the parametric variable, $\gamma_h = \gamma_1 = 10^7$, $\gamma_h = \gamma_{ho} = 0.9 \times 10^9$, and $\gamma_h = \gamma_2 = 10^{11}$; vert. scale, 10 dB/div.; horiz. scale, 1.2514 GHz/Div..

and climbs rapidly in an oscillatory but irregular manner with the peaks reaching approximately +30 dB. What is happening here is that the errors in $\hat{E}_2(n)$ and $\hat{E}_1(n)$ dominate $\hat{D}(n,0)$ above 4.6 GHz. DFT inversion of $D(n,0)$ most assuredly would provide a very unrecognizable result compared to the impulse response of a causal physical system. But keep in mind that we don't know the details of the exact value $\left|H(n)\right|$, in this case $\left|\hat{S}_{21}(n)\right|$. It is true that we know the network is a low-pass filter that nominally cuts-off at 3.5 GHz, but we will not use that information in the deconvolution, since in general we would not know it.

Upon increasing γ_h to $\gamma_1 = 10^7$, we see that the filter is beginning to suppress the irregular response. The response corresponding to $\gamma_{ho} = 0.9 \times 10^9$ is the optimum one as defined earlier (just before (77)); note that the -3 dB response occurs very close to 2.8 horizontal divisions (3.5 GHz). Upon increasing γ_h to $\gamma_2 = 10^{11}$ we see that the filter now strongly alters the smooth portion of the response, which in turn means that the filtering is altering components belonging to the exact result $\hat{H}(n)$, in addition to those of the errors.

Turning to the corresponding time-domain deconvolution results, Fig. 14 shows the real (upper) and imaginary (lower) parts of $\hat{d}_{hc}(k,\gamma_h)$, (77), for $\gamma_h = 0$, which are the results when filtering is not employed; in other words, the direct result given by $\hat{e}_2(k)(1/*)\hat{e}_1(k)$. Clearly, this is a useless result. The optimum results, $\hat{d}_h(k,\gamma_{no})$ and $\hat{d}_{hI}(k,\gamma_{ho})$, $\gamma_{ho} = 0.9 \times 10^9$, are shown in Fig. 15. In Fig. 16, $\hat{d}_h(k,\gamma_n)$ is compared to the optimum case $\hat{d}_h(k,\gamma_{ho})$ for the under-filtered case $\gamma_h = 10^7$ and the over-filtered case $\gamma_h = 10^{11}$. The γ_h dependence of the standard deviation (78) for the imaginary part of $d_{hc}(k,\gamma)$ is shown in Fig. 17, where the minimum value γ_{ho} equals 0.9×10^9. Finally, in Figs. 14 and 15, the scale of the $\hat{d}_{hI}(k,\gamma_h)$ are three orders-of-magnitude less than for the real part $\hat{d}_h(k,\gamma_h)$ (as the imaginary parts should be much smaller than those of their respective real parts.)

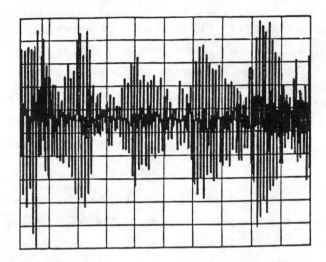

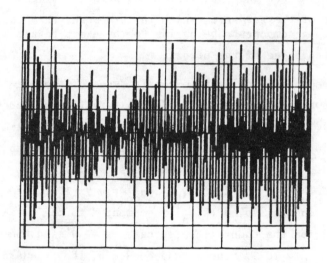

Fig. 14.   The deconvolution result without filtering, $\gamma_h = 0$.
No. of points 512; horiz. scale, 0.999 ns/div. (Upper
fig.) The real part $\hat{d}_h(k,\gamma_h)$: min. and max., -0.559
and 0.449; vert. scale, 0.100/div. (Lower fig.). The
imaginary part $\hat{d}_{hI}(k)$: min. and max., -0.302 x 10^{-3}
and 3.23 x 10^{-3}; vert. scale, 0.626 x 10^{-4}/div.; mean
value, 0.144 x 10^{-11}; std. dev., 0.111 x 10^{-3}.

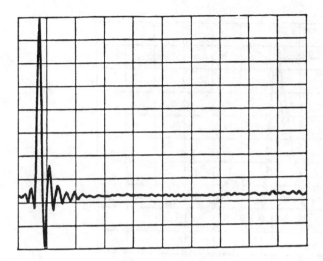

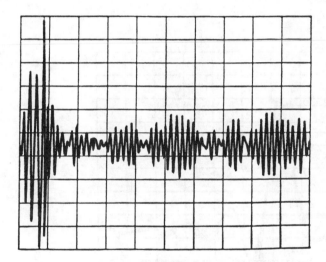

Fig. 15. The deconvolution result with optimum filtering, γ_{ho} = 0.9 x 10^9. No. of points 512; horiz. scales, 0.999 ns/ div. (Upper fig.) The real part $\hat{d}_{ho}(k,\gamma_{ho})$: min. and max., -0.0404 and 0.138; vert. scale, 0.0179/div. (Lower fig.) The imaginary part $\hat{d}_{hI}(k,\gamma_{ho})$: min. and max., -0.642 x 10^{-4} and 0.791 x 10^{-4}; vert. scale, 0.143 x 10^{-4}/div.; mean value, 0.209 x 10^{-12}; std. dev., 0.137 x 10^{-4}.

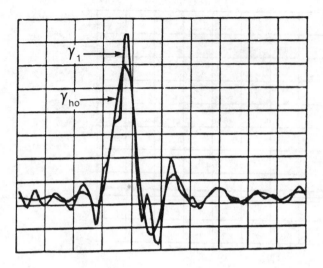

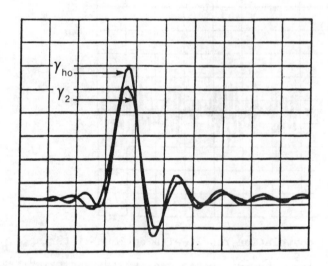

Fig. 16.   Comparison of the optimum deconvolution result $\hat{d}_h(k,$
           $\gamma_{ho})$, $\gamma_1 = 10^7$, and that (Lower fig.) of the overfil-
           tered case, $\gamma_2 = 10^{11}$. Vert. scales, 0.0242/div.; min.
           and max., -0.0552 and 0.187; horiz. scales, 0.2 ns/
           div.; No. of points each waveform, 512.

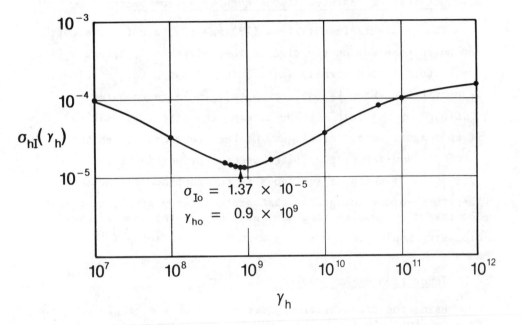

Fig. 17. The dependence of the standard deviation, $\sigma_{hI}(\gamma_h)$,
of the imaginary part in the deconvolution results,
$\hat{h}_c(k)$, on the degree of filtering, γ_h. The minimum
standard deviation, σ_{Io}, is 1.39 x 10^{-5} which occurs
for γ_{ho} = 0.9 x 10^9.

6.   IMPULSE RESPONSE OF THE REGULARIZATION FILTER

6.1 Overview

In this section we will show how to obtain the impulse response of a regularization filter and will apply the method to
the optimum regularization filter obtained in the measurement
problem of Section 5. Along the way as we develop this method we
will gain an insight into the nature of the processes which produces the imaginary part of the deconvolution result and the
minimization of it.

6.2 Qualitative Features of the Regularization Filter

The regularization filters (73) and (74) are filters which depend in some way on the signals they filter. $\hat{X}_1(n)$ and $\hat{H}_1(n)$ are factors in the signals which $\hat{R}_h(n,\gamma_h)$ and $\hat{R}_x(n,\gamma_x)$ filter, respectively. When $\left|\hat{X}_1(n)\right|$ or $\left|\hat{H}_1(n)\right|$ is large compared to $\gamma_h\left|\hat{C}(n)\right|^2$ or $\gamma_x\left|\hat{C}(n)\right|^2$, respectively, the filter is weak and it approaches unity for all n. On the other hand, when the $\gamma_i\left.\hat{C}(n)\right.^2$ dominates, the filter becomes proportional to $1/\{\gamma_i\left|\hat{C}(n)\right|^2\}$ which is analogous to the continuous function $1/\omega^4$, the fourth-order integrator that produces very strong smoothing. The insertion-impulse response case (85) is the same as the ordinary impulse response case with $\hat{E}_i(n)$ replacing $\hat{X}_1(n)$.

6.3 Generation of An Impulse $\hat{\delta}(k)$

Using the deconvolution program, we enter a single waveform data set for both the $\hat{x}_1(k)$ and $\hat{y}_1(k)$ waveforms. Here we will use the $\hat{e}_1(k)$ sequence of Section 5. The DFT of $\hat{e}_1(k)$ is $\hat{E}_1(n)$, and $\left|\hat{E}_1(n)\right|$ is shown in Fig. 18. We then have:

$$\hat{H}(n) = \hat{E}_1(n)/\hat{E}_1(n) = 1, \tag{90}$$

which corresponds to zero dB, Fig. 19. The inverse DFT yields the complex response, Fig. 20,

$$\hat{h}_c(k) = \hat{\delta}(k) + j\,\hat{h}_I(k). \tag{91}$$

$\delta(k)$ equals unity for k=0 and zero elsewhere, i.e., 0.999...to the eighth place, and 0.000... to the eighth place (single precision, for the minicomputer used here) as indicated in Fig. 20.

6.4 The Minimum in σ_I

$\hat{h}_c(k)$ (91) is the result of an inverse DFT of 512 numbers which are just about exactly equal to each other, out to about nine-places (0.999...ninth place). The maximum value of the imaginary part is eight orders smaller than that of the real part, while the minimum value of both the real and imaginary are

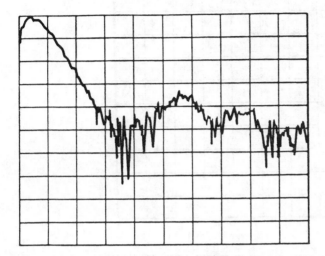

Fig. 18. The magnitude $\hat{E}_1(n)$ of the DFT of the reference waveform $\hat{e}_1(k)$, Fig. 12 (upper). Vert. scale, 10 dB/div.;
horiz. scale, 2.5027 GHz/div.; No. of points, 512.

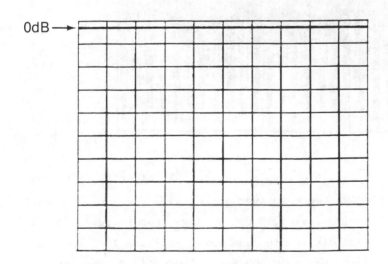

Fig. 19. The quotient $\hat{E}_1(n)/\hat{E}_1(n)$ equals unity. Vert. scale,
10 dB/div.; horiz. scale, 2.5027 GHz/div.; No. of
points, 512.

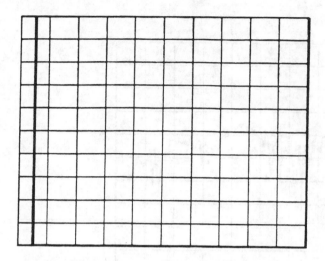

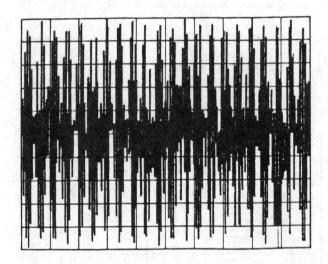

Fig. 20. $\hat{h}_c(k)$, the complex inverse DFT of unity (Fig. 19 only
          shows the magnitude of $\hat{H}_c(n)$). No. of points, 512;
          horiz. scales, 0.999 ns/div. (upper fig.) The real
          part $\hat{\delta}(k)$: min. and max., -0.215 x 10^{-8} and 1.00;
          vert. scale, 0.100/div. (Lower fig.) The imaginary
          $\hat{h}_I(k)$: min. and max., -0.197 x 10^{-8} and 0.196 x 10^{-8};
          vert. scale, 0.393 x 10^{-9}/div.; mean, 0.141 x 10^{-14};
          std. dev., 0.949 x 10^{-9}.

also eight orders smaller. The statistical parameters for the imaginary part are also very small, with a mean of 0.14×10^{-14} and a standard deviation 0.9×10^{-7}.

The imaginary part is present because of computation errors in the computer's computations; this is a situation which always exists [19]. The accuracy of FFT algorithms are commonly tested by performing the FFT on a given sequence, subsequently inversing FFT, and then noting the error which is evidenced by a non-zero imaginary part. But what we observe here is that the smallest error fluctuation occurs for an N-point $[FFT]^{-1}$ operating on N identical numbers, which gives, $\sigma_I = 10^{-9}$. On the other hand, for N different numbers, σ_I is approximately larger by a factor of N^2, which in the present case is $(512)^2$; thus $\sigma_I = 10^{-9} \times N^2 = 2.62 \times 10^{-4}$. This is the order of the magnitude of σ_I seen in Fig. 17 at both extremes, $\gamma = 0$ (no filtering) and $\gamma = 10^{12}$ (extreme filtering).

As filtering is increased, the standard deviation, σ_I, passes through a broad shallow minimum which indicates that the filter is acting on a noise process contained in the signal $\hat{H}_3(n) = \hat{E}_2(n)/\hat{E}_1(n)$ (cf. Eqs. 54, 56, and 85). Because the filter itself is an operation which can generate a set of N-different numbers upon which the inverse FFT operates, the minimum is shallow. Its effect on σ_I is slight for small γ, approaching that of the delta function $\hat{\delta}(k)$.

6.5 Excitation of $\hat{r}_i(k,\gamma_i)$ by $\hat{\delta}(k)$

The impulse response of the regularization filter is given by the real part of the complex result

$$\hat{r}_{ci}(k,\gamma_i) = [DFT]^{-1}\{R_i(n,\gamma_i)\}, \qquad (92)$$

$$= \hat{r}_i(k,\gamma_i) + j\,\hat{r}_{iI}(k,\gamma_i) \qquad (93)$$

To use the deconvolution computer program to obtain the regularization filter impulse response for a given γ_i (93), we use the manual version which allows the operator to select γ_i. We proceed in the following way. Since

$$\hat{D}_h(n,\gamma_h) = \frac{\hat{Y}_1(n)}{\hat{X}_1(n)} R_h(n,\gamma_h) \ , \text{ and} \tag{94}$$

$$\hat{D}_x(n,\gamma_x) = \frac{\hat{Y}_1(n)}{\hat{H}_1(n)} R_x(n,\gamma_x) \ , \tag{95}$$

let

$$\hat{Y}_1(n) = \begin{cases} \hat{X}_1(n) \text{ when } \hat{R}_h(n) \text{ is desired} \\ \hat{H}_1(n) \text{ when } \hat{R}_x(n) \text{ is desired} \end{cases} \tag{96}$$

We then have for either case

$$\hat{D}_i(n,\gamma_i) = \hat{R}_i(n,\gamma_i), \tag{97}$$

and the deconvolution result gives the regularization filter impulse response (93),

$$\hat{r}_{ci}(k,\gamma_i) = \hat{r}_i(k,\gamma_i) + j \ \hat{r}_{iI}(k,\gamma_i). \tag{98}$$

Figures 21 through 23 show typical results for (93) using the regularization filter from Section 5 for two values of γ_h, 10^3 and 0.9×10^9, the latter being the optimum γ_{ho}. We now consider the details of these two in pulse responses, $\hat{r}_h(k,\gamma_h)$, and their respective transfer function magntiudes, $\left| \hat{R}_h(n,\gamma_h) \right|$.

In Fig. (21) the two transfer function magnitudes are shown; note that when $\gamma_h = 10^3$, the magnitude contains many values close to unity (0 dB), while for the optimum value, $\gamma_{ho} = 0.9 \times 10^9$, $\left| \hat{R}_h(n,\gamma_{ho}) \right|$ contains many values different than unity and is down about 3 dB at 2 divisions (5 GHz). Correspondingly, we see that σ_I of the imaginary part of the impulse response $\hat{r}_{hI}(k,\gamma_h)$ increases as γ is increased, as does their mean values. Furthermore, in the real parts $\hat{r}_h(k,\gamma_h)$ we see that for $\gamma_h = 10^3$, the response is still rather close to a delta function with a noisy base line, while for $\gamma_{ho} = 0.9 \times 10^9$ the response has departed from that of a delta function and has an even-function symmetry about $k = 0$. The ringing even-response is characteristic of a sharp cut-off real DFT domain filter.

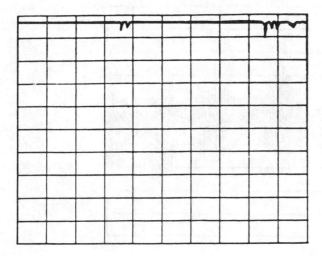

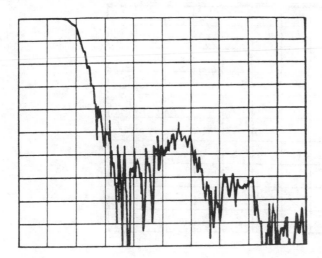

Fig. 21. Typical transfer function magnitudes for the regulari-
          zation filter $\hat{R}_h(n,\gamma_h)$. No. of points, 512; vert.
          scales, 10 dB/div.; horiz. scales, 2.5027 GHz/div.
          (Upper fig.) $\gamma_h = 10^3$; (Lower fig.) $\gamma_{ho} = 0.9 \times 10^9$.

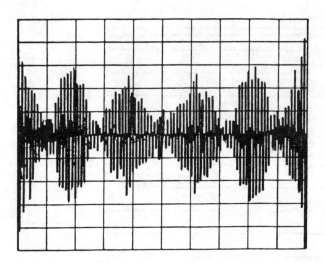

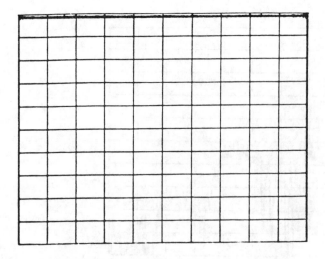

Fig. 22.   Complex inverse DFT of the regularization filter for
$\gamma_h = 10^3$.   No. of points, 512; horiz. scales, 0.999
ns/div.   (Upper fig.) real part $\hat{r}_h(k,\gamma_h)$:   min. and
max., -0.9977×10^{-2} and 0.967; vert. scale, 0.0986.
(Lower fig.) Imaginary part $\hat{r}_h(k,\gamma_h)$:   min. and max.,
-0.688×10^{-4} and 0.688×10^{-4}; vert. scale, $0.137 \times$
10^{-4}/div.; mean, -0.284×10^{-11}; std. dev., $0.181 \times$
10^{-4}.

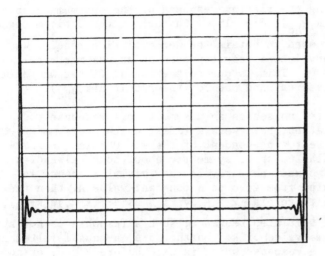

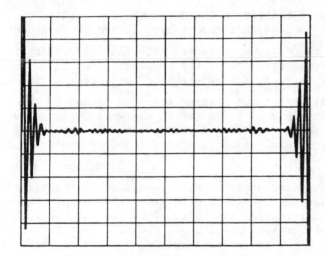

Fig. 23. Complex inverse DFT of the regularization filter for the optimum value γ_{ho} = 0.9 x 10^9. No. of points, 512; horiz. scales, 0.999 ns/div. (Upper fig.) Real part $\hat{r}_h(k,\gamma_{ho})$: min. and max., -0.0350 and 0.195; vert. scale, 0.0230/div. (Lower fig.) Imaginary part $\hat{r}_{hI}(k,\gamma_{ho})$: min. and max., -0.536 x 10^{-3} and 0.536 x 10^{-3}; vert. scale, 0.107 x 10^{-3}/div.; mean, -0.108 x 10^{-9}; std. dev., 0.856 x 10^{-4}/

Upon comparing the ringing response of the optimum deconvol-
ution result, $\hat{d}_h(k, \gamma_{ho})$, Fig. 15, with that of the optimum regularization filter used to obtain the deconvolution result, we
see that the deconvolution result possesses an unsymmetrical
ringing about k = 0. This is due to the causal filter whose impulse is being measured, and has for the result $\hat{d}(k, \gamma_{ho})$.

A causal filter possesses an unsymmetrical response to a
symmetrical excitation; here we are referring to the symmetry of
the excitation itself, independent of the k = 0 axis. For example, a rectangular pulse is symmetric about some value of k.
When applied to a causal network, if the initial response is
f(k)u(k), say rising from zero to a constant value A, then the
subsequent return to zero at $k = k_1$ varies as $A-f(k-k_1)u(k-k_1)$,
where u(k) is the unit step sequence; such a response is not symmetrical about some value of k. Furthermore, a causal filter has
an impulse or pulse response which is zero before k = 0, or some
other k when the response is delayed from k = 0.

Returning our attention to the deconvolution result,
$\hat{d}_h(k\gamma_{ho})$, Fig. 15, notice that the ringing period is different
before and after the maximum response. The initial ringing is
due to the non-causal response of the regularization filter impulse response, $\hat{r}_h(k, \gamma_{ho})$, Fig. 23a, while the larger period of
ringing immediately after the maximum response is due to the
actual cut-off frequency (3.5 GHz) of the network being measured.
In this case the regularization is very strong and some of the
features due to the regularization filter response appear in the
deconvolution result.

With respect to the need for strong regularization filtering, the experiments of Section 5 were made using a pulse generator whose spectrum (Fig. 18) was about 20-dB down at the cut-off
frequency of the insertion network, and thus about 10 dB above
the noise peaks. Also, the measurements were purposely made in
the presence of electromagnetic interference which intermittently
caused severe time jitter and amplitude variation during data
acquisition; this, in turn, decreased the effective bandwidth of
the measurement system. The results reported here are not typical of the results usually obtainable by the APMS [18] in which
the regularization filtering is relatively weak due to the use of
larger generator bandwidths and very small interference effects.
Thus, the regularization has a small unnoticeable effect on the
resultant deconvolution waveform; Section 8 presents a typical
example. Summarizing, strong regularization possesses a regularization filter impulse response whose duration is not negligible

compared to the impulse response or signal response being measured; thus some regularization filtering features appear in the deconvolution result, $\hat{d}_i(k, \gamma_i)$, (81) or (82).

7. THE EFFECTIVE BANDWIDTH OF A MEASUREMENT SYSTEM

7.1 Overview

The ability of an impulse response measurement system to obtain the impulse response of a given network is dependent upon measurement bandwidth. Here we determine the bandwidth of the measurement system used to obtain the low-pass network insertion impulse response of Section 5. To do so, we acquire two independent sets of data for $x_1(k)$. If the two sets of data were identical, then when they are considered to be the input and output data sets for the deconvolution algorithm, the deconvolution result would be unity in the DFT domain, while in the time domain it would be a delta function. Thus the measurement system would possess an infinite bandwidth. In practice, due to measurement errors, the DFT response will not be constant, nor will a delta function appear in the time domain. The measurement system includes the pulse generator, the time domain measurement instrument, and all error sources; the latter includes electronic equipment noise and drift, and electromagnetic environment effects (electromagnetic interference).

7.2 The Measurement Method

In general, a statistical approach must be used to characterize the measurement system bandwidth. To do so, we would acquire a relatively large number of independent data sets. Next, we would apply some statistical averaging strategy in the time and/or DFT domains to pairs of data in such a way that we would end up with just one set of data in the DFT domain representing a statistically derived ratio of data-pairs. Finally, we would apply the inverse DFT and obtain an impulsive response whose departure from a delta function is a measure of the measurement system bandwidth.

7.3 The Experiment

Here we acquired two data-sets by two independent data acquisitions, i.e., we acquired $\hat{x}_1(k)$ and $\hat{x}_2(k)$, Fig. 24, each data set being an ensemble average of 100 waveforms. The DFT of each

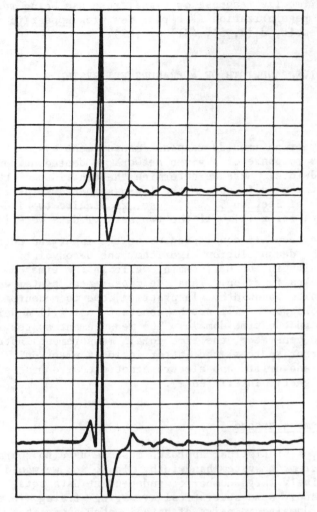

Fig. 24. Two independent acquisitions of the signal $\hat{x}(k)$. No.
of points, 512; horiz. scales, 0.999 ns/div. The spectra magnitudes of the two waveforms are approximately
given by $\hat{E}_1(n)$, Fig. 18. (Upper fig.) $\hat{x}_1(k)$: min.
and max., -0.607 x 10^4 mV and 0.210 x 10^5 mV; vert.
scale, 0.271 x 10^4 mV/div.; signal (max voltage)-to-
noise (RMS volts) ratio, 57.5 dB. (Lower fig.) $\hat{x}_2$:
min. and max., -0.661 x 10^4 mV and 0.214 x 10^5 mV;
vert. scale 0.280 x 10^4 mV/div., signal (max. voltage)
-to-noise (RM volts) ratio, 52.1 dB.

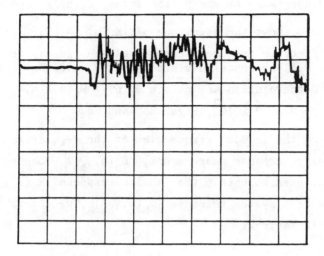

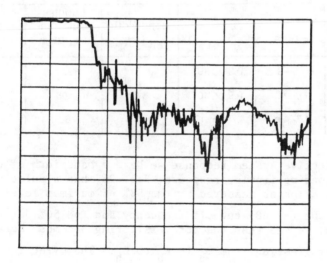

Fig. 25. The discrete transfer function magnitudes. Number of points, 512; vert. scales, 10 dB/div.; horiz. scales, 2.5027 GHz/div. (Upper fig.) $\hat{H}_b(n)$. (Lower fig.) $\hat{D}_b(n, \gamma_{bo})$, $\gamma_{bo} = 4.79 \times 10^7$.

waveform and the ratio of their transforms, $\hat{X}_1(n)/\hat{X}_2$, were computed; Fig. 25a shows the magnitude of $\left|\hat{H}_b(n)\right| = \left|\hat{X}_2(n)/\hat{X}_1(n)\right|$. Notice that discrete frequency components are closely reproduced out to about 2.4 horizontal divisions (-3 dB, 6.0 GHz). Applying regularized deconvolution we, obtain $\hat{D}_b(n) = \hat{H}_b(n)R(n,\gamma_{bo})$, and $d_b(n,\gamma_{bo})$, $\gamma_{bo} = 4.79 \times 10^7$, Figs. 25b and 26.

Comparing the (- 3dB) frequencies of the regularized results for the impulse response measurement, $\left|\hat{D}_h(n,\gamma_{ho})\right|$, the regularization filter used to obtain the impulse response $\left|\hat{R}_h(n,\gamma_{ho})\right|$, and the effective system response $\left|\hat{D}_b(n,\gamma_{bo})\right|$, we obtain the results shown in Table 2.

Table 2. Effective 3-dB bandwidths and 50% level impulse durations for the impulse response measurement.

Response	Figure	$\gamma_{o,i}$	Bandwidth f_i, GHz	Impulse Duration t_i, ps.	$t_i f_i$
$\left\|\hat{D}_h(n,\gamma_{ho})\right\|$	1.5–3	0.9×10^9	3.5	174	0.609
$\left\|\hat{R}_h(n,\gamma_{ho})\right\|$	1.6–4b	0.9×10^9	5.0	108	0.54
$\left\|\hat{D}_b(n,\gamma_{bo})\right\|$	1.7–2b	$4.79 = 10^7$	6.0	87	0.52

For the effective system response we have a bandwidth of 6 GHz. The corresponding impulse response, Fig. 26, has a pulse duration t_b of about 87 ps as measured at the 50% of maximum level. The product of the (-3 dB) cut-off frequency and the 50% duration is 0.52. Also, we see that the total filtering becomes increasingly stronger, progressing from $\hat{D}_b$ to $\hat{R}_h$ to $\hat{D}_h$. The effective cut-off for $\hat{D}_h$ is at the limit, for all practical purposes, because the product $t_h f_h$ is approximately 0.6. The limiting cut-off characteristic would be the real rectangular frequency function. In the continuous frequency domain the rectangular function is given by

$$f(\omega) = \begin{cases} 1.0, & -0.5\omega_1 \le \omega \le 0.5\omega_1 \\ 0, & \text{elsewhere.} \end{cases} \tag{99}$$

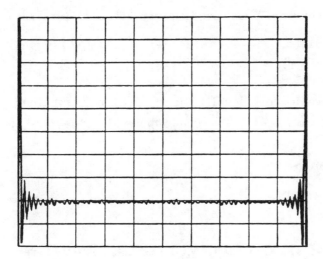

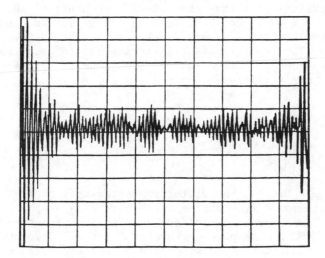

Fig. 26. The complex deconvolution results $\hat{d}_{bc}(k)$ for the effective bandwidth measurement. No. of points, 512; horiz.
scales, 0.999 ns/div; $\gamma_{bo} = 4.79 \times 10^7$. (Upper fig.)
$d_b(k, \gamma_{bo})$: min. and max., -0.0553 and 0.233; vert.
scale, 0.0289/div. (Lower fig.) $\hat{d}_{bI}(k, \gamma_{bo})$: min. and
max., -0.216×10^{-3} and 0.203×10^{-3}; vert. scale,
0.203×10^{-3}/div.; mean -0.585×10^{-11}; std. dev.,
0.388×10^{-4}.

whose inverse Fourier Transform is given by the (sin x)/x pulse,

$$f(t) = \frac{\omega}{\pi} \frac{\text{Sin}(\omega_1 t)}{\omega_1 t} .$$ (100)

When ω_1 is approximately equal to 0.6π, $f(t)$ is very close to 50% of the maximum level. Thus the 50% level pulse duration t_1 follows from

$$\omega_1 t_1/2 = 0.6\pi ,$$ (101)

and

$$t_1 f_1 = 0.6.$$ (102)

Consequently, we see that $\left| \hat{D}_h(n, \gamma_{ho}) \right|$ is the result of very strong filtering. Also, the optimum filter has not significantly altered the physical fact that the cut-off frequency of the low-pass filter is 3.6 GHz. However, because the effective bandwidth of the measurement system is only 1.7 times that of the network being measured, the regularization filtering has to be strong, and its effects appear in the deconvolution results with approximately a (sin x)/x variation.

8   MEASUREMENT OF A PULSE WAVEFORM

8.1 Overview

In this application, the distorting effect of the measurement system is removed from the measured data of a step-like waveform. The experiment was performed using the present-day version of the APMS. The pulse generator is an NBS Reference Waveform Pulse Generator whose step-like output has a nominal 10-90% transition duration of 50 ps with a 50-ohm load.

8.2 Measurement Method

The measurement system is shown in Fig. 27. The measured or observed waveform is $\hat{y}(k)$; it is desired to obtain an estimate of the generator waveform $\hat{x}(k)$. The sequence length for both waveforms is 1024. To estimate $\hat{x}(k)$, we use the modeled impulse response $\hat{h}_1(k)$ of the APMS [22, 23] in the regularized deconvolution (82),

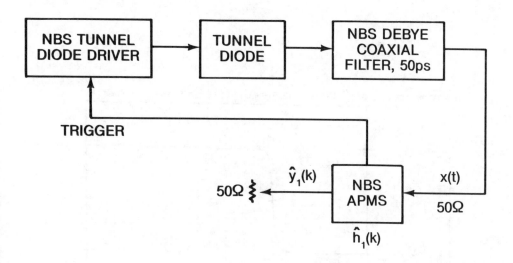

Fig. 27. Measurement method for pulse generator output x(t).

$$\hat{d}_x(k,\gamma_{xo}) = \hat{y}_1(k)(1/*)\hat{h}_1(k). \tag{103}$$

$$= [DFT]^{-1}\left\{\frac{\hat{Y}_1(n)}{\tilde{H}_1(n)}\hat{R}_x(n,\gamma_{xo})\right\} \tag{104}$$

8.3 The Experiment

Using the measurement system of Fig. 27 we acquire $\hat{y}_1(k)$ in the following way. The time at which a sample is taken is held constant for ten repetitions of the repetitive signal, and the resultant 10 samples are averaged in an analog integrator circuit. Consequently, to acquire a given 1024 sequence representing the pulse waveform, 10,240 repetitions of the pulse are required. Each acquired 1024 sequence is stored until 100 sequences are obtained, and then the ensemble of 100 sequences are averaged; the result is the acquired signal $\hat{y}_1(k)$, Fig. 28; the detailed waveform parameter values are an integral part of the present APMS display as seen in Figs. 28 through 32.

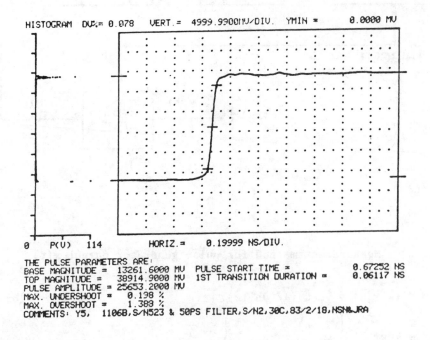

HISTOGRAM DU%= 0.078   VERT.= 4999.9900MV/DIV.   YMIN =   0.0000 MV

0 P(V) 114   HORIZ.=   0.19999 NS/DIV.

THE PULSE PARAMETERS ARE:
BASE MAGNITUDE =   13261.6000 MV   PULSE START TIME =   0.67252 NS
TOP MAGNITUDE =   38914.9000 MV   1ST TRANSITION DURATION = 0.06117 NS
PULSE AMPLITUDE = 25653.2000 MV
MAX. UNDERSHOOT =   0.198 %
MAX. OVERSHOOT =   1.389 %
COMMENTS: Y5, 1106B,S/N523 & 50PS FILTER,S/N2,30C,83/2/18,NSN&JRA

Fig. 28.   The acquired waveform $\hat{y}_1(k)$. The modes of the ampli-
           tude histogram are used to determine the zero and 100%
           levels of the waveform. The vertical scale has been
           multiplied by 100 (pulse amplitude is actually 256.53
           mV). No. of points, 1024.

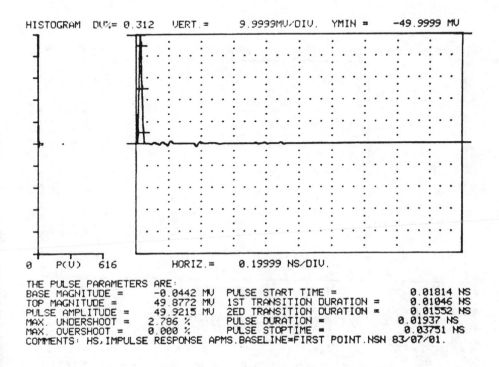

HISTOGRAM DU%= 0.312   VERT.=   9.9999MV/DIV.   YMIN =   -49.9999 MV

0   P(V)   616   HORIZ.=   0.19999 NS/DIV.

THE PULSE PARAMETERS ARE:
BASE MAGNITUDE =   -0.0442 MV   PULSE START TIME =   0.01814 NS
TOP MAGNITUDE =   49.8772 MV   1ST TRANSITION DURATION =   0.01046 NS
PULSE AMPLITUDE =   49.9215 MV   2ED TRANSITION DURATION =   0.01552 NS
MAX. UNDERSHOOT =   2.786 %   PULSE DURATION =   0.01937 NS
MAX. OVERSHOOT =   0.000 %   PULSE STOPTIME =   0.03751 NS
COMMENTS: HS,IMPULSE RESPONSE APMS.BASELINE=FIRST POINT.NSN 83/07/01.

Fig. 29.   The modeled impulse response of the NBS AMPS, $\hat{h}_1(k)$.
           Pulse duration defined at the 50% level. No. of
           points, 1024.

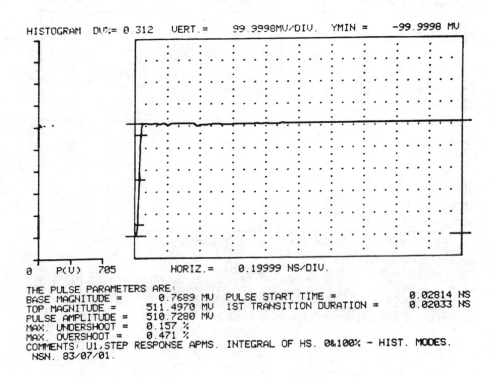

HISTOGRAM DUX= 0.312   VERT.=   99.9998MV/DIV. YMIN =   -99.9998 MV

0   P(V)   705   HORIZ.=   0.19999 NS/DIV.

THE PULSE PARAMETERS ARE:
BASE MAGNITUDE =   0.7689 MV PULSE START TIME =   0.02814 NS
TOP MAGNITUDE =   511.4970 MV 1ST TRANSITION DURATION =   0.02033 NS
PULSE AMPLITUDE =   510.7280 MV
MAX. UNDERSHOOT =   0.157 %
MAX. OVERSHOOT =   0.471 %
COMMENTS: U1,STEP RESPONSE APMS. INTEGRAL OF HS. 0&100% - HIST. MODES.
NSN. 83/07/01.

Fig. 30.   The modeled step response of the NBS APMS, $\hat{U}_1(k)$.

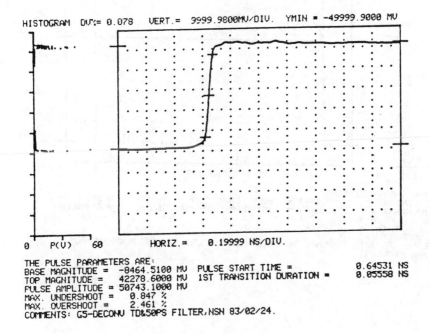

HISTOGRAM DV%= 0.078   VERT.= 9999.9800MV/DIV. YMIN = -49999.9000 MV

0   P(V)   60   HORIZ.=   0.19999 NS/DIV.

THE PULSE PARAMETERS ARE:
BASE MAGNITUDE = -8464.5100 MV   PULSE START TIME =   0.64531 NS
TOP MAGNITUDE   = 42278.6000 MV   1ST TRANSITION DURATION =   0.05558 NS
PULSE AMPLITUDE = 50743.1000 MV
MAX. UNDERSHOOT =   0.847 %
MAX. OVERSHOOT =   2.461 %
COMMENTS: G5-DECONV TD&50PS FILTER, NSN 83/02/24.

Fig. 31.   The deconvolution estimate for the pulse generator output $\hat{x}(k)$. The vertical scale is referred to the open
circuit generator output and is multiplied by 100.

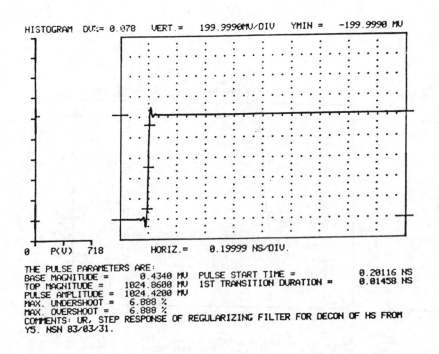

Fig. 32.   The step response of the deconvolution regularization
          filter, $U_R(k, \gamma_{xo})$.

The APMS modeled impulse and step responses, $\hat{h}_1(k)$ and $\hat{u}_1(k)$, are shown in Figs. 29 and 30, respectively [20]. Applying the regularized deconvolution (82), we obtain the deconvolution estimate, $\hat{d}_x(k,\gamma_o)$, for the pulse generator output $x(t)$, Fig. 31.

Next, we obtain the impulse response of the regularizing filter using the method in Section 6.5, and integrate (sum) the result to obtain the step response of the filter, $\hat{U}_R(k,\gamma_{xo})$, Fig. 32.

The 10-90% transition duration of the acquired waveform $\hat{y}_1(k)$ is 61.17 ps, while that of the deconvolved result is 55.58 ps. Notice that the ringing on the regularization filter step response, $\hat{U}_R(k,\gamma_{xo})$, is not evidenced in the deconvolution result, $\hat{d}(k,\gamma_{xo})$; this is because the filtering required to make the estimate (perform the deconvolution) is relatively light.

We can now make an estimate of the bounds on the 10-90% transition duration. We define τ_b as the upper bound due to the deconvolution operation; we wish to calculate τ_b.

To do so, we will use the relation between the transition durations given by

$$\tau_t^2 = \tau_1^2 + \tau_2^2 + \ldots + \tau_n^2 \tag{105}$$

which is valid for Gaussian transitions and is a good approximation for other relatively smooth transitions [23]. First of all, we have the following transition durations from Figs. 28, 30, 31, and 32, respectively,

τ_y = 61.17 ps, for the observed waveform

τ_h = 20 ps, for the measurement instrument

τ_d = 55.58 ps, for the deconvolution result

τ_u = 14.58 ps, for the regularization filter

We have for the upper bound,

$$\tau_b = \{\tau_d^2 + \tau_u^2\}^{1/2}, \tag{106}$$

$$= 57.46 \text{ ps.}$$

The lower bound is the deconvolution result transition duration, τ_d. Thus, we have for the estimated transition duration

$$\Delta\tau_x = \left| 57.46 - 55.58 \right|$$

$$= 1.88 \text{ ps}.$$

Because we could never obtain a regularized waveform estimate for the exact pulse generator output $x(t)$ without filtering, we can only say that the generator waveform transition duration is

$$\tau_x = (\tau_d + \frac{\Delta\tau_x}{2}) \pm \frac{\Delta\tau_x}{2}. \tag{107}$$

$$\tau_x = 56.52 \pm 0.94 \text{ ps}.$$

Also, notice that $[(\tau_y^2 - \tau_h^2 - \tau_u^2)^{1/2}$ gives the value 55.93 ps, which is very close to the value for the deconvolution result of 55.46 ps. Thus, the relation (105) is an accurate estimator for the waveforms $\hat{y}_1(k)$, $\hat{h}_1(k)$, $\hat{d}_x(k)$, $\hat{U}_R(k)$.

9.   SUMMARY AND CONCLUSIONS

9.1   Summary

Our objective in this paper has been to bring out the ideas of ill-posedness and to give examples of applications to pulse measurement problems which require deconvolution. This class of problems arises when we attempt to remove the effects of a pulse source and/or a measurement system on observed pulse waveforms.

The removal of these effects is accomplished by computation; and in light of this, the term "software-correction" is a fitting synonym for "deconvolution".

We have presented three kinds of examples: (a) simulation experiments to illustrate that the origin of ill-posedness resides in inexact data; (b) the removal of measurement system effects applied to pairs of actual experimental waveform data in impulse response measurements; and (c) the removal of the effects of a measurement instrument's impulse response from the observed pulse waveform experimental data using an instrument model.

9.2 Conclusions

By virture of what has been presented here we can conservatively say the following:

a. Very small errors in experimental data lead to an ill-posed deconvolution problem.

b. The solution to an ill-posed problem is a solution which is in the neighborhood of the true solution, and, as such, criteria must be given a priori as to when a solution is an estimate whose error bounds may not be known.

c. Solution of an ill-posed problem requires some form of regularization or filtering to achieve a physically and/or mathematically acceptable result; the regularization may be implemented in the time domain or in a transform domain, or, conceivably, simultaneously in both domains.

d. In general, the selection of a solution method is arbitrary; thus a method may be selected that utilizes physical and/or mathematical constraints which are appropriate for the problem at hand.

APPENDIX A

For further study the reader may refer to references [1] and [13].

A.1 Fourier Transforms Continuous and Discrete

The continuous Fourier Transform (CFT) of the continuous function $f(t)$ is given by

$$F(j\omega) = \int_{-\infty}^{\infty} f(t)\, e^{-j\omega t}\, dt, \tag{A-1}$$

while the inverse transform $(CFT)^{-1}$ of $F(j\omega)$ is given by

$$f(t) = \frac{1}{2\pi} \int_{-\infty}^{\infty} F(j\omega)\, e^{j\omega t} d\omega. \tag{A-2}$$

The Discrete Fourier Transform (DFT) of the sequence or discrete function $\hat{f}(kT)$ is given by

$$\hat{F}(n\Omega) = \sum_{k=0}^{N-1} \hat{f}(kT) \; e^{-j\Omega Tkn}, \; n = 0,1,2,\ldots,N-1; \qquad (A-3)$$

while the inverse transform $(DFT)^{-1}$ of $\hat{F}(n\Omega)$ is given by

$$\hat{f}(kT) = \frac{1}{N} \sum_{n=0}^{N-1} F(n\Omega) \; e^{j\Omega Tkn}, \; k = 0,1,2,\ldots,N-1$$

T is the time interval between the sequence values in the discrete time domain, while Ω is the frequency interval in the discrete frequency domain and is given by

$$\Omega = 2\pi/NT , \qquad\qquad\qquad\qquad\qquad\qquad (A-5)$$

where N is the number of terms in both the sequences (A-3) and (A-4). With regard to notation, for brevity we ordinarily write $\hat{f}(k)$ for $\hat{F}(kT)$ and $\hat{F}(n)$ for $\hat{F}(n\Omega)$. Also, the "T" in CFT and DFT may mean transform or transformation depending upon the context of usage. The function F and the sequence F are transforms while the corresponding integral and sum relations are transformations.

A.2 Parseval's Relations

For a continuous function Parseval's relation is

$$\int_{-\infty}^{\infty} f^2(t) \; dt = \frac{1}{2\pi} \int_{-\infty}^{\infty} \left| F(j\omega) \right|^2 \; d\omega , \qquad (A-6)$$

while for discrete function or sequence we have

$$\sum_{k=0}^{N-1} \hat{f}^2(k) = \frac{1}{N} \sum_{n=0}^{N-1} \left| \hat{F}(n) \right|^2 \qquad\qquad (A-7)$$

A.3 The Derivative and Difference Transforms

For a continuous function the transform of a derivative operator operating on the function is given by

$$(j\omega)^m \ F(j\omega) = [CFT] \left\{ \frac{d^m}{dt^m} f(t) \right\} , \qquad (A-8)$$

where m is the order of the derivative. Thus the derivative operator transforms to the multiply factor, $(j\omega)^m$, i.e., the CFT is

$$(j\omega)^m = \int_{-\infty}^{\infty} \frac{d^m}{dt^m} d^{-j\omega t} \ dt \qquad (A-9)$$

For discrete functions, the transform corresponds to the transformation of a backward difference (BD) sequence operator operating on the sequence, e.g.,

$$(BD)^1 * \hat{f}(k) = [1,-1,0,0,\ldots,0] * \hat{f}(k) \qquad (A-10)$$

$$(BD)^2 * \hat{f}(k) = [1,-2,1,0,0,\ldots, 0] * \hat{f}(k) , \qquad (A-11)$$

for the first and second backward difference operators. There are $(N-2)$ and $(N-3)$ zeroes in $(BD)^1$ and $(BD)^2$, respectively. For the difference operator we have

$$\hat{C}(n) \equiv [DFT] \ (BD)^2 = \sum_{k=0}^{N-1} [1,-2,1,0,0,\ldots,0]e^{-j\Omega Tkn}$$

$$n=0,1,1,\ldots,N-1. \qquad (A-12)$$

$$= 1 - 2e^{-j\frac{2\pi n}{N}} + e^{-j\frac{\pi 4n}{N}} , \quad n=0,1,2,\ldots,N-1$$

$$n=0,1,1,\ldots,N-1 . \qquad (A-13)$$

and

$$\left| \hat{C}(n) \right|^2 = \hat{C}(n) \ \hat{C}^*(n) \qquad (A-14)$$

$$= \left[1 - 2e^{-j\frac{2\pi n}{N}} + e^{-j\frac{4\pi n}{N}} \right] \left[1 - 2e^{j\frac{2\pi n}{N}} + e^{+j\frac{4\pi n}{N}} \right]$$

$$n = 0,1,2,\ldots, N-1. \qquad (A-15)$$

$$= 6 - 8 \cos \frac{2\pi n}{N} + 2 \cos \frac{4\pi n}{N}; \quad n = 0,1,2,\ldots,N-1 \qquad \text{(A-16)}$$

$$= 16 \sin^4 \frac{\pi n}{N} ; \quad n = 0,1,2,\ldots, N-1. \qquad \text{(A-17)}$$

REFERENCES

1. C. D. McGillem and G. R. Cooper, Continuous and Discrete Signal and System Analysis, Holt, Rinehart, and Winston, New York, 1974, Chapter 4.

2. Ibid, p. 167.

3. Ibid, p. 130.

4. Ibid, p. 179.

5. W. L. Gans and N. S. Nahman, "Continuous and Discrete Fourier Transforms of Steplike Waveforms," IEEE Trans. Instru. and Meas., Vol. IM-31, No. 2, June 1982, pp. 97-101.

6. P. H. Van Cittert, "On the Influence of the Aperature-slit on the Intensity Distribution of Spectral Lines," Z. Physik, Vol. 69, pp. 298-307 (1931). (In German).

7. Op. Cit [1], pp. 170-188.

8. Ibid, p. 277.

9. N. S. Nahman and M. E. Guillaume, "Deconvolution of time domain waveforms in the presence of noise," National Bureau of Standards Technical Note 1047, October 1981, 115 pages. National Bureau of standards, Boulder, CO. 80303, pp. 90-93.

10. A. V. Tikhonov and V. Y. Arsenin, Solutions to Ill-posed Problems, John Wiley and Sons, Wash., D.C., 1972.

11. S. Twomey, "The Application of Numerical Filtering to the Solution of Integral Equations Encountered in Indirect Sensing Measurements", J. Franklin Institute, Vol. 269, Feb. 1965, pp. 95-108.

12. B. R. Hunt, "Deconvolution of Linear Systems by Constrained Regression and Its Relationahip to the Wierner Theory", IEEE Trans. on Auto. Control, AC-17(5), October 1979, pp. 703-705.

13. A. V. Oppenheim and R. W. Schafer, Digital Signal Processing, Prentice Hall, New Jersey, 1977.

14. Op. Cit. [9], pp. 28-33.

15. Op. Cit. [10], p. 148.

16. M. E. Guillaume and J-C. Bizeul, Extract Annales Des Télécommunications, tome 64, nos. 3-4, Mars-Avril 1981, pp. 1/8-8/8.

17. N. S. Nahman and M. E. Guillaume, "Some Results Using the Guillaume-Nahman Automated Deconvolution Method," Paper A4-1, January 6, 1983, U. S. National Radio Science Meeting (URSI), Jan. 5-7, 1983, Boulder, CO 80303.

18. W. L. Gans, "Present Capabilities of the NBS Automatic Pulse Measurement System," IEEE Trans. Instru. meas., IM-25, No. 4, Dec. 1976, pp. 384-388.

19. Op. Cit. [13], p. 409.

20. J. R. Andrews (NBS) developed the presently used impulse response model shown in Fig. 29. It is a scaled version of an earlier model [21,22], but also includes reflection effects peculiar to the specific AMPS sampling head.

21. S. M. Riad and N. S. Nahman, "Modeling of the Feed-through Wideband Sampling-head," Digest of the 1979 IEEE-MTT-Society International Microwave Symposium, Ottawa, Canada, June 27-29, 1978.

22. S. M. Riad, "Modeling of the HP-1430A Feed-through Wide-band (28-ps) Sampling-head", IEEE Trans. Instru. & Meas., Vol. IM-31, No. 2, June 1982, pp. 110-115.

23. G. E. Valley, Jr. and H. Wallman, Eds., "Vacuum Tube Amplifiers", Vol. 18, Radiation Laboratory Series, McGraw-Hill, Inc., New York, 1948, pp. 77-78.

FIBER OPTIC LINKS FOR DATA ACQUISITION, COMMUNICATION, AND CONTROL

George Chandler

Los Alamos National Laboratory
Albuquerque, NM
USA

1. INTRODUCTION

The era of fiber optics communications began in 1970 when Kapron, Keck, and Maurer of Corning Glass Works [1] produced the first fiber with low loss, less than 20 db per kilometer. Today fiber optics communication links are commonplace, particularly in the telephone industry, where they are replacing coaxial cable links for long distance communications; and increasingly in the data communications field for digital data transfer and computer networking.

Fiber optics links have also found increasing application in experimental physics, particularly where electromagnetic interference is a problem. Glass fibers cabled in plastic protective layers can replace copper wires in diagnostics and other applications near experiments with large induced electric fields, or high voltages. Elaborate shielding and grounding schemes often associated with diagnostics, control, and data communication cabling may be obviated. Optical fibers are usually smaller and lighter than the coaxial cables they often replace, so that many physical restraints are removed. A variety of fiber types is available, offering bandwidths from 20 MHz-km to several hundred MHz-km. Finally, the cost of fiber optic cable and the associated hardware makes the cost of fiber optic links comparable to the cost of hardwired links. Table 1 gives typical costs for the necessary components.

Table 1. Typical Component Costs.

COMPONENT	COST	UNIT
Fiber	$1 $2	meter
Connectors (SMA)	$5 - $15	ea.
TO-18 mount panel connector	$4	ea.
LED	$2 - $20	ea.
Photodiode	$8 - $200	ea.
Packaged analog link	$500 - $1200	pair
Packaged digital link	$500 - $1000	pair

2.   TYPES OF FIBERS [2]

The simplest model of light propagation in an optical fiber
is that of total internal reflection at an interface between two
transparent materials of slightly different indexes of refraction. The core of the fiber has a slightly higher index than
the clad. The clad may be glass, doped to produce the different
index (glass-on-glass), or it may be a plastic (plastic-clad
silica, PCS). In early fibers it was simply air. There are some
plastic fibers being produced, of increasingly high quality, but
not yet approaching the quality of glass. In graded-index fibers, the index of refraction of the core is not constant but is
graded radially at a rate determined by the properties desired.
The gradient is usually but not always parabolic, i.e., varies
as the radius squared. If the core of a step-index fiber is made
small enough it will propagate only one mode, and this is called
single-mode fiber. Fig. 1 illustrates the basic types of fiber.

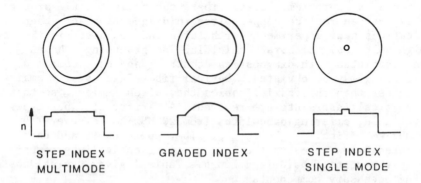

| STEP INDEX | GRADED INDEX | STEP INDEX |
| MULTIMODE | | SINGLE MODE |

Fig. 1.   Sketches of the cross sections and index-of-refraction
          profiles of three basic fiber types.

Commercially available fiber types are now narrowed down to a few, the properties of which are summarized in Table 2.

3.   FABRICATION

There is a large number of processes for producing optical fibers. Two are illustrated in Figs. 2 and 3. An early technique is called the double-crucible process. Two concentric crucibles contain molten glasses of slightly different index; a fiber is drawn directly from the melt (Fig. 2).

Most fibers today are made from preforms; one technique for producing them, the chemical vapor deposition (CVD) process, is shown in Fig. 3. A silica tube, which will become the cladding, is rotated in a glass lathe with a torch which can be moved along its length. A mixture of gases reacts inside the tube, depositing oxides of silica, germanium, or boron in desired ratios, on the inside of the tube. The ratios are varied to achieve the desired radial distribution of the index of refraction. The completed tube is collapsed into a solid rod or preform which is then transported to a drawing tower to be pulled into a fiber which will have the same index profile.

Table 2.   Typical Properties of Commercial Fibers

Type	Core Microns	Clad Microns	Wavelength nm	Attenuation $d\beta/Km$	Bandwidth MHz	NA*	Price $/m	
Plastic clad	200-300	Various	820-900	5-40	20	.4	1-1.5	Radiation Resistance
silica (PCS)	400-1500	Various	200-900	-	-	-	3-25	
Step Index Multimode	100	140	850	4-10	200-100	.25	.9-1.5	
	50	125	850	3-8	40-200	.2	.6-1	
Graded Index	50	125	850	3-5	400-800	.2	.6	
	50	125	1300	1-3	400-1200	.2		
	50	125	1550	1-2	400-1200	.2		
	100	140	850	3-5	100-400	.25	.9	
	100	140	1300	1.5-2	200-400	.25		
Single Mode	3-4	80-125	633	7-12	1000	.1	1-2	
	5-6	80-125	820	5-10	2000	.1	1.2	
	8-11	80-125	1300	.6-4	20,000	.1	1-3	

* Numerical Aperture

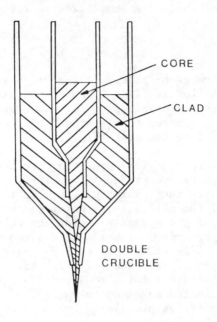

Fig. 2.   Sketch of the double-crucible process for drawing opti-
          cal fiber.

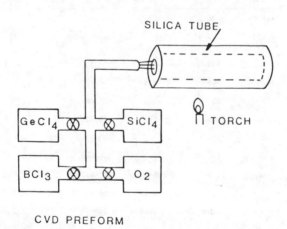

Fig. 3.   Sketch of the CVD process for producing preforms to be
          drawn into optical fiber.

4.   ATTENUATION

Fig. 4 is a sketch of attenuation-vs-wavelength for a hypothetical fiber, illustrating qualitatively the contributing
factors and the spectral regions where they are important. The
short wavelength attenuation in modern fibers is dominated by
Rayleigh scattering. The peaks at about 720, 950, and 1400 nm
are due to absorption by hydroxl ions. Above 1400 nm the infrared absorption of the glass dominates. There are two important windows, around 850 nm (first window) and 1300 nm (second
window).

The first window is important because emitters of the GaAs
and AlGaAs families operate there, and most of the short-range
communications work and fiber are designed for the first window.
Silica has a zero in its material dispersion around 1300 nm so
that high bandwidth is possible in the second window. That point
is also fortuitously near the theoretical minimum of attenuation,
where the Rayleigh scattering and infrared absorption of silica
cross over.

Ions other than hydroxyl produce attenuation in glass. The
most troublesome include ions of iron, copper, manganese, and
chromium; effective techniques have been developed to remove
these from the glass.

5.   DISPERSION

Pulse dispersion is the temporal widening of a pulse with
distance travelled along the fiber. This is explained by four
mechanisms. First, material dispersion is familiar to optics
students: the variation of the index-of-refraction of a medium
with wavelength. The sources used to excite fibers have a finite
spectral width, from a few nanometers for laser diodes to a few
tens of nanometers for light-emitting diodes. The spectral components from a pulsed source in a dispersive material therefore
travel at different velocities, causing the pulse to spread. In
silica the material dispersion has a zero around 1300 nm, and
much device development work is being done to exploit this fact
for high-bandwidth communications.

Second, modal dispersion is explained by describing the propagation of light in a dielectric waveguide in terms of modes,
similar in concept to the modes in a metallic waveguide or a
laser cavity. Each mode has a different velocity, hence a pulse
from a source which excites several modes will spread as it propagates. The development of single-mode fiber for long distance
communication is intended to remove the difficulty of modal dispersion.

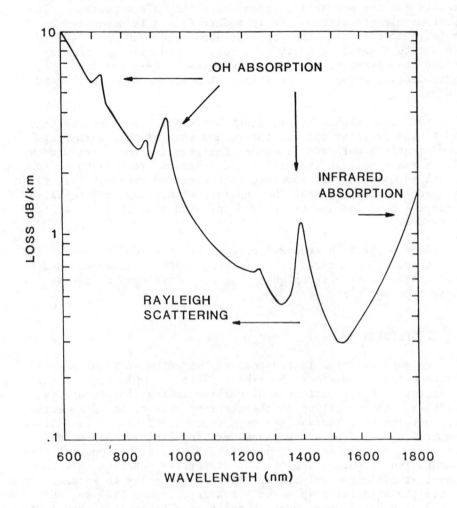

Fig. 4. Sketch of attenuation-vs-wavelength for a hypothetical
          fiber.

Third, polarization dispersion arises because the single, lowest-order mode is actually degenerate: it is a sum of two orthogonal states of polarization which can propagate at different velocities if there is birefringence in the fiber.

Finally, waveguide dispersion is due to the wavelength dependence of the propagation constants of each mode; thus even in a single-mode fiber with zero material dispersion the finite bandwidth of the source causes pulse spreading.

6. NUMERICAL APERTURE

For design purposes one of the most important parameters of an optical fiber is the numerical aperture, NA. The ray-tracing model based on total internal reflection is illustrated in Fig. 5. The angle ϕ_{crit} is the angle of the input ray which, after refraction at the air-core interface, arrives at the core-clad interface at the critical angle, above which all rays are totally reflected. We define:

$$\text{Sin}(\phi_{crit}) \equiv NA = (n_1^2 - n_2^2)^{1/2} \simeq n_1(2\Delta n)^{1/2} \tag{1}$$

where n_1, n_2 are the indices of the core and the clad, respectively, and $\Delta n = n_1 - n_2$.

For design purposes, only light that is incident on the fiber end at angles less than $\sin^{-1}(NA)$ will contribute to useful signal at the output of the fiber. In addition, the designer should be aware that there is a range of angles (or modes) near the critical angle which do, in fact, propagate for varying distances in the fiber so that the measured numerical aperture of a fiber may depend on its length, for lengths less than a few hundred meters. These marginal modes are highly attenuated, and they can also be stripped with certain techniques. The measurement of fiber properties such as attenuation and numerical aperture is considerably complicated by this situation.

7. LOSS BUDGET [3]

This is a process of accounting for the power from a source as it propagates through an optical system, encountering various loss mechanisms. Table 3 lists the major sources of loss.

One often designs a system backwards, starting at the receiver. The desired bit-error rate, or signal-to-noise ratio at the receiver, determines the optical power that must be delivered

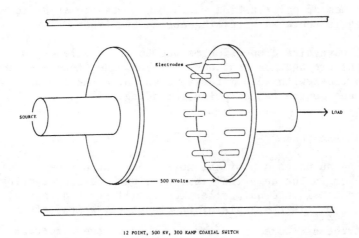

12 POINT, 500 KV, 300 KAMP COAXIAL SWITCH

Fig. 5. Sketch of a simple ray-tracing model of optical fibers,
         illustrating the definition of numerical aperture.

Table 3.   Loss Mechanisms.

JUNCTURE	LOSS MECHANISM	TYPICAL LOSS
LED-fiber	NA mismatch separation core/source size	70-98%
Fiber-fiber	NA mismatch end separation misalignment core size mismatch angular misalignment	.1 - 3 db
Fiber-detector	separation core/detector size	< 1 db
Fiber	attenuation	1 - 40 dB/km

by the fiber. The major loss mechanism in most short systems of
interest to diagnosticians is at the input. Connectors and
splices usually offer low loss but can be unpredictable, i.e.,
not repeatable, and play havoc especially with simple analog
systems. There should be enough loss margin to allow for connector variations and aging of the emitter.

8.   SOURCES AND DETECTORS

Most links useful for diagnostics and data communications
use light-emitting diodes, LED's, for the source. Higher bandwidth (hundreds of Megahertz) or long distances (Kilometers) may
require laser diodes. Most commercially available LED's operate
in the first window, and inject from ten microwatts to a milli-
watt into the fiber, depending on the particular geometry and
fiber. Most detectors are silicon PIN photodiodes. Certain
applications may require avalanche photodiodes (APD's), a sort of
solid-state photomultiplier with gain of up to several hundred.
These are difficult to bias and operate reliably, but if properly
used can be less noisy and faster than PIN photodiodes. They are
available packaged with bias and amplifier circuitry. LED's have
risetimes from 2 to 20 nsec., laser diodes from < 0.5 nsec, PIN
photodiodes 1 to 10 ns, and APD's 0.5 ns to 5 ns. There are also
available relatively slow LED's (microsecond risetimes), and pho-
totransistors and photo-Darlingtons having risetimes from tens to
hundreds of microseconds.

Light output with respect to LED current is typically linear
from one to five percent over the entire current range. They can
be more linear than that over a restricted range, and opto-elec-
tronic feedback can be used to linearize an LED or laser diode
transmitter. Photodiodes are inherently linear over a wide range
of photocurrent. Quality LED's and photodiodes in TO46, TO18, or
similar packages compatible with fiber optic component mount connectors are priced from ten to thirty dollars. Laser diodes
start around fifty dollars and go into the thousands.

9.   CONNECTORS AND SPLICES

Possibly the most popular connectors in use today are adap-
tations of the microwave SMA type connectors. These are available in a variety of prices depending on the precision of the
hole location, the method of attaching fibers, and the material.
Prices range from five to twenty dollars in single quantities.
Early versions required gluing the fiber into the connector, with
a crimp to fasten the body to the cable, followed by a polishing
process. Assembly time including polishing might be fifteen to

twenty minutes. Recently manufacturers have produced "epoxyless" SMA connectors and cleaving techniques that minimize fabrication and nearly eliminate the polishing step, so that assembly time can be as little as five minutes. These connectors show typical loss figures of 2 to 4 db, and one percent or so repeatability.

Splices and bulkhead-mounted component adapters are priced in the five-dollar range. Compatible components are readily available; the most common packages are the T046, T018, and a similar package with the same diameter.

Much more sophisticated connectors with low loss (.1 db) are marketed; prices range into the hundreds of dollars. Connectors for single-mode fibers are beginning to appear. Several companies market fusion splicers, which connect two fibers by aligning them and fusing them in place; such splices exhibit losses of .05 db and less. These splicers cost upwards to five thousand dollars and are aimed at the telecommunications market.

10. DIGITAL LINKS

The rate at which errors occur in digital data communication links is primarily a function of the signal-to-noise ratio (SNR) at the receiver. In a so-called baseband system, i.e., a simple on-off digital signaling system, the Bit-Error-Rate (BER), is approximated by:

$$P_e \simeq \text{erfc}\left(\frac{SNR}{2}\right) \tag{2}$$

where the assumption is made that the comparator threshold is set at half the signal "one" voltage; SNR is peak-signal-to-rms noise, and erfc(x) is the error function

$$\text{erfc}(x) \equiv \frac{1}{(2\pi)^{1/2}} \int_x^\infty e^{-y^2/2} \, dy \tag{3}$$

Fig. 6 plots BER as a function of SNR for this approximation. The prediction of error rate in signaling systems is complex; see reference 4 for an entry to the subject.

To maintain a specified BER a baseband link needs to have an automatic gain control (AGC) circuit in the receiver to keep the threshold adjusted properly. AGC circuits need occasional transitions to operate, and this often leads to more sophisticated signaling systems such as bi-phase modulation because of the need to transmit long streams of one's or zero's. Synchronous digital systems may encounter many other sources of errors such as skew

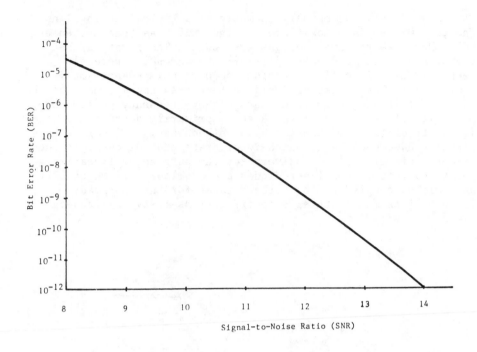

Fig. 6.   Plot of bit-error-rate as a function of peak signal-to-
          rms noise for a baseband signaling system with threshold
          at one-half of peak.

or asymmetric waveforms. The calculation of BER for systems,
other than the one described, is considerably more complicated
than Equation 2.

   Many commercial links are available; most claim BER's in the
range 10^{-8} to 10^{-9}. Short-haul links with bandwidths of 5 to
10 MHz cost in the low hundreds of dollars; higher bandwidths of
20 to 50 MHz sell in the $1000 - $2000 range. A wide variety of
packages are available, from stand-alone to rack-mounted, to PC-
mounted units. Several manufacturers sell integrated circuits
which combine optical devices and logical driving circuitry in a
TO46/TO18 package that mounts in an SMA connector, so that one
has, for example, TTL in and TTL out of the panel-mounted connectors. The components for a 5-Mbit, 10^{-9}-BER link so constructed
would cost less than fifty dollars exclusive of fiber and power
supplies.

11. ANALOG LINKS

Fig. 7 is a conceptual schematic of a typical analog link. The transmitter is a Los Alamos design which has been tested up to 10 MHz; two receiver designs are shown. The first is the ubiquitous transimpedance amplifier; the second, a recently developed circuit called a bootstrapped transimpedance amplifier [5,6]. The bootstrap design effectively cancels most of the diode capacitance, permitting much higher frequency performance. Linearity of designs such as these is primarily dependent on the linearity of the LED; these can be as good as 1%. Low frequency circuits have been built which incorporate photodiodes into the feedback circuits of LED drivers for the purpose of linearizing the transmitter; some laser diodes are available [7] which have photodiodes coupled to the "other" facet for this purpose. Table 4 lists manufacturer's specifications on typical analog links.

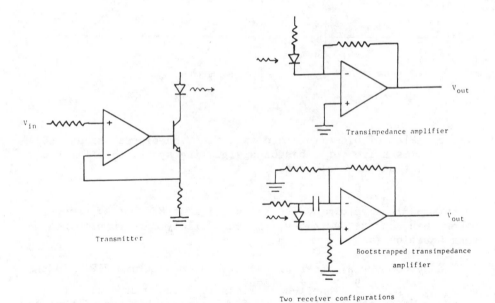

CONCEPTUAL SCHEMATIC OF AN ANALOG LINK.

Fig. 7. Typical analog link.

Table 4.   Properties of some Commercial Analog Links.

MANUFACTURER	MODEL	BANDWIDTH	DYNAMIC RANGE	LINEARITY
ITT	T6211/6261	10 Hz-20 MHz	25dB	distortion 30 dB
LeCroy	FAT3HS	15 Hz-10 MHz	SNR 48dB	1%
	FAS4	15 Hz-30 MHz	---	1%
MERET	MDL271-4	DC-2 MHz	40 dB	.1%
	MDL275-4	DC-10 MHz	26 dB	.1%
	MDL278-4	DC-25 MHz	20 dB	.1%
	MDL24702	DC-500 kHz	60 dB	.05%
OPTELECOM	3100TAA	6 Hz-6 MHz	SNR 48 dB	1%
MATH	XA/RA-1000	DC-5 MHz	50 dB	1%
DYNAMIC MEASUREMENTS	5700	DC-1 MHz	SNR 40 dB	.2%

12.   DYNAMIC RANGE

The dynamic range of an analog system is the ratio, expressed in decibels, of the lowest and the highest signal levels that can be passed through the system while meeting some specified set of criteria. Applied to analog links, that generally means the ratio between the highest signal the system will pass, while preserving the linearity and bandwidth specifications, and the tangential noise. Some manufacturers specify SNR instead of dynamic range.

13.   LINEARITY

This is illustrated in Figure 8. The linearity figure most often quoted is simply the maximum deviation from a straight line divided by the amplitude of the maximum signal transmitted. This figure can be related to harmonic distortion; at least one manufacturer of analog links measures harmonic distortion with a spectrum analyzer and converts that result to a linearity specification. Linearity should be measured over the specified bandwidth and input amplitude range of the link.

14.   BANDWIDTH

Receivers are generally the limiting factor.

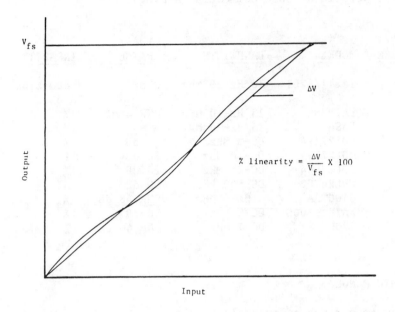

ILLUSTRATING THE DEFINITION OF LINEARITY

Fig. 8.   Illustrating the definition of linearity.

15.   FM LINKS

These are based on voltage-to-frequency converter (VFC) circuits, and exhibit linearity on the order of .1% or better. Bandwidth is limited; commercially available links operate at data bandwidths of 1 MHz or less. Design tradeoffs are dynamic range and bandwidth.

16.   CONTROL LINKS

To realize the full potential of fiber optics in severe electromagnetic environments, designers should consider controlling the auxiliary equipment with optics, especially if the machine is to be controlled by a computer. On a large machine the challenge is economic; for example, the next generation of reversed field-pinch at Los Alamos is expected to have some 3000 control functions. To spend several hundred dollars per link, as would be required using presently available commercial hardware, would be prohibitively expensive. We are scouring the market-place for inexpensive components and fiber, and expect to be able to implement the optical links for between $50 and $100 each, including the cost of the fiber.

The distribution of timing pulses in a pulsed power device is likewise feasible; at Los Alamos we use laser-diode driven links with receivers powered by air-driven generators [8] for this purpose, achieving timing resolution of the order of 10 nsec.

17. DIAGNOSTICS APPLICATION

Figure 9 shows an experiment under construction at this time. A 12-point high voltage switch for 300 kA was designed for the High Density Z-Pinch [9]. The experimenters need to know how the current is distributed among the 12 electrodes, but be-

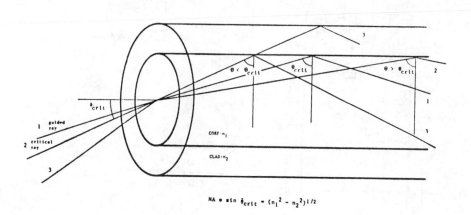

Fig. 9. Sketch of 12-point, 500-kV, 300-kA switch.

cause the electrode structure goes to 500 kV when the switch
is fired, no conventional diagnostic was feasible. The light
emitted from the discharge is also not useful as a measurement
of the currents, although timing information is available. The
solution being implemented is shown in Figure 10. An LED is
mounted near each electrode with a current-limiting resistor
which forms a loop, coupling flux from the electrode current.
The light output is coupled into a fiber and to a fast photodiode
and amplifier for display on an oscilloscope. The combination
has a response time of about 10 nanoseconds, and is expected to
be near 2 nanoseconds when new components on order are received.

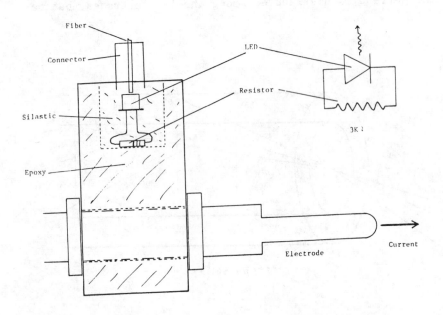

Fig. 10. Sketch of LED and loop for measuring electrode current.

REFERENCES

1. F. P. Kapron, D. B. Keck, and R. D. Maurer, "Radiation
    Losses in Glass Optical Waveguides," Appl. Phys. Lett. 17,
    pp 423-425 (1970).

2. J. E. Midwinter, Optical Fibers for Transmission, Wiley,
    1979.

3. H. Gempe, "Applications of Ferruled Components to Fiber
    Optic Systems," Application Note AN-804, Motorola Semicon-
    ductor Products Inc., P.O. Box 20912, Phoenix AZ 85036, USA.

4. S. D. Personick, "Design of Receivers and Transmitters for
    Fiber Systems," Fundamentals of Optical Fiber Communica-
    tions, M. K. Barnoski, editor, Academic Press, 1981.

5. M. Abraham, "Design of Butterworth-type Transimpedance and
    Bootstrap-transimpedance Preamplifiers for Fiber-Optic
    Receivers," IEEE Trans. on Circuits and Systems CAS-29,6,
    p. 375-382 (1982).

6. Application Note, "Using the CLC 103 AI in Photodiode De-
    tector Applications," Comlinear Corp., 2468 E. 9th St.,
    Loveland CO 80537 USA.

7. RCA Solid State Div., Lancaster PA 17604 USA.

8. G. Chandler, A. Brousseau, R. Hall, and R. Gribble, "Use
    of Fiber Optics to Eliminate Ground Loops and Maintain
    Shielding Integrity on a Controlled Fusion Experiment,"
    Proc. Los Alamos Conference on Optics '79, SPIE Vol. 190,
    1979.

9. J. Hammel, D. W. Scudder, and J. S. Shlachter, "Recent Re-
    sults on Dense Z-Pinches," to be published in Nuclear In-
    strumentation Methods.

A HIGH SPEED MULTI-CHANNEL DATA RECORDER

J. Chang, J. Foesch, and C. Martinez

Sandia National Laboratories
Albuquerque, NM
USA

1.   INTRODUCTION

Data acquisition requirements for large pulsed power accelerators have become increasingly difficult to fulfill. There are two contributing factors: one is the requirement of high band width (> 1 GHz) analog data recording, and the other is the need for a large number of recorders.

For example, PBFA I [1] has thirty-six separate lines, with each containing a Marx bank, intermediate storage capacitor, gas switch, pulse-forming line, water switch, and magnetically insulated transmission line. If 5 to 10 monitors for each line are necessary to diagnose all its elements, then to study the power flow for the whole accelerator 180 to 360 data channels are needed. A resource commitment of this magnitude is obviously difficult to fulfill. Furthermore, an additional number of these data channels are needed for diode experiments, and some of these must have bandwidths in excess of 1 GHz in order to observe the ultrafast diode events. Present oscilloscope technology, including digital oscilloscopes, could not economically meet these requirements. Usually, due to budgetary limitations, only wave forms from a highly selected irreducible set of monitors are recorded. However, these may be supplemented by a multi-channel peak amplitude detector [2] if the knowledge of the complete wave form is not essential and provided only the time-of-arrival of the peak amplitude is desired. This approach to overcome the data acquisition difficulty obviously is not ideal and certainly leaves much to be desired. An alternate approach is to develop an economical system that would allow a more extensive coverage of the data channels.

In this paper, we present a new approach based on optical
fiber and electrooptical streak camera technologies in developing
an economical, high-speed, multi-channel, data recording system.

2.   SYSTEM DESCRIPTION

The approach is to use an electrooptical streak camera to
record as many separate channels of optical analog signal, with
as much dynamic range, as possible. This approach requires that
input signals be optical analog in nature and electric analog
input signals need to be first converted by an electrooptical
transmitter interface.

The system, as represented by the schematic shown in Fig. 1,
consists of an optical fiber input array, an electrooptical
streak camera, a digital readout system, and a data-handling computer. The input signal could be from either an optical analog
sensor or an electrooptical transmitter interface driven by an
electric analog signal. The optical analog signals are coupled
into separate fibers of the optical fiber array and the array, in
turn, is lens-coupled into the streak camera. The streak camera
records these channels of data as separate streaks. A SIT Vidicon TV camera is coupled to the output of the streak camera and
records these streaks in digital form on a 256 x 256 x 8-bit target array. In this target, the pixel elements are arranged into
256 pixel columns of 256 pixels each. This digital format is
accessed by a digital computer which performs the necessary data
handling and processing functions so that data from each channel
can be plotted as final output in the familiar amplitude-vs-time
format.

A prototype system has been assembled with off-the-shelf and
commercially available components. The fiber optics input array
uses the Western Electric [3] optical fiber cable. This cable is
roughly 1.25 cm in diameter and has 144, 50 μm-core and 124-μm
cladding, multi-mode graded index fibers. The 144 optical fibers
are divided into groups of 12, with each group of 12 fibers combined into a flat ribbon. The 12 ribbons form the core of the
cable and it is jacketed by a layer of polyethylene. Outside
this layer are two layers of steel-wire strengthening members
separated by another layer of polyethylene. Finally, an outside polyethylene jacket encases the entire assembly to form the
finished cable that is both durable and environmentally inert.
The cable is rated at ~ 400 MHz-km and 2-3 dB/km. The ribbon-
ized format makes this cable ideally suited for the linear array
arrangement required for the streak camera input.

⊓IGH SPEED MULTICHANNEL DATA RECORDING SYSTEM

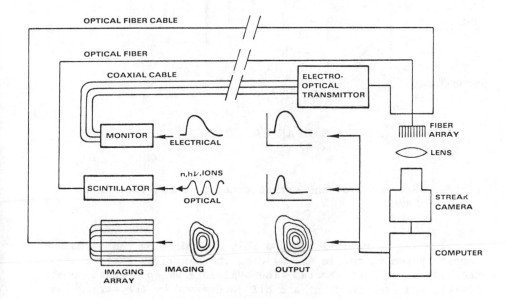

OPTICAL FIBER CABLE

OPTICAL FIBER

COAXIAL CABLE

ELECTRO-OPTICAL TRANSMITTOR

MONITOR

ELECTRICAL

FIBER ARRAY

LENS

SCINTILLATOR

n,hν,IONS

OPTICAL

STREAK CAMERA

IMAGING ARRAY

IMAGING

OUTPUT

COMPUTER

Fig. 1.   Schematic of the multi-channel data recorder shown with a number of possible applications.

The electrooptical streak camera (Hamamatsu C-1370) [4], as shown in Fig. 2, is a device that renders a temporally varying optical analog signal into a spatially varying optical analog signal. To transform the time dimension to the space dimension, the streak camera first converts the optical input signal with a photocathode into an electron signal. A set of accelerating and focusing electrodes then focuses the electrons through a set of electrostatic deflection plates onto a phosphor screen. As the deflection plates are activated by an applied voltage ramp, the electrons are swept down the phosphor screen. In this manner, the time-dependent input amplitude is displaced on the phosphor screen as a spatially varying output light signal. To record these light streaks on the phosphor screen in digital form, a SIT Vidicon TV camera with 256 x 256 x 8-bit target array is used. The digital output from the Vidicon is output to an LSI 11/23-based computer system for data processing.

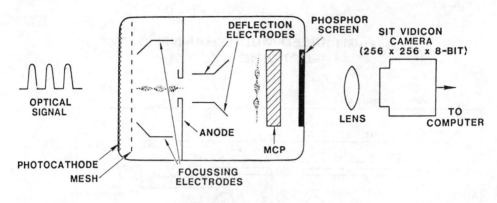

STREAK CAMERA

Fig. 2.   Schematic of the streak camera and SIT Vidicon TV
Camera.

The system capability is highly dependent on the streak
camera system which, in the case of the Hamamatsu unit, contains
the Vidicon readout camera. The available sweep speeds, sensitivity, and dynamic range are all determined by the streak camera.
With the particular model used, sweep rates of 10, 20, 50, 100,
200, 500 ns are available. The streak camera has a manufacturer-
specified sensitivity of at least $6 \times 10^{-15} J/mm^2$ at the photocathode and dynamic range of $\geq 50 + 1$ [5]. Since the streak
camera has a time resolution of ~ 30 ps, potentially an analog
bandwidth ≥ 10 GHz is available. However, in practice the operational band width is limited by the optical fiber, or the transmitter interface, if the initial signal is electrical. The band
width limitation associated with the fiber is ~400 MHz-km.
Therefore, for laboratory applications where transmitter interface is necessary, the transmitter band width becomes the limiting factor. State-of-the-art transmitters today are in the
200-MHz to 1-GHz range.

3.   SYSTEM TEST RESULTS

The total prototype system occupies two, 5-foot (1.53-m)
tall instrument racks and a portion of a 4' x 8' (1.22 m x
2.44 m) optical table. A 30-m fiber cable is used in this
system.

To determine the channel capacity of this system, a simple test was carried out to determine the number of pixel columns one channel of streak occupies. Once this quantity is determined, it is straightforward to estimate the number of channels 256 pixel columns can accommodate.

A light pulse train is inputed into a single fiber of the linear input array. The pulse train consists of a series of 250-ps wide light pulses separated by 4 ns. Such a pulse train is shown in Fig. 3. The light intensity is adjusted to just single input channel is made to occupy the maximum number of pixel columns. A profile is taken across a light pulse near mid screen and is plotted in Fig. 4. The maximum number of pixel columns occupied by a single channel is 5. Based on this measurement, approximately 50 channels could be inputed into a single streak camera. With this input density, the per channel cost is only 1/5 to 1/10 of the cost of conventional oscilloscopes.

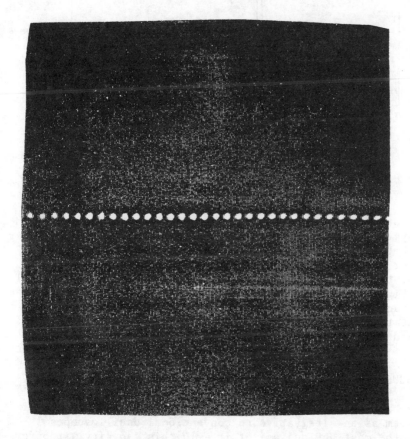

Fig. 3.   Single pulse train, 250-ps pulser, separated by 4 ns.

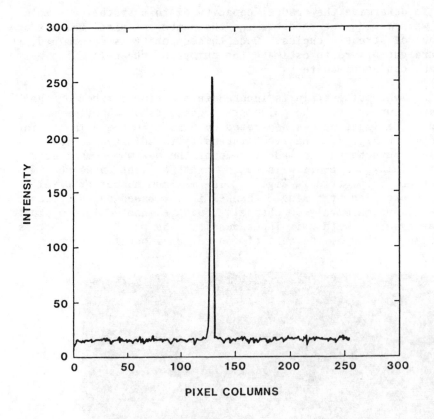

Fig. 4.   Profile across light pulse near mid-screen.

The other system characteristic that needs to be considered
is the system dynamic range. By inspection, the center-to-center
spacing between channels cannot be any closer than 5 pixel
columns or the cross talk between channels would be significant.
As it is, the amplitude at 5 pixel columns away from the channel
center (Fig. 5) is near 2% of the saturation level. From this, a
dynamic range of 50:1 is derived. Based on these test results,
we expect to be able to input 50 channels of separate data per
system and to achieve a dynamic range of up to 50:1.

4.   DISCUSSION AND CONCLUSION

The high-speed multi-channel data recording application of
this system as an alternative to conventional oscilloscopes is
probably the most obvious use of it. In this capacity, it offers
economy, large numbers of channels, and compactness.

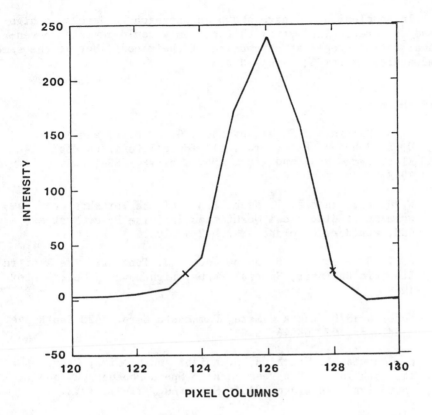

Fig. 5. Profile (expanded).

The system can be placed in a room measuring 18 m^3. and can
be used in two modes. It can be integrated into existing data
acquisition systems strictly as a replacement for oscilloscopes.
In this mode, each channel has a transmitter, permitting interfacing with existing coaxial cables and electric analog monitors.
In other words, it can be retrofitted into existing data acquisition systems to provide upgrade. The disadvantage of this mode
of operation is the limited band width imposed by the transmitter.

The other mode of operation is to use optical analog monitors and optical fibers directly to acquire data. The optical
analog monitors are an interesting topic in themselves and suf-
fice is to say that various electrooptic and magnetooptic effects
in optical components, as well as in optical fibers, could be
developed as monitors for voltage and current. In this mode of
operation, the ultimate performance and bandwidth could be
achieved with this system.

In conclusion, we have shown an approach to develop a high-speed data recording system that not only could replace conventional oscilloscopes at a fraction of their cost, but at the same time achieve multi-GHz bandwidth.

REFERENCES

1. T. H. Martin, J. P. VanDevender, G. W. Barr, and
    D. L. Johnson, Proc. 3rd Intl. Topical Conf. on High Power
    Elect. Beam Res. and Tech., Novosibirsk, USSR, July 3-6,
    1979.

2. W. B. Boyer and E. L. Neau, Proc. of the Workshop on Meas-
    urement of Electrical Quantities in Pulse Power Systems,
    NBS, Boulder, Colorado, March 2-4, 1981.

3. F. T. Dozelsky, R. B. Sprow and F. J. Topolski, The Western
    Electric Engineer, Special Issue, Lightwave, p. 81, Winter
    1989.

4. Model C-1370 Streak Camera, Hamamastu Corp., 420 South Ave.,
    Middlesex, NJ   08846, U.S.A.

5. E. Inuzuka, Y. Tsuchiya, M. Koishi and M. Miwa, Proc. of
    the 15th Int's. Congress on High Speed Photography and
    Photonics, San Diego, CA, U.S.A., Aug. 21-27, 1982.

AN ITERATIVE DECONVOLUTION ALGORITHM FOR THE RECONSTRUCTION OF HIGH-VOLTAGE IMPULSES DISTORTED BY THE MEASURING SYSTEM

K. Schon

Physikalisch-Technische Bundesanstalt,
Braunschweig
FRG

1.   INTRODUCTION

In high voltage (HV) techniques, the measurement of fast, HV impulses is usually done by means of resistive or damped capacitive voltage dividers. Of course, the low-voltage output signal of the divider should be an accurate portion of the HV impulse. But due to the large dimensions of HV dividers and test circuits, especially for voltages in the megavolt range, the transfer characteristic of the divider is influenced by stray capacitances and inductances. On the low-voltage side of the divider, only a distorted signal can therefore be measured, and errors in the evaluation of the peak value, the front time, the voltage-time area, or other significant values of the HV impulse may occur.

In this paper, a numerical method for the reconstruction of HV impulses which are distorted by the transfer characteristic of the divider is described. Other correction methods are given in Refs [1, 2, 3].

2.   NUMERICAL SOLUTION OF THE CONVOLUTION INTEGRAL

Considering the divider to be a linear system with step response $g(t)$, its output signal $y(t)$ can be calculated for any input signal $x(t)$ using the convolution (Duhamel's) integral:

$$y(t) = \int_{o}^{t} x(\tau)\, g'(t-\tau) \tag{1}$$

where

$$g'(t-\tau) = \frac{dg(t-\tau)}{dt} \tag{2}$$

and $g(0) = 0$.

It is generally impossible to solve Eq. (1) analytically for $x(t)$, which in our case is the HV impulse to be measured. A numerical solution of the integral is obtained by the application of the trapezoidal rule [4]:

$$y_k = \frac{\Delta t}{2} \left[x_o g'_k + x_k g'_o + 2 \sum_{i=1}^{k-1} x_i g'_{k-i} \right] \tag{3}$$

for $k = 1, 2, 3, \ldots N$,

where Δt is the sample interval length,

N is the total number of sample intervals, and

x_k, y_k, g'_k are the discrete values at times $t = k\Delta t$.

In most cases, $x_0 = 0$, i.e., the input signal starts with zero voltage. Eq. (3) can be easily solved for the input signal x_k ($x_0 = 0$):

$$x_k = \frac{2}{g'_0} \left[\frac{y_k}{\Delta t} - \sum_{i=1}^{k-1} x_i g'_{k-i} \right] \tag{4}$$

But Eq. (4) was found to be generally not applicable [4]. The calculation of x_k strongly depends on the knowledge of the correct value g'_0 and on the absense of disturbances which, in practice, may result from measuring errors and electromagnetic interference. An example of the application of Eq. (4) is given in Fig. 1a for an ideal first-order system. From $y(t)$ and $g(t)$, the input signal values x_k are calculated for $\Delta t = 5$ ns and $N = 200$, using Eq. (4). By comparing x_k with the true input signal which is a triangular impulse, the relative error of the numerical reconstruction was found to be less than 1% in the peak region (Fig. 1a). However, further investigations on systems of higher order showed that much higher errors occured, which could not be tolerated.

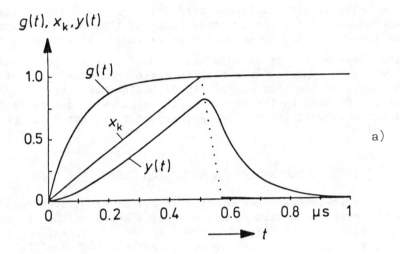

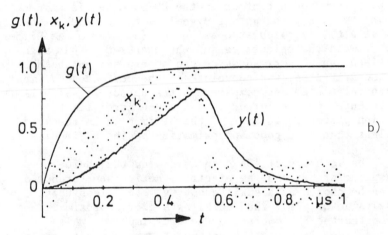

Fig. 1.   Reconstruction of the input signal of an ideal first
          order system by application of the direct deconvolution
          algorithm
          g(t) - step response
          y(t) - output signal
          x_k - reconstructed values of the input signal
          a) exact values of g(t) and y(t)
          b) y(t) superposed by small disturbances

     In order to reduce the errors, the sampling time Δt was
decreased and N increased, but then the calculation tended to
produce instabilities and oscillations [5].   In Fig. 1b, the influence of random disturbances of less than 0.5% relative to the

peak value of y(t) is illustrated. Though the disturbances
superposed on y(t) can scarcely be seen, they exert a tremendous
influence on the deconvolved input signal: the values x_k are
widly scattered. In this rather simple example, the reconstruction may be improved if smoothing techniques were used but higher
harmonics of the true signal itself would then also be eliminated. Summing up the results, Eq. (4) as well as any other
direct deconvolution method is regarded to be not applicable in
practice.

3.   ITERATIVE DECONVOLUTION ALGORITHM (IDA)

The iterative deconvolution algorithm (IDA) overcomes most
of the problems mentioned above. It is based on the iterative
application of Eq. (3), and proved to be very useful and accurate, and its application posed no problems [4]. In the first
step, a very rough contour of the HV impulses is assumed and its
convolute, i.e., the output signal of the divider with step response g(t), is calculated using Eq. (3). From the difference,
δy, between the calculated and the measured output signal, corrections can be derived which are superposed on the HV impulse.
In the next step, the convolute of the corrected HV impulse is
again calculated and compared to the output signal. If its
difference, δy, is larger than a given error limit, ε, the correction procedure to improve the HV impulse is repeated. If δy
is less than ε, a good approximation of the HV impulse is found.

The accuracy of IDA depends now on the choice of ε. If the
measuring errors and the electromagnetic interference are small,
ε can be chosen as a small quantity, e.g., less than 1% of the
peak value of y(t), thus enabling a good approximation of the
reconstructed HV impulse to the true impulse, which is not known.
In practice, fast digital recorders may be conveniently used for
the measurement of the output signal and the step response. As
the digitizer errors are randomly distributed [6], they can be
reduced by carefully applying smoothing techniques to the recorded signals.

The accuracy of IDA was tested for ideal systems. Since
g(t) and y(t) were analytically known, ε could be chosen very
small, and the input signals were calculated with an error of
less than 1%.

In Fig. 2, an application example is given for a system
which consisted of low-voltage RCL elements and was screened.
The step response is shown in Fig. 2a. The signals were recorded by an 8-bit, 50-MHz digitizer. The reconstructed input

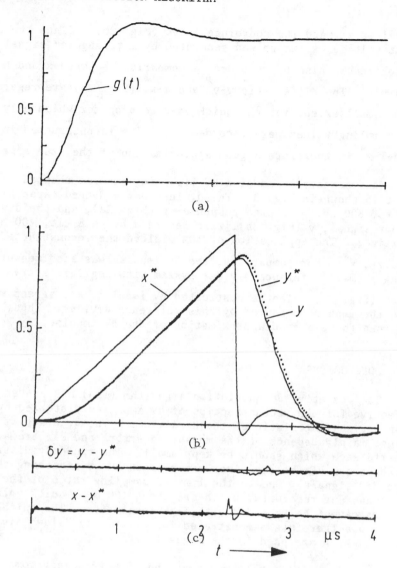

Fig. 2. Reconstruction of a low-voltage impulse by application
         of the iterative deconvolution algorithm (IDA)
         a) step response g(t) of the RLC circuit
         b) y(t) - output signal
              x^* - reconstructed input signal
              y^* - convolute of x^*
         c) differences between the measured and calculated out-
            put and input signals.

signal is x*, and its convolute is y* (Fig. 2b). Since the true
input voltage, x, which was generated by a triangular signal generator, could also be recorded, a comparison between x and x* was
possible. The differences y-y* and x-x* (Fig. 2c) are regarded
to be small except for the quick voltage drop for which only a
few sampling values were recorded [7]. The reconstructed input
signal x* is therefore a good approximation of the true signal x.

The reconstruction of a HV lightning impulse chopped in the
front is shown in Fig. 3. The divider was a damped capacitive
one for 800 kV. The step response, g (Fig. 3a), and the divider
output signal, y (Fig. 3b), were recorded by an 8-bit, 100-MHz
digitizer. The application of IDA yielded the reconstructed HV
impulse, x*. For comparison, the HV impulse was also measured
using a small reference divider system with negligible transfer
error (Fig. 3d). The reconstructed HV impulse, x*, agrees well
with the impulse measured by this reference divider. Thus, x* is
regarded to be a good approximation of the HV impulse.

4.   CONCLUSIONS

The iterative deconvolution algorithm (IDA) has proved to be
effective for the reconstruction of HV impulses distorted by the
transfer characteristic of the HV divider. The accuracy of this
method mainly depends on the measuring errors and electromagnetic
interference which should be kept small. The randomly distributed errors of the digitizer can be reduced by smoothing the
recorded signals. Due to the limited sampling rates of the fastest transient recorders, which are about 100 MHz, quick voltage
changes cannot be recorded with high resolution. The suitability
of IDA was therefore demonstrated for relatively "slow" divider
systems which are used in the megavolt range.

Sometimes it is believed that the high-frequency components
of a signal, which are lost by the imperfect measuring system,
could be regained by calculations using a suitable deconvolution
method. This would enable the use of cheap HV dividers with poor
transfer characteristics. Theoretically this is true only for
ideal systems and undisturbed signals. But in practice, small
high-frequency components will be covered by the disturbances,
and a real improvement of the measurement by calculations cannot
be expected.

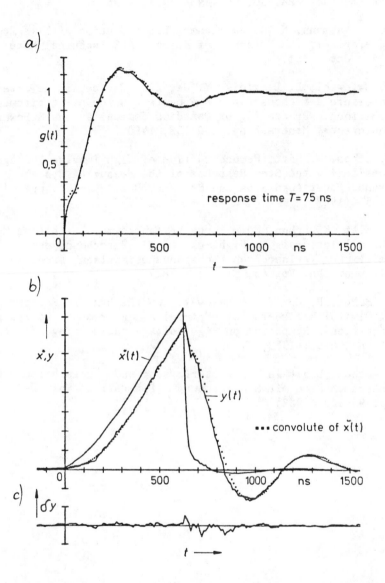

Fig. 3. Reconstruction of a high-voltage impulse by application
of the iterative deconvolution algorithm (IDA)

REFERENCES

1. S. E. Kiersztyn, Numerical Correction of HV impulse deformed
    by the Measuring System. IEEE Trans. on Power Apparatus and
    Systems, Vol. PAS-99, pp. 1984-1991, (1980).

2. N. S. Nahmann, M. E. Guillaume, Deconvolution of Time Domain
    Waveforms in the Presence of Noise. NBS Technical Note
    1047, Oct. 1981.

3. R. Malewski; B. Langlois, Y. Gervais, Correction de erreurs
    de mesure des chocs rapides a`haute tension par traitement
    numerique. Proceeding of Canadian Communications & Power
    Conference, Montreal pp. 150-153, 1980.

4. K. Schon, W. Gitt, Reconstruction of High Impulse Voltages
    Considering the Step Response of the Measuring System. IEEE
    Trans. Power Apparatus and Systems, Vol. PAS-101, pp. 4147-
    4155, (1982).

5. W. Gitt, K. Schon, Numerische Verfahren zur Entfaltung des
    Duhamel-Integrals im Hinblick auf die Berechnung des
    zeitlichen Verlaufs von Hochspannungsimpulsen, Arch, f.
    Elektrotechn. Vo. 63, p. 317 (1981).

6. K. Schon, H. Korff, R. Malewski, On the Dynamic Performance
    of Digital Recorders for HV Impulse Measurement, Fourth In-
    ternational Symposium on High Voltage engineering, No. 65.06
    Athens, Sept. 1983.

7. K. Schon, Digitale Messwerterfassung und - verarbeitung bei
    Untersuchungen mit Stosspannung. PTB-Bericht Vol. E-21
    pp. 99-109 (1982).

TEST METHODS FOR THE DYNAMIC PERFORMANCE OF FAST DIGITAL RECORDERS

K. Schon

Physikalisch-Technische Bundesanstalt,
Braunschweig
FRG

1. INTRODUCTION

Digital waveform recorders have been in use for measurements of harmonic and transient signals in all areas of science and technology for at least a decade. The most convenient feature of these instruments is their ability to store the quantized data in a digital memory which enables computerized data to be acquired. In the last five years, a variety of types have been developed by various manufacturers, but up to now, there are no generally accepted specifications or standards on how to test the accuracy of ultra-fast digitizers with sampling frequencies of about 100 MHz. Several test methods have been proposed by the manufacturers, but these differ widely from each other, and even for the same test method, the results depend on numerous test parameters. It is often, therefore, impossible to compare the test results for digitizers from different manufacturers.

This paper deals with the dynamic errors of fast digitizers, and the various test methods known so far are discussed. It should be noted that the errors originating in the analog devices of the digitizers are not usually covered by these test methods.

2. IDEAL QUANTIZATION

The relative maximum quantization error of an ideal N-bit digitizer for a full scale signal is given by:

$$e_q = \pm\ 0.5\ LSB = \pm\ 2^{-(N+1)} \tag{1}$$

where LSB is the least significant bit. For an 8-bit ideal
digitizer, $e_q = \pm\ 0.2\%$. When evaluating the peak value of a fast
signal we must take into consideration an additional sampling
error, e_s, in the case where the sampling is not exactly at the
maximum [1]. Within the limits given by Eq. (1), the individual
error values, δ_{ik}, of an ideal digitizer at the sampling times
$k\Delta t$ are uniformly distributed. It can be shown that the standard
deviation of this uniform frequency distribution is:

$$\sigma_i = \sqrt{\frac{1}{m-1} \sum_{k=1}^{m} \delta_k^2} = \frac{1}{3}\ \delta_{i_{max}} \tag{2}$$

$$= \frac{1}{\sqrt{12}}\ LSB = 0.29\ LSB$$

This relationship is valid for every signal frequency.

3.   TEST METHODS FOR REAL DIGITIZERS

     Real digitizers behave much less well than ideal ones when
recording fast signals. It is well known that their dynamic
accuracy decreases more or less dramatically with higher frequencies. Most of the test methods proposed are well known in
the field of analog measuring techniques. A reasonable way to
test fast digitizers is to apply a well-defined test signal to
the input terminals and to evaluate the digital output signal
[2, 3, 4]. It should, however, be noted that the output signal
exists only at the discrete sampling times. Interpolation may
be possible if the sampling frequency is much higher than the
significant signal frequency, thus excluding the aliasing effect.
A critical overview of the different test methods is given in the
following sections.

3.1 Fourier Analysis

     Figure 1 shows the result of the Fourier analysis for an
8-bit, 100-MHz digitizer. An ideal digitizer would show only a
single line in the spectrum for every sinusoidal test signal and
a negligible wideband noise due to the small quantization noise.
The real, i.e., imperfect, digitizer produces an increasing number of harmonics, subharmonics, and quantization noise for higher

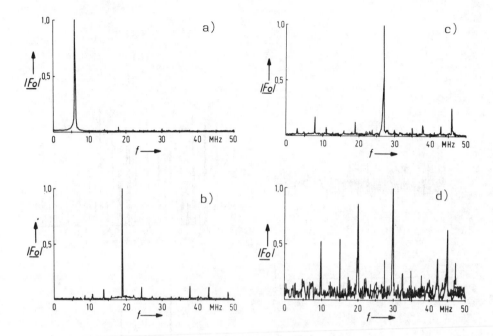

Fig. 1. Spectrum $|F_o|$ of the digitizer output data for sinusoidal
input voltages of frequency f_o.

    a) f_o = 6 MHz   b) f_o = 19 MHz

    c) f_o = 27 MHz   d) f_o = 45 MHz

frequencies of the sinusoidal test signal. In Fig. 1d, the
amplitude of the fundamental oscillation, f_o = 45 MHz, is even
less than that of the subharmonics which are the alias components
of the higher harmonics, nf_o [3].

The question arises of how to summarize the test results
given in the numerous spectra. As regards analog amplifiers, it
would be possible to define and calculate a distortion factor
dependent on the frequency. However, this is not adequate for
characterizing the dynamic performance of digitizers with a mode
of operation quite different from that of analog instruments.

3.2 Probability Density Function

This test method is illustrated in Fig. 2 for the same recorder (8-bit, 100-MHz) as above, applying free-running, full-
scale triangular signals. The frequency-of-code-occurrence in

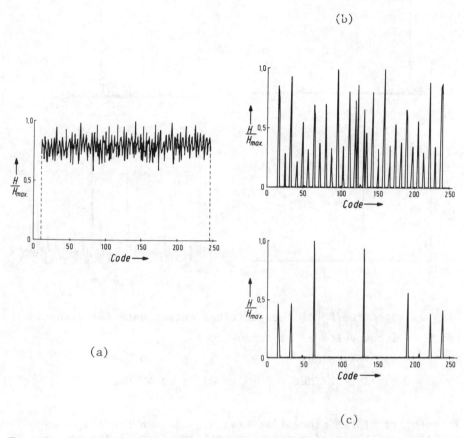

Fig. 2. Probability density function H/H_{max} of the digitizer
output data for free-running triangular input voltages
of slope s.

a) $s = 0.1 \ \mu s^{-1}$ b) $s = 6 \ \mu s^{-1}$ c) $s = 22 \ \mu s^{-1}$

the digital record is plotted versus the code number from 10 to
245. For an ideal digitizer, the probability density function
of free-running triangular voltages is constant. For real digitizers, this is approximately true only of signals with low
slopes (Fig. 2a). For higher slopes, it can be clearly seen in
Figs. 2b and 2c that more and more codes fail to appear. This
indicates a severe failure of the digitizer resulting in a considerable loss of amplitude resolution. For $s = 22 \ \mu s^{-1}$, only 7
codes seem to be present, which is the resolution of a 4-bit

digitizer. The information given in these diagrams may be summarized in a diagram "remaining-bit-number-versus-ramp-slope".
The test can also be made with sinusoidal test signals, but a
more complex probability density function must then be considered [3, 4].

3.3 Curve-Fitting Using Sinusoidal Test Voltage

The principle of the curve-fitting test for sinusoidal test
signals is described in more detail in Refs. [1] and [5]. This
test gives global information on the digitizer's dynamic performance. The output data of the recorder for the sinusoidal input
voltages are best-fitted by an ideal sinusoidal signal with regard to frequency, amplitude, offset, and phase (Fig. 3a). The
differences, δ_k, i.e., the digitizer errors at the discrete sampling times, $k\Delta t$, are calculated in Fig. 3b. Since the differences, δ_k, are discrete values, their standard deviation in LSB
can be taken to characterize the real digitizer [1, 2]:

$$\sigma_r = \sqrt{\frac{1}{m-1} \sum_{k=1}^{m} \delta_k^2} \qquad (3)$$

Whereas the errors, δ_k, of an ideal digitizer are uniformly distributed, the error histogram of the real digitizers tested by
us can be approximately described by a normal distribution
(Fig. 3c).

For a 6-bit and three 8-bit digitizers, Fig. 4 (curve a)
shows σ_r versus the signal frequency, f, for full-scale sinusoidal test voltages. For the 6-bit digitizer (curve c), σ_r is
given in the equivalent LSB of an 8-bit digitizer. At low signal
frequencies, σ_r of the curves a, c, and d is approximately equal
to σ_i of the corresponding ideal digitizers, but curve b is not
better than an ideal 6-bit digitizer. Whereas σ_i of the ideal
digitizer is constant, σ_r of the real digitizers increases more
or less with frequency, especially that of curve a. The largest
errors of most digitizers generally occur at the steepest part of
the sinusoidal test voltage, at zero crossing. The increase in
the standard deviation with sine frequency indicates a loss of
the digitizer's nominal N-bit resolution. In Refs. [1, 2] the
remaining effective bits were defined as:

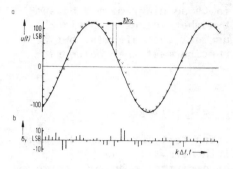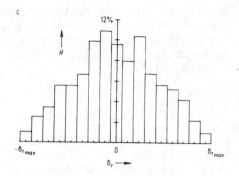

Fig. 3.   Curve-fitting test with full-scale sinusoidal test
          voltages.
          a) Output data of the digital recorder (crosses) and
              best-fit sinewave
          b) Difference δ_{rk} between the sampling values and the
              best-fit sine wave at the sampling times $k\Delta t$
          c) Error histogram of δ_{rk}

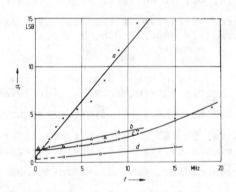

Fig. 4.   Standard deviation, σ_{rk}, of the digitizer errors, δ_{rk},
          versus frequency, f, of full-scale sinusoidal test
          signals
          Curve a - Transient recorder (8 bits, 10 ns)
          Curve b - Digital oscilloscope (8 bits, 20 ns)
          Curve c - Transient recorder (b bit, 2 ns)
                    σ_r is given in LSB of an 8-bit digitizer
          Curve d - Programmable digitizer (8 bit, 5 ns)

$$EB(f) = N - \log_2\left[\frac{\sigma_r(f)}{\sigma_i}\right], \tag{4}$$

with σ_i given by Eq. (2) and σ_r evaluated by Eq. (3). In Ref. [4], the definition of the effective bits is related to the rms value of the errors instead of the standard deviation, but it so happens that both definitions yield the same EB values. The advantage of the definition given in Eq. (4) is its profound mathematical background related to the standard deviation.

The graphs "effective-bits-versus-the-signal frequency" can be seen in Fig. 5 for the four digitizer models. Due to the different test parameters, e.g., the signal amplitude, they may differ from those given by the manufacturers. Examples are given in Fig. 6 of full and half-scale sine signals. The higher EB values for a half-scale sine signal can be explained by the smaller digitizer errors due to the reduced slope of the sinusoid at zero crossing. On the other hand, the resolution of the digitizer is only half the value compared with that of a full-scale signal, a fact which is not considered in this diagram and in Eq. (4). Furthermore, σ_r and EB depend on the sample numbers taken for the curve-fitting program.

The test result for a 9-bit waveform recorder is summarized in Fig. 7. This recorder is different from the other digitizers in that it consists of an analog oscilloscope system where the signal is stored on the screen and a digital read-out device. The EB characteristic therefore depends on the setting of the oscilloscope time base, which is 4 ns/division and 2 ns/division

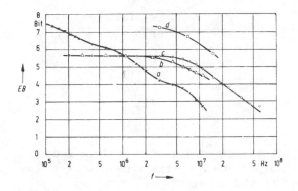

Fig. 5. Effective bits, EB, versus frequency, f, of full-scale test sinusoidal signals a, b, c, d (see Fig. 4).

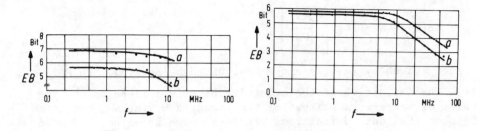

Fig. 6. Effective bits, EB, versus, f, for two digitizer models
         a) 8-bit, 50-MHz digitizer
             Curve a - half-scale input signal
             Curve b - full-scale input signal
         b) 6-bit, 500-MHz digitizer
             a, b (as in 6a)

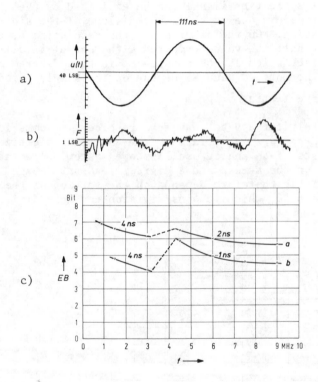

Fig. 7. Curve-fitting test of a 9-bit waveform recorder
         a) Data output and best-fit sine signal
         b) Error of the recorder
         c) Effective bits, EB, versus signal frequency, f
             Curve a - half-scale input signal
             Curve b - full-scale input signal

for the higher frequencies (Fig. 7c). The great advantage of this instrument is its high bandwidth, making possible the digital recording of signals with frequency components in the 100-MHz range. The EB value is then presumably as high as that for the lower frequencies in Fig. 7c.

3.4 Curve-Fitting Using Ramp Voltages

The σ_r or EB characteristics for sinusoidal test voltages being known, a rough classification of the various digitizers on the market may be possible. For more detailed information on the dynamic performance of digitizers, the use of ramp or triangular test voltages with constant slope over the full input range was proposed in Refs. [1, 2]. The test procedure is the same as that for sinusoidal test voltages described above. Figure 8a shows the digitizer output data and the best-fit ramp with slope s, Fig. 8b the instantaneous differences, δ_{rk}, and Fig. 8c the error histogram, which is approximately of the Gaussian type. In Fig. 9, σ_r-versus-slope is given for the two 8-bit digitizer models a and b. For low slopes, model a operates almost as an ideal digitizer with $\sigma_i = 0.29$ LSB. This model is therefore qualified for the measurement of peak values but not for quick voltage changes. Model b exhibits rather large digitizer errors for low slopes which originate from internal noise effects. Even if the input terminals are short-circuited, the output is noisy between ±3 LSB. For slopes higher than s = 0.6 μs^{-1}, model b is better than model a. From σ_r, the effective-bit characteristic is calculated (Fig. 10).

A knowledge of σ_r-versus-slope makes possible the estimation of the probable maximum digitizer error, δ_{rmax}, for any input signal [1]. For any slope of the signal, the corresponding σ_r value can therefore be taken from Fig. 9 and multiplied by the factor 3 to obtain δ_{rmax} within the confidential level of 99.7%. This information on the probable error may be of great advantage and cannot be obtained by the other test methods. Of course, the individual instantaneous digitizer error cannot be predicted by any method.

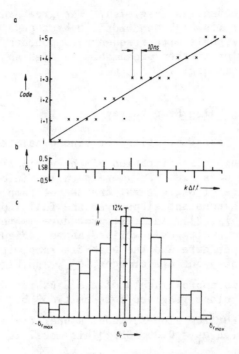

Fig. 8.   Curve-fitting test with full-scale ramp test voltages
          a)   Output data (crosses) and best-fit ramp
          b)   Instantaneous digitizer error at $k\Delta t$
          c)   Error histogram

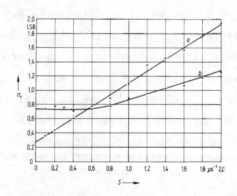

Fig. 9.   Standard deviation, σ_r, versus slope, s, of full-scale
          ramp test voltages
          (a, b see Fig. 4)

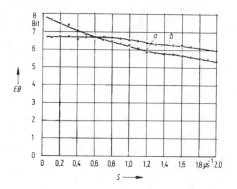

Fig. 10.   Effective bit, EB, versus slopes, s, of full-scale ramp
           test voltages
           (a, b see Fig. 4)

3.5 Aperture Uncertainty

     This test method states the time inaccuracy ("time jitter")
of an A/D conversion [6]. The aperture uncertainty produces an
amplitude error which increases at higher slew rates, du/dt, of
the signal (Fig. 11). For two hypothetical 8-bit, 100-MHz,
digital recorders with aperture uncertainties of 2 ns and 3 ns,
respectively, the digitizer errors were calculated for sine voltages and expressed by the remaining effective bit number, EB,
versus frequency (Fig. 12, curves b and c). For comparison,
curve a in Fig. 12 represents the graph "EB-versus-f" of the
digitizer model a, shown in Fig. 5. Obviously, the real digitizer error cannot be described by the aperture uncertainty.

3.6 Comparison with a Standard Digitizer

     A comparison between the digitizer to be tested and a stan-
dard digitizer with negligible or well-known errors would be very
convenient and effective because any reproducbile test signals -
continuous ones or impulses - could be used. This may be important because the different performances of digital recorders for
continuous and discontinuous sine test voltages have been reported in [7]. A 10-bit standard digitizer of at least 200 MHz
would be required for comparison, but such a digitizer has not
yet been constructed.

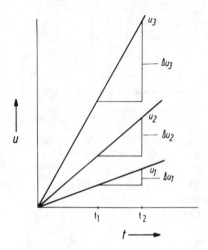

Fig. 11.   Aperture uncertainty, t_2-t_1, and signal error, Δu, for
           different slew rates, du/dt.

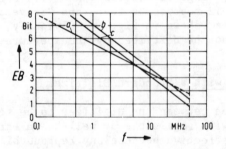

Fig. 12.   Effect of the aperture uncertainty on the calculation
           of EB-versus-f of an 8-bit, 100-MHz digitizer.
           Curve a - Measured
           Curve b - Calculated for an aperture uncertainty of
                     2 ns
           Curve c - as b for 3 ns

4.   CONCLUSIONS

The digitizers on the market reveal a loss of resolution and accuracy when recording fast signals. Considering all the pros and cons, the test method described in Section 3.4, using ramp or triangular test voltages, is regarded as the most informative one. A knowledge of the standard deviation, σ_r, versus slope of the ramp test signal (Fig. 9) enables the probable maximum digitizer error, $3 \cdot \sigma_r$, for any other input signal, to be estimated, since all digitizers tested showed a random error distribution. It is, however, not possible to predict the true instantaneous error with this or any other method. The only objection to this test method may be the difficulty in generating the fast ramp or triangular voltages as precisely as needed.

From the mathematical point of view there is no need to use sinusoidal signals as is usually done for analog systems whose transfer characteristic can be calculated for any signal from Fourier analysis and the convolution integral. However, this does not apply to digital systems. The graphs "EB-versus-f" given in Fig. 5 only enable a rough comparison to be made of the digitizers for sinusoidal voltages, but a prediction of the errors when other signals are recorded is not possible.

REFERENCES

1. K. Schon, H. Korff, R. Malewski, "On the dynamic performance of digital recorders for HV impulse measurements". Fourth Int. Symp. on High-Voltage Eng.

2. K. Schon, H. Korff, "Test methods for digitizers." Working document (Kind 2) presented at the meeting of CIGREWG 33.3 in Cairo, Sept. 1983.

3. B. E. Peetz, A. S. Muto, J. M. Neil, "Measuring waveform recorder performance". HP-Journal Vol. 33 No. 11, pp 21-29, (1982).

4. P. Wiesendanger, "Automatische, digitale Aufzeichnung und Auswertung von transienten Signalen in der Hochspannungstechnik." Doctoral Thesis ETH Zurich, 1977.

5. N. A. Robin, B. Ramirez, Capture fast waveforms accurately with a 2-channel programmable digitizer. Electronic Design, Vol. 28, No. 3, (1980).

6. Application Note from Le Croy, "Aperture Uncertainty", Nov. 1980.

7. T. M. Souders, D. R. Fach, "Measurement of the transient versus stead-state response of waveform recorders". Proceedings of the Waveform Recorder Seminar, NBS Special Publication 634, June 1982.

GROUNDING AND SHIELDING

ELECTROMAGNETIC TOPOLOGY FOR THE ANALYSIS AND DESIGN OF COMPLEX ELECTROMAGNETIC SYSTEMS

Carl E. Baum

U.S. Air Force Weapons Laboratory
Kirtland Air Force Base
Albuquerque, New Mexico
USA

1.   INTRODUCTION

The emerging fundamental basis for analysis and design of complex electromagnetic systems, for protection against undesirable propagation of electromagnetic signals through the system, is known as electromagnetic topology. While this may be viewed as the basis for grounding and shielding, it is more general than that.

There have been many papers written on electromagnetic topology and the reader may acquaint himself with the variety of topics by perusing a basic bibliography presented in Table 1.1. For convenience this bibliography is divided into two parts. Part I covers what may be called qualitative EM topology. Here the emphasis is on the descriptive aspects of EM topology, including the topological diagrams, dual graphs (interaction sequence diagrams), and system design options and rules that can be developed. Part II covers what may be called quantitative EM topology. Here the development begins from transmission line network theory and goes on to the BLT equation and various kinds of bounds that can be developed for system response using norm concepts in conjunction with interaction or scattering matrices and operators. Of course these two aspects are not mutually exclusive and some papers consider both aspects. However, each is assigned to one of these categories for convenience.

Many of these papers have been published in the open literature, and so it is not necessary to reproduce that material here. However, there is one very fundamental paper which is not so readily available. This paper, "On the Analysis of General

Multiconductor Transmission-Line Networks" is included in the
bibliography as [8]. It is the beginning of the quantitative
type of EM topology and states the BLT equation which is exploited in subsequent papers in a more general EM context. It is
felt that this paper is of sufficient importance that it is
published herewith to make it available to the general scientific
and engineering community.

Table 1.1   Basic Bibliography - EM Topology
            Part 1:   Qualitative EM Topology

1. Baum, C. E., "How to Think About EMP Interaction", Proc.
    1974 Spring FULMEN Meeting (FULMEN 2), pp. 12-23,
    April 1974.
2. Baum, C. E., "The Role of Scattering Theory in Electromag-
    netic Interference Problems", in P.L.E. Uslenghi (ed.),
    Electromagnetic Scattering, Academic Press, 1978.
3. E. F. Vance, "On Electromagnetic Interference Control", In-
    teraction Note 380, October 1979, and IEEE Trans. Electro-
    magnetic Compatibility, Nov. 1980, pp. 319-328.
4. C. E. Baum, "Multiple Electromagnetic Topologies for De-
    scribing Various Aspects of Electronic Systems and Various
    System States", Proc. 1980 Spring FULMEN Meeting (FULMEN 4),
    March 1980, p. 33.
5. K. S. H. Lee (ed.), "EMP Interaction: Principles, Tech-
    niques, and Reference Data", EMP Interaction 2-1, December
    1980, AFWL-TR-80-402 (available from NTIS).
6. C. E. Baum, "Sublayer Sets and Relative Shielding Order in
    Electromangetic Topology", Interaction Note 416, April 1982,
    and Electromagnetic, Oct-Dec 1982, pp. 335-354.
7. C. E. Baum, "Topological Considerations for Low-Frequency
    Grounding and Shielding", Interaction Note 417, June 1982,
    and Electromagnetics, April-June 1983, pp. 145-157.

            Part 2:   Quantitative EM Topology (including transmis-
                      sion-line network theory, BLT equation, and
                      norms)

8. C. E. Baum, T. K. Liu, and F. M. Tesche, "On the Analysis of
    General Multi-conductor Transmission-Line Networks", Inter-
    action Note 350, November 1978.
9. C. E. Baum, "Electromagnetic Topology: A Formal Approach to
    the Analysis and Design of Complex Electronic Systems", In-
    teraction Note 400, September 1980, and Proc. EMC Symposium,
    Zürich, 1981, pp. 209-214.

10. A. K. Agrawal and C. E. Baum, "Bounding of Signal Levels at Terminations of a Multiconductor Transmission-Line Network", Interaction Note 419, April 1983.

11. F. C. Yang and C. E. Baum, "Use of Matrix Norms of Interaction Supermatrix Blocks for Specifying Electromagnetic Performance of Subshields", Interaction Note 427, April 1983.

12. C. E. Baum, "Black Box Bounds", Interaction Note 429, May 1983, and Proc. EMC Symposium, Zurich, 1981.

13. C. E. Baum, "Bounds on Norms of Scattering Matrices", Interaction Note 432, June 1983.

14. C. E. Baum, "Some Bounds Concerning The Response of Linear Systems with a Non-Linear Element", Interaction Note 438, June 1984.

15. C. E. Baum, "On The Use of Electromagnetic Topology for the Decomposition of Scattering Matrices for Complex Physical Structures," Interaction Note to be published.

16. F. M. Tesche, "Topological Concepts for Internal EMP Interaction," IEEE Trans. Antennas and Propagation, Jan 1978, pp. 60-64, and IEEE Trans. EMC, Feb 1978, pp. 60-64.

* * * * * * * * * *

Transmission-line theory has been with us for quite some time [1]. Its impact on communication technology should be obvious. However, its expression in one-dimensional scalar form for a single voltage-current pair has rather limited application to modern complex electronic systems. In the analysis of EMP interaction with electronic systems, transmission-line theory is commonplace [2]; however, its practical use is still in a rudimentary form, usually being applied in a one-dimensional scalar form as in the case of a coaxial cable or some simple approximation of a more complex system in one-dimensional scalar form [3].

Recent investigations have considered multiconductor transmission lines as an extension of transmission-line theory applicable to complex systems problems such as involved in EMP and EMC [4,5]. However, one should recognize that such models are still quite simplistic in the context of the total system response in typical cases. This paper addresses the problem of networks of such multiconductor transmission lines.

Basic to the analysis of transmission-line networks is the network topology based on junctions and tubes, each tube being a representation of a multiconductor transmission line, and tube terminations (including connections to other tubes) occurring at junctions. So first we consider the network topology and the associated interconnection matrices which will be used to construct the network equation. This transmission-line network topology is compared to other kinds, such as those used for lumped-element networks and electromagnetic scatterers.

Next the equations describing a single tube or multiconductor transmission line are considered. The problem is reduced to a first-order matrix differential equation through the introduction of a combined voltage vector which is a special linear combination of the voltage and current vectors. This equation is readily solved for given boundary conditions at the tube ends and source vectors along the tube. The propagation matrix is assumed diagonalizable and the resulting eigenmodes and eigenvalues are used to give representations of the various parameters describing the tube and its response.

The remainder of the paper then integrates the result of a single tube with the scattering matrices of the junctions using the interconnectivity of the transmission-line network topology and its associated wave indexing. This forms the overall equation of the multiconductor transmission-line network, referred to as the BLT equation [6,7]. For this purpose it is useful to introduce the concept of supermatrices, or matrices of matrices, and corresponding supervectors. This separates the indices in a manner which associates different indices with different physical aspects of the multiconductor transmission-line network and its associated topology. In addition, the supermatrices correspond to a symmetric partioning of matrices in a manner which makes them block sparse.

The supermatrices used in this paper can also be referred to as dimatrices corresponding to a single level of partition resulting in two pairs of indices or subscripts to describe the dimatrix elements. This concept has already been generalized to higher order partitions, and hence higher order supermatrices, in applications concerning the topology of complex electromagnetic scatters [8]. So when the reader has waded through the present tome he can cheerfully contemplate that more is to come (or he may need cheering up for the same reason).

2. TOPOLOGY

Network topology is a generic name given to the topological properties of a network. It is studied widely in lumped circuit theory to gain reliable knowledge concerning the number of independent equations in a circuit of arbitrary structural complexity [9]. For electromagnetic problems, a more general type of topology is required to describe the three-dimensional volumes and surfaces that are generally associated with the scatterers. Indeed, the scatterer topology has been introduced by Baum [10, 11] and Tesche [12], and has found useful applications in considering practical shielding and grounding problems [13, 14].

For transmission-line networks, the topological description falls between that of lumped circuits and that of scatterers. Similar to a lumped circuit, there are well-defined material paths along which energy propagates, and there are positions in the transmission-line network where energy is distributed according to Kirchhoff's laws. Unlike a lumped circuit, energy may be coupled between the material paths, and the path characteristics (length, geometry, etc.) alter the ways energy propagates. Energy sources may also be induced along the lengths of the paths. These latter properties which are attributed to the distributed nature of transmission lines are more similar to those of scatters. In fact, it is more appropriate to describe the transmission-line behavior as wave phenomena.

The specialized topological description of a transmission line network has been called the transmission-line topology [11]. In the following, we first review the concept of circuit graphs and circuit topology. The transmission-line topology is then described. A brief outline of other related topologies is also included.

2.1 Graphs

The basic elements of a graph are edges and vertices. They are termed differently in specific types of topology, and are summarized in Table 2.1 under Section 2.6.

A graph is thus made up of a set of vertices which are inter-connected by edges. It is structured to represent, in a simplified form, the electrical connections and/or signal flow paths of the network or system.

Often, for a complicated network, one cannot represent the network comprehensively by a single graph. Parts of the network are then represented by subgraphs. A subgraph fulfills all requirements of being a graph, but is limited to represent only part of the network. An example is that of a multiconductor transmission-line network where a network graph is used to show the transmission-line connections using tube and junction representations, but the detailed electrical connections within junctions are represented by separate subgraphs [14].

On the other hand, a network graph may be a subset to a graph which represents a larger system. The latter is called a supergraph. In the previous example, the transmission-line network graph is a supergraph of that of the junctions and tubes.

The concepts of supergraph, graph, and subgraph define a hierarchical order of representing a system. However, the naming of super- and sub- are only relative when compared to a smaller or larger part of the network.

2.2 Electrical Circuits

The most common network topological concepts have been applied to lumped circuits. The construction of the network graph and its associated development of cut sets and tie sets are assumed to be familiar to the reader. In transmission-line networks, often the junctions contain lumped elements, or the transmission-line discontinuities may be accurately modeled by lumped circuits at low frequencies [15, 16, 17, 18]. Hence, in transmission-line network analysis, it is often necessary to include lumped circuit analysis. We summarize the essential points in the following.

The basic elements of a lumped circuit network graph are branches and nodes. A branch is a component part of a circuit characterized by two terminals to which connections can be made. A node is formed where two or more branches are connected. The graphic symbols for branches and nodes are, respectively, lines and dots. The nodes in a circuit are numbered and the nth node is denoted by N_n. The total number of nodes is N_N. The branch connecting nodes N_n and N_m is labeled $B_{n,m}$. Hence, $B_{n,m}$ and $B_{m,n}$ denote the same branch. The total number of branches is N_B. If between a node pair there is more than one branch, then by parallel combinations one can reduce this to a single branch.

As an example, a circuit graph is shown in Fig. 1a, representing a four-node, five-branch circuit. The branches are numbered by double subscripts using the rule outlined above.

It is useful to introduce three topology matrices which define the topological structure of a graph. They describe node-node (or node interconnection), node-branch and branch-branch (or branch interconnection) connections. These matrices contain somewhat redundant information and usually only one of them is sufficient to specify the associated graph. However, the last type (branch-branch) does not give unique node numbering.

For a circuit with N_N nodes, the node-node interconnection matrix $(C_{n,m})_{N-N}$ is an $N_N \times N_N$ matrix. The elements are defined by:

$n \neq m$	$C_{n,m;N-N} = 1$ if nodes N_n and N_m are connected
	$= 0$ if no connection
$n = m$	$C_{n,n;N-N} = 1$ to denote self connection

$$(1)$$

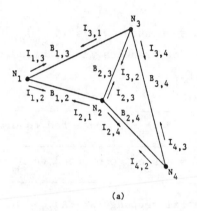

(a)

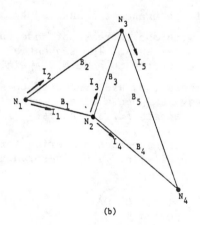

(b)

Fig. 1. A circuit graph with (a) double-subscripted branch
numbers, (b) single-subscripted branch numbers.

Note that the node-node interconnection matrix is symmetric, i.e.,

$$C_{n,m;N-N} = C_{m,n;N-N} \tag{2}$$

For example, in Fig. 1, the node-node matrix is

$$(C_{n,m})_{N-N} = \begin{pmatrix} 1 & 1 & 1 & 0 \\ 1 & 1 & 1 & 1 \\ 1 & 1 & 1 & 1 \\ 0 & 1 & 1 & 1 \end{pmatrix} \tag{3}$$

Similarly, the node-branch matrix $(C_{n,m})_{N-B}$ is $N_N \times N_B$, where N_B is the total number of branches in the circuit. The elements $C_{n,m;N-B}$ are defined by

$C_{n,m;N-B} = 1$ if node N_n is connected to branch B_m
$C_{n,m;N-B} = 0$ if N_n is not connected to B_m

$$\tag{4}$$

This matrix is in general rectangular instead of square. A single-subscripted denotation of the branches is necessary for matrix manipulations. The numbering of the branches starts at node N_1 for the branch going to the node with the lowest node number. The rest of the branches connected to node N_1 are numbered consecutively with the increase of node numbers they are connected to at the other ends. The sequential numbering continues for branches connected to node N_2, again according to the increase of node numbers they are connected to at the other ends. Here, branches already assigned a branch number are not renumbered. This process repeats for all nodes.

The single-subscripted branch numbering for the graph in Fig. 1a is depicted in Fig. 1b. The node-branch matrix is given by

$$(C_{n,m})_{N-B} = \begin{pmatrix} 1 & 1 & 0 & 0 & 0 \\ 1 & 0 & 1 & 1 & 0 \\ 0 & 1 & 1 & 0 & 1 \\ 0 & 0 & 0 & 1 & 1 \end{pmatrix} \tag{5}$$

The branch-branch interconnection matrix $(C_{n,m})_{B-B}$ is an $N_B \times N_B$ matrix describing branch-to-branch connections. The elements are defined by

$n \neq m$	$C_{n,m;B-B}$ = number of connections between the ends of branches B_n and B_m
	= 0 implies no connections
	= 1 implies one end of each branch is connected
$n = m$	$C_{n,m;B-B}$ = 0 excluding branches with both ends connected to the same node

(6)

Note that this matrix is symmetric, i.e.,

$$C_{n,m;B-B} = C_{m,n;B-B} \tag{7}$$

The branch-branch matrix corresponding to the graph in Fig. 1b is

$$(C_{n,m})_{B-B} = \begin{pmatrix} 0 & 1 & 1 & 1 & 0 \\ 1 & 0 & 1 & 0 & 1 \\ 1 & 1 & 0 & 1 & 1 \\ 1 & 0 & 1 & 0 & 1 \\ 0 & 1 & 1 & 1 & 0 \end{pmatrix} \tag{8}$$

Note that the node-node matrix specifies the connections for a given set of nodes. The node-branch matrix specifies the connection for a given set of nodes and branches. Either of these two matrices can regenerate the graph with the same node and branch numbering. This may be much more difficult for the branch-branch matrix. However, all the matrices are derived from the given network (graph) consisting of nodes and branches.

The Kirchhoff's laws can be written down comprehensively. One defines voltages $\tilde{V}_n(s)$ at each node N_n, and currents $\tilde{I}_{n,m}(s)$, leaving N_n to N_m along the branch $B_{n,m}$, with the condition

$$\sum_m \tilde{I}_{n,m} = 0 , \quad \tilde{I}_{n,n} = 0 \tag{9}$$

The voltages and currents are related by

$$\tilde{V}_n - \tilde{V}_m = \tilde{Z}_{n,m} \tilde{I}_{n,m} - \tilde{V}_{n,m}^{(s)} \tag{10}$$

where $\tilde{V}_{n,m}^{(s)}$ is a voltage source along branch $B_{n,m}$ (and increases from n to m), and $\tilde{Z}_{n,m}$ is some impedance (assumed linear) on the same branch. In a degenerate case the branch current might be specified by a current source. For a closed loop, $\tilde{V}_n - \tilde{V}_n \equiv 0$, and (10) becomes

$$\tilde{Z}_{n,m_1} \tilde{I}_{n,m_1} + \tilde{Z}_{m_1,m_2} \tilde{I}_{m_1,m_2} + \ldots + \tilde{Z}_{m_i,n} \tilde{I}_{m_i,n}$$

$$- (\tilde{V}_{n,m_1}^{(s)} + \tilde{V}_{m_1,m_2}^{(s)} + \ldots + \tilde{V}_{m_i,n}^{(s)}) = 0 \tag{11}$$

Appropriate applications of (9) and (11) to a given circuit yield the network equations. There are many forms of network equations which are derived according to the cut sets or tie sets chosen [9]. It is not intended to go into this subject here, but instead one form of the network equations for the example illustrated in Fig. 1 is given. Currents are labeled by single subscripts in the same way as the single-subscripted branch numbers. Application of (9) to N_1, N_2, and N_3 yields

$$\begin{pmatrix} 1 & 1 & 0 & 0 & 0 \\ -1 & 0 & 1 & 1 & 0 \\ 0 & -1 & -1 & 0 & 1 \end{pmatrix} \cdot \begin{pmatrix} \tilde{I}_1 \\ \tilde{I}_2 \\ \tilde{I}_3 \\ \tilde{I}_4 \\ \tilde{I}_5 \end{pmatrix} = (0_n) \tag{12}$$

For the mesh containing branches B_1, B_2, B_3, and the mesh containing B_3, B_4, B_5, (11) becomes

$$\begin{pmatrix} \tilde{Z}_{1,2} & -\tilde{Z}_{1,3} & \tilde{Z}_{2,3} & 0 & 0 \\ 0 & 0 & \tilde{Z}_{2,3} & -\tilde{Z}_{2,4} & \tilde{Z}_{3,4} \end{pmatrix} \cdot \begin{pmatrix} \tilde{I}_1 \\ \tilde{I}_2 \\ \tilde{I}_3 \\ \tilde{I}_4 \\ \tilde{I}_5 \end{pmatrix} = \begin{pmatrix} 1 & 1 & 1 & 0 & 0 \\ 0 & 0 & 1 & 1 & 1 \end{pmatrix} \cdot \begin{pmatrix} \tilde{V}_{1,2}^{(s)} \\ \tilde{V}_{1,3}^{(s)} \\ \tilde{V}_{2,3}^{(s)} \\ \tilde{V}_{2,4}^{(s)} \\ \tilde{V}_{3,4}^{(s)} \end{pmatrix} \tag{13}$$

For dimensional consistency, using an arbitrary impedance $\tilde{Z}_{ref}$, combination of (12) and (13) gives one form of the network equation, viz.,

$$
\begin{pmatrix}
\tilde{Z}_{ref} & \tilde{Z}_{ref} & 0 & 0 & 0 \\
-\tilde{Z}_{ref} & 0 & \tilde{Z}_{ref} & \tilde{Z}_{ref} & 0 \\
0 & \tilde{Z}_{ref} & -\tilde{Z}_{ref} & 0 & \tilde{Z}_{ref} \\
\tilde{Z}_{1,2} & -\tilde{Z}_{1,3} & \tilde{Z}_{2,3} & 0 & 0 \\
0 & 0 & \tilde{Z}_{2,3} & -\tilde{Z}_{2,4} & \tilde{Z}_{3,4}
\end{pmatrix}
\cdot
\begin{pmatrix}
\tilde{I}_1 \\ \tilde{I}_2 \\ \tilde{I}_3 \\ \tilde{I}_4 \\ \tilde{I}_5
\end{pmatrix}
=
\begin{pmatrix}
0 & 0 & 0 & 0 & 0 \\
0 & 0 & 0 & 0 & 0 \\
0 & 0 & 0 & 0 & 0 \\
1 & 1 & 1 & 0 & 0 \\
0 & 0 & 1 & 1 & 1
\end{pmatrix}
\cdot
\begin{pmatrix}
\tilde{V}^{(s)}_{1,2} \\ \tilde{V}^{(s)}_{1,3} \\ \tilde{V}^{(s)}_{2,3} \\ \tilde{V}^{(s)}_{2,4} \\ \tilde{V}^{(s)}_{3,4}
\end{pmatrix}
\qquad (14)
$$

The currents are readily obtained by inverting the 5x5 matrix.

It is the purpose of this paper to present similar network equations for tranmission-line networks. These equations are complicated by the wave nature of the voltages and currents, and their dependence on positions and modal properties.

2.3 Transmission-Line Networks

Concepts similar to those of circuit topology are developed for transmission-line networks to help summarize the network configurations, to define topology matrices, and to set up network equations.

The basic elements of the transmission-line network graphs are tubes and junctions. A tube is a collection of wires characterized by two ends to which electrical connections can be made. A junction is where wires terminate. Usually a bundle of wires is considered as a tube which may be terminated by a circuit. Branching of a bundle of wires can be considered as a tube divided into a few tubes with the position of branching as a junction within which only direct electrical connections occur.

The graphic symbols for tubes and junctions are respectively "parallel" lines and circles. The vth junction is denoted as J_v for $v = 1,\ldots,N_J$, where N_J is the number of junctions in the network. For transmission-line networks, it is possible to have more than one tube between two junctions. The pth tube between J_v and $J_{v'}$ is labeled $T^{(p)}_{v,v'}$. If there is only one tube between J_v and $J_{v'}$, the simplified notation $T_{v,v'}(=T^{(1)}_{v,v'})$ is often used.

Each tube can be characterized by two sets of waves: the forward traveling wave and the backward traveling wave. The waves on the pth tube between junctions J_v and $J_{v'}$ are labeled $W_{v,v'}^{(p,+)}$ and $W_{v,v'}^{(p,-)}$, where $W_{v,v'}^{(p,+)}$ travels from J_v to $J_{v'}$, and $W_{v,v'}^{(p,-)}$ travels from $J_{v'}$ to J_v. Thus, $W_{v,v'}^{(p,+)} = W_{v,v'}^{(p,-)}$. There are thus two waves traveling in opposite directions on a given tube.

As an example, a transmission-line graph is shown in Fig. 2a. There are four junctions and six tubes. The tubes and waves are numbered with double subscripts according to the rules outlined above. The parallel tubes and the self tube* are unique for transmission-line networks as there are no corresponding elements for the circuits. Topology matrices similar to those used for lumped circuits can be defined here involving junctions, tubes, and waves. Specifically, there are six useful interconnection matrices: junction-junction (or junction interconnection), junction-tube, tube-tube (or tube interconnection), junction-wave, wave-wave (or wave interconnection), and tube-end-wave.

For a transmission-line network with N_J junctions, the junction-junction interconnection matrix $(t_{v,v'})_{J-J}$ is an $N_J \times N_J$ matrix. The elements are defined as:

$$
\begin{array}{|c|l|}
\hline
v \neq v' &
\begin{aligned}
&t_{v,v';J-J} = \quad \text{number of tubes connecting} \\
&\qquad\qquad\qquad \text{junctions } J_v \text{ and } J_{v'} \\[6pt]
&t_{v,v';J-J} = 0 \quad \text{implies no connection between} \\
&\qquad\qquad\qquad\quad\ \text{the two junctions}
\end{aligned} \\
\hline
v = v' &
\begin{aligned}
&t_{v,v;J-J} = 1 \quad \text{denotes self connection (since} \\
&\qquad\qquad\qquad\quad \text{a junction is always connected} \\
&\qquad\qquad\qquad\quad \text{to itself)} \\[6pt]
&t_{v,v;J-J} > 1 \quad \text{denotes existence of self} \\
&\qquad\qquad\qquad\quad\ \text{tubes} \\[6pt]
&\qquad\qquad = 1 + 2 \times \text{(number of self tubes)}
\end{aligned} \\
\hline
\end{array}
\tag{15}
$$

*A self tube is one that has both ends terminating in the same junction.

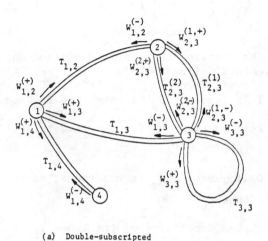

(a) Double-subscripted

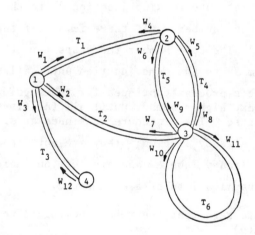

Fig. 2.   A transmission line graph with (a) double-subscripted
          numbering and (b) single-subscripted numbering.

For the example in Fig. 2a, the junction-junction matrix is

$$
(t_{v,v'})_{J-J} = \begin{pmatrix} 1 & 1 & 1 & 1 \\ 1 & 1 & 2 & 0 \\ 1 & 2 & 3 & 0 \\ 1 & 0 & 0 & 1 \end{pmatrix} \tag{16}
$$

Similarly, the junction-tube interconnection matrix $(t_{v,n})_{J-T}$ has elements defined by:

$$t_{v,n;J-T} = 1 \text{ if junction } J_v \text{ is connected to tube } T_n$$

$$= 2 \text{ if junction } J_v \text{ is connected to self} \tag{17}$$

tube T_n

$$= 0 \text{ if junction } J_v \text{ is not connected to}$$

tube T_n

The single-subscripted denotation of the tubes T_n is useful for matrix manipulations. The numbering system is similar to that for branches in the circuit topology. Consecutive numbering starts at the first tube linking junction J_1 to the junction that has the lowest node number. (If there is a self tube at J_1 this would be first.) The tube number increases for other parallel tubes going from J_1 to the same junction until all these tubes are labeled. This process continues for tubes going to the junction with the next higher number until all functions connected to J_1 are exhausted. The procedure continues at J_2 except for tubes which are already labeled; they are not repeated (i.e., tubes $T_{1,2}^{(p)}$ are already labeled and are left out here). The process continues until all tubes are numbered.

One may note here that the tube labeling is not oriented, i.e., $T_{v,v'}^{(p)} = T_{v',v}^{(p)}$.

For the example in Fig. 2b, the junction-tube matrix is:

$$
(t_{v,n})_{J-T} = \begin{pmatrix} 1 & 1 & 1 & 0 & 0 & 0 \\ 1 & 0 & 0 & 1 & 1 & 0 \\ 0 & 1 & 0 & 1 & 1 & 2 \\ 0 & 0 & 1 & 0 & 0 & 0 \end{pmatrix} \tag{18}
$$

The tube-tube interconnection matrix $(t_{n,m})_{T-T}$ is defined as follows:

$n \neq m$	$t_{n,m;T-T}$	= number of connections between the ends of tube T_n and tube T_m
		= 0 implies no connections
		= 1 implies one end of each tube is connected
		= 2 implies either (i) two parallel tubes or (ii) one is a self tube
		= 4 implies both are self tubes
$n = m$	$t_{n,n;T-T}$	= 0 for a simple tube (normal situation)
		= 2 for a self tube

(19)

The $(t_{n,m})_{T-T}$ matrix for the case of Fig. 2b is:

$$(t_{n,m})_{T-T} = \begin{pmatrix} 0 & 1 & 1 & 1 & 1 & 0 \\ 1 & 0 & 1 & 1 & 1 & 2 \\ 1 & 1 & 0 & 0 & 0 & 0 \\ 1 & 1 & 0 & 0 & 2 & 2 \\ 1 & 1 & 0 & 2 & 0 & 2 \\ 0 & 2 & 0 & 2 & 2 & 2 \end{pmatrix} \qquad (20)$$

Each tube is also characterized by two waves. The junction-wave matrix $(t_{v,u})_{J-W}$ describes the waves that are incident or reflected from a junction. One defines:

$$t_{v,u;J-W} = \begin{cases} 1 \text{ if wave } W_u \text{ is leaving junction } J_v \\ \quad \text{(transmitted and/or reflected)} \\ \\ 1 \text{ if wave } W_u \text{ is entering junction } J_v \\ \quad \text{(incident)} \\ \\ 0 \text{ if junction } J_v \text{ is not associated with} \\ \quad \text{wave } W_u \\ \\ 2 \text{ if wave } W_u \text{ is on a self tube} \end{cases} \qquad (21)$$

The single-subscripted denotation of a wave, W_u, is numbered similar to that of a tube. However, there are two waves on a tube, oriented to propagate in opposite directions. Thus, numbering starts at junction J_1 for a wave leaving J_1 on tube T_1 until all tubes are exhausted. Numbering continues at junction J_2 for all tubes (in ascending tube numbers), again for waves leaving J_2. This is repeated for all junctions. This results in different numbering as compared to the tubes. In fact, for N_T tubes, there are N_W waves given by

$$N_W = 2 N_T \tag{22}$$

The junction-wave matrix for the example in Fig. 2b is:

$$(t_{v,u})_{J-W} = \begin{pmatrix} 1 & 1 & 1 & 1 & 0 & 0 & 1 & 0 & 0 & 0 & 0 & 1 \\ 1 & 0 & 0 & 1 & 1 & 1 & 0 & 1 & 1 & 0 & 0 & 0 \\ 0 & 1 & 0 & 0 & 1 & 1 & 1 & 1 & 1 & 2 & 2 & 0 \\ 0 & 0 & 1 & 0 & 0 & 0 & 0 & 0 & 0 & 0 & 0 & 1 \end{pmatrix} \tag{23}$$

The wave-wave matrix $(W_{u,v})$ describes interconnection of waves (via junctions). Elements are defined by:

$W_{u,v} = 1$ if wave W_v scatters into wave W_u, i.e., if W_v is connected to W_u via a junction into which W_v is incoming and W_u is outgoing.

$$W_{u,v} = \begin{cases} 1 \text{ for a self tube} \\ 0 \text{ otherwise (normal situation).} \end{cases} \tag{24}$$

For the example in Fig. 2b, $(W_{u,v})$ is:

$$(W_{u,v}) = \begin{pmatrix} 0 & 0 & 0 & 1 & 0 & 0 & 1 & 0 & 0 & 0 & 0 & 1 \\ 0 & 0 & 0 & 1 & 0 & 0 & 1 & 0 & 0 & 0 & 0 & 1 \\ 0 & 0 & 0 & 1 & 0 & 0 & 1 & 0 & 0 & 0 & 0 & 1 \\ 1 & 0 & 0 & 0 & 0 & 0 & 0 & 1 & 1 & 0 & 0 & 0 \\ 1 & 0 & 0 & 0 & 0 & 0 & 0 & 1 & 1 & 0 & 0 & 0 \\ 1 & 0 & 0 & 0 & 0 & 0 & 0 & 1 & 1 & 0 & 0 & 0 \\ 0 & 1 & 0 & 0 & 1 & 1 & 0 & 0 & 0 & 1 & 1 & 0 \\ 0 & 1 & 0 & 0 & 1 & 1 & 0 & 0 & 0 & 1 & 1 & 0 \\ 0 & 1 & 0 & 0 & 1 & 1 & 0 & 0 & 0 & 1 & 1 & 0 \\ 0 & 1 & 0 & 0 & 1 & 1 & 0 & 0 & 0 & 0 & 1 & 0 \\ 0 & 1 & 0 & 0 & 1 & 1 & 0 & 0 & 0 & 1 & 0 & 0 \\ 0 & 0 & 1 & 0 & 0 & 0 & 0 & 0 & 0 & 0 & 0 & 0 \end{pmatrix} \tag{25}$$

Another matrix of interest is the tube-end-wave interconnection matrix. We denote by the index r in $J_{v;r}$ the tube ends reaching the junction J_v; then, for a wave entering J via $J_{v;r}$, it is labeled as $J_{v;r,-}$, and for a wave leaving J_v via $J_{v;r}$, it is labeled $J_{v;r,+}$.

The tube-end-wave interconnection matrix $(t_{r,u})_{v;E-W}$ is defined as follows:

$$
t_{r,u;v;E-W} = \begin{cases} 0 \text{ if wave } W_u \text{ does not connect to } J_v \text{ via} \\ \quad \text{end } J_{v;r} \\ -1 \text{ if wave } W_u \text{ enters } J_v \text{ via } J_{v;r} \text{ (i.e.,} \\ \quad J_{v;r,-}) \\ 1 \text{ if wave } W_u \text{ leaves } J_v \text{ via } J_{v;r} \text{ (i.e.,} \\ \quad J_{v;r,+}) \end{cases} \quad (26)
$$

Thus, for junction J_3 of Fig. 2b, a new illustration is depicted in Fig. 3. Here the tube-end-wave interconnection matrix is:

$$
(t_{r,u})_{v;E-W} = \begin{pmatrix} 0 & -1 & 0 & 0 & 0 & 0 & 1 & 0 & 0 & 0 & 0 & 0 \\ 0 & 0 & 0 & 0 & -1 & 0 & 0 & 1 & 0 & 0 & 0 & 0 \\ 0 & 0 & 0 & 0 & 0 & -1 & 0 & 0 & 1 & 0 & 0 & 0 \\ 0 & 0 & 0 & 0 & 0 & 0 & 0 & 0 & 0 & -1 & 1 & 0 \\ 0 & 0 & 0 & 0 & 0 & 0 & 0 & 0 & 0 & -1 & 1 & 0 \end{pmatrix} \quad (27)
$$

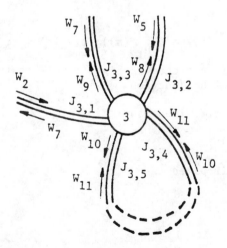

Fig. 3.   Tube-end labeling for Junction J_3 .

Note that the junction-wave interconnection matrix is formed by

$$t_{v,u;J-W} = \sum_{r=1}^{r_v} \left| t_{r,u;v;E-W} \right| \tag{28}$$

where

$r_v \equiv$ maximum value or r

$\quad\quad = $ number of tube ends at J_v $\tag{29}$

$|| \equiv$ absolute value

2.4 Equivalent Circuits of Functions and Tubes

Most junctions of interest either consist of physical lumped circuits or are transmission-line discontinuities modeled by lumped circuit elements [16, 17, 18, 19]. Topological descriptions of a junction can thus be similar to those for a lumped circuit, as outlined in Section 2.2.

Junctions can be classified according to their complexities. The simplest one involves only one tube terminated by an imped-ance network (including sources). This includes the special cases of open-circuited and short-circuited terminations. The voltage-current relation is given by

$$(\tilde{V}_n(s)) + (\tilde{V}_n^{(s)}(s)) = (\tilde{Z}_{n,m}(s)) \cdot [(\tilde{I}_n(s)) + (I_n^{(s)}(s))] \tag{30}$$

The dual relationship is

$$(\tilde{I}_n(s)) + (\tilde{I}_n^{(s)}(s)) = (\tilde{Y}_{n,m}(s)) \cdot [(\tilde{V}_n(s) + (\tilde{V}_n^{(s)}(s))] \tag{31}$$

The configuration is depicted in Fig. 4; note that current is taken positive into the junction.

For the short-circuited case

$$(\tilde{V}_n(s)) = (0_n) \tag{32}$$

and for the open-circuited case

$$(\tilde{I}_n(s)) = (0_n) \tag{33}$$

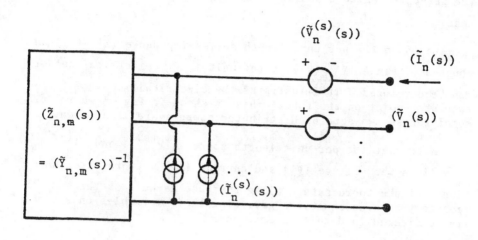

Fig. 4.   Terminating Network.

A more general type of junction is one that connects to many tubes, and connections between wires of the tubes are by direct electrical contacts. This type is extremely relevant in modeling the case of branching. As is well known, branches connected at a node have equal voltages and the sum of currents leaving the node is zero. Thus, each connecting point within the junction is a node and the above voltage and current relations (9-11) apply [15].

More general forms of junctions are multitube junctions. While the equations take the same form as in (34, 35), these equations need to be partitioned according to the different tubes (and associated waves) that connect to the junction. This partitioning is considered in some detail in Section 6.0; it utilizes the topology matrices discussed in the previous subsection.

A tube is characterized by its physical construction and geometry. These are in turn described by the per-unit-length quantities such as the per-unit-length impedance matrix $(\tilde{Z}'_{n,m}(s))$ and the per-unit-length admittance matrix $(\tilde{Y}'_{n,m}(s))$, for the general case, or the per-unit-length inductance matrix $(L'_{n,m})$ and the per-unit-length capacitance matrix $(C'_{n,m})$ for the lossless case.

Based on the per-unit-length series impedance $(\tilde{Z}'_{n,m}(s))$ and shunt admittance $(\tilde{Y}'_{n,m}(s))$, a per-unit-length electrical network can be developed. It illustrates the electrical model of the tube at a point on the tube. This is shown in Fig. 5 where for completeness two sets of distributed sources are also shown.

Note that the per-unit-length sources $(\tilde{V}_n^{(s)\,'})$ and $(\tilde{I}_n^{(s)\,'})$, as well as the voltage $(\tilde{V}'_n)$ and current $(\tilde{I}'_n)$ vectors, are functions of the coordinate z along the tube as well as the complex frequency s. The equations governing these variables on a tube are considered in detail in Section 3.

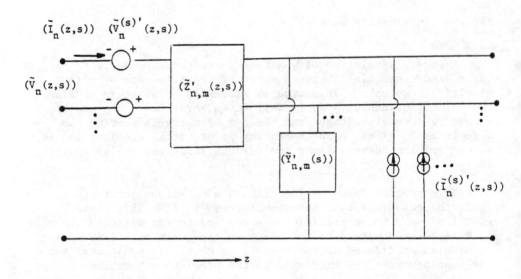

Fig. 5.   The per-unit-length model of a multi-conductor transmission line.

2.5 Other Related Topologies

The development of scatterer topology and the hierarchical
scatterer topology is useful in dealing with scattering and
penetration problems [10, 11, 12]. For an aircraft, missile,
or other systems, there are many cable bundles enclosed inside
the walls of the system. The use of hierarchical scatterer
topological concepts is well-suited to aid in the solution of
these kinds of problems. Here one deals with the problem layer-
by-layer, dividing it into many subproblems of coupling, propagation, and penetration. The details are treated in the above-
cited references and are not described here.

2.6 Topology Summary

The various types of topologies and their associated quantities are summarized in Table 2.1, which was first presented in
ref. [11].

As mentioned earlier, the hierarchical scatterer problem
can be divided into the following subproblems corresponding to
the electromagnetic processes associated with each layer (principal volume) in the transport of signals into the system:

a. Coupling. This relates the response of each system
layer to the electromagnetic fields coming from the layer external to the one under consideration. Quantitatively coupling can
be identified with the source terms in the equations used to describe the response of the layer of interest.

b. Propagation. This deals with the distribution of signals within the layer of interest in the system. It is concerned
with the operator (integral, differential, etc.) in the equations
describing the response of the layer as well as the resulting
response itself.

c. Penetration. This deals with the excitation of the signals in the next layer (going to the interior). Specifically,
penetration is concerned with the conversion of the response
within a layer into an appropriate set of parameters which can be
used for the coupling process in the next layer. It is then the
transition from one layer to the next.

In the hierarchical decomposition of a system, one or more of the
layers of interest may be represented as transmission line networks, in which case the above breakdown is relevant to transmission-line problems. The above breakdown within a layer is
summarized in Table 2.2

Table 2.1. Various Topologies

	Topology	Basic Topological Quantity and Symbol	Interconnecting Topological Quantity and Symbol	Diagramatic form of Topology
1	Graph (Generalized)	Vertex	Edge	
2	Circuit	Node N_n ●	Branch $B_{n,m}^{(T)}$ ———	●——●
3	Transmission Line	Junction J_n ◯	Tube $T_{n,m}^{(T)}$ ═══	◯══◯
4	Scatterer	Volume V_n	Surface $S_{n,m}$	
5	Hierarchical Scatterer	Principal Volume $(V)_\lambda^{(\lambda')}$	Principal Surface $(S)_\lambda^{(\lambda')}$ (closed but sometimes in more than one part)	
		Elementary Volume $V_{\lambda,\tau}^{(\lambda')}$	Elementary Surface $S_{1,\tau_1;\lambda_2,\tau_2}^{(\lambda_1',\lambda_2')}$ (usually open but sometimes closed)	and

Table 2.2. Hierarchical Decomposition of a System

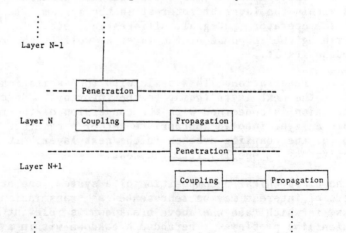

3.   PROPAGATION ON AN N-WIRE TRANSMISSION LINE TUBE

In this section the phenomena of waves propagating on a
single section, N-wire transmission-line tube are considered.
An N-wire transmission line is one that consists of N conductors
and a reference which may be infinity or ground. As will be derived later, such a system has N modes of propagation.

The equations governing the voltage and current propagation
on the N-wire transmission line, i.e., the generalized transmission equations, are the current change equation

$$\frac{d}{dz}(\tilde{I}_n(z,s)) = -(\tilde{Y}'_{n,m}(s))\cdot(\tilde{V}_n(z,s)) + (\tilde{I}_n^{(s)'}(z,s)) \tag{34}$$

and the voltage change equation

$$\frac{d}{dz}(\tilde{V}_n(z,s)) = -(\tilde{Z}'_{n,m}(s))\cdot(\tilde{I}_n(z,s)) + (\tilde{V}_n^{(s)'}(z,s)) \tag{35}$$

where

z = position along the tube

$(\tilde{I}_n(z,s))$ = current vector at z

$(\tilde{V}_n(z,s))$ = voltage vector at z $\hspace{2cm}$ (36)

$(\tilde{Y}'_{n,m}(s))$ = per-unit-length shunt admittance matrix

$(\tilde{Z}'_{n,m}(s))$ = per-unit-length series impedance matrix

$(\tilde{I}_n^{(s)'}(z,s))$ = per-unit-length shunt current source vector

$(\tilde{V}_n^{(s)'}(z,s))$ = per-unit-length series voltage source vector.

It is noted that all vectors are of dimension N, and all matrices
are N x N. The per-unit-length equivalent circuit has been given
in Figure 3.

There are a few ways of solving (34) and (35). One could
reduce the equations to a second-order differential equation in
either the voltage vector or the current vector, or one could express the voltage-current relations in terms of a transmission
supermatrix [19]. Still another way is to solve for the unknown
propagation vectors that are associated with the waves. Here,
the derivation is in terms of a yet undefined combined voltage.

This approach, as will be illustrated later, has many definite advantages over other methods.

3.1 Combined Voltage Equation

Pre-multiplying (34) by a matrix $(\tilde{A}_{n,m}(s))$ which is N x N and non-singular, and adding (35), then

$$\frac{d}{dz} [(\tilde{A}_{n,m}(s)) \cdot (\tilde{I}_n(s)) + (\tilde{V}_n(s))]$$

$$= [- (\tilde{A}_{n,m}(s)) \cdot (\tilde{Y}'_{n,m}(s)) \cdot (\tilde{V}_n(s))$$

$$- (\tilde{Z}'_{n,m}(s)) \cdot (\tilde{I}_n(s))] + [(\tilde{A}_{n,m}(s)) \cdot (\tilde{I}_n^{(s)}{}'(s))$$

$$+ (\tilde{V}_n^{(s)}{}'(s))] \tag{37}$$

Defining the following quantities:

$$(\tilde{V}_n(z,s))_q \equiv (\tilde{V}_n(z,s)) + (\tilde{A}_{n,m}(s)) \cdot (\tilde{I}_n(z,s)) \tag{38}$$

and

$$(\tilde{V}_n^{(s)}{}'(z,s))_q \equiv (\tilde{V}_n^{(s)}{}'(z,s)) + (\tilde{A}_{n,m}(s)) \cdot (\tilde{I}_n^{(s)}{}'(z,s)) \tag{39}$$

(37) becomes

$$\frac{d}{dz} (\tilde{V}_n(s))_q = - (\tilde{C}_{n,m}(s)) \cdot (\tilde{V}_n(s))_q + (\tilde{V}_n^{(s)}{}'(s))_q \tag{40}$$

where

$$(\tilde{C}_{n,m}(s)) \cdot (\tilde{V}_n(z,s))_q = (\tilde{A}_{n,m}(s)) \cdot (\tilde{Y}'_{n,m}(s)) \cdot (\tilde{V}_n(z,s))$$

$$+ (\tilde{Z}'_{n,m}(s)) \cdot (\tilde{I}_n(z,s)) \tag{41}$$

Using (41) and definition (38), one obtains

$$(\tilde{C}_{n,m}(s)) = (\tilde{A}_{n,m}(s)) \cdot (\tilde{Y}'_{n,m}(s)) \tag{42}$$

and

$$(\tilde{C}_{n,m}(s)) \cdot (\tilde{A}_{n,m}(s)) = (\tilde{Z}'_{n,m}(s)) \tag{43}$$

Equating $(\tilde{A}_{n,m}(s))$ in (42) and (43)

$$(\tilde{C}_{n,m}(s)) \cdot (\tilde{Y}'_{n,m}(s))^{-1} = (\tilde{C}_{n,m}(s))^{-1} \cdot (\tilde{Z}'_{n,m}(s)) \tag{44}$$

i.e.,

$$(\tilde{C}_{n,m}(s))^2 = (\tilde{Z}'_{n,m}(s)) \cdot (\tilde{Y}'_{n,m}(s)) \tag{45}$$

One can also write

$$(\tilde{C}_{n,m}(s)) = [(\tilde{Z}'_{n,m}(s)) \cdot (\tilde{Y}'_{n,m}(s))]^{1/2} \tag{46}$$

which has many values. One may define a principal value (or matrix)* $(\tilde{\gamma}_{c_{n,m}}(s))$, i.e.,

$$(\tilde{\gamma}_{c_{n,m}}(s)) = \text{principal value of } [(\tilde{Z}'_{n,m}(s)) \cdot (\tilde{Y}'_{n,m}(s))]^{1/2} \tag{47}$$

$(\tilde{\gamma}_{c_{n,m}}(s))$ is called the propagation matrix. Expressing (40) in this new form, one obtains

$$[(1_{n,m}) \frac{d}{dz} + q(\tilde{\gamma}_{c_{n,m}}(s))] \cdot (\tilde{V}_n(z,s))_q = (\tilde{V}_n^{(s)'}(z,s))_q$$

$$1_{n,m} = \begin{cases} 1 \text{ for } n=m \\ 0 \text{ for } n \neq m \end{cases} \tag{48}$$

$q = \pm 1$ for forward- and backward-traveling combined N-vector waves, respectively.

Essentially, $(\tilde{C}_{n,m}(s))$ has been restricted to $q(\tilde{\gamma}_{c_{n,m}}(s))$ in (38)

From (42), (43), and (47)

$$(\tilde{A}_{n,m}(s)) = q(\tilde{\gamma}_{c_{n,m}}(s)) \cdot (\tilde{Y}'_{n,m}(s))^{-1}$$

$$= q(\tilde{\gamma}_{c_{n,m}}(s))^{-1} \cdot (\tilde{Z}'_{n,m}(s)) \tag{49}$$

---

* see definition under Section 3.2.

Thus, $(\tilde{A}_{n,m}(s))$ is a characteristic of the transmission line and
has the dimensions of an impedance. Define

$$(\tilde{Z}_{c_{n,m}}(s)) \equiv (\tilde{Y}_{c_{n,m}}(s)) \cdot (\tilde{Y}'_{n,m}(s))^{-1}$$

$$= (\tilde{Y}_{c_{n,m}}(s))^{-1} \cdot (\tilde{Z}'_{n,m}(s)) \qquad (50)$$

which is called the characteristic impedance matrix. Now,
definitions (38) and (39) are rewritten to be

$$(\tilde{V}_n(z,s))_q \equiv (\tilde{V}_n(z,s)) + q(\tilde{Z}_{c_{n,m}}(s)) \cdot (\tilde{I}_n(z,s))$$

$$\qquad (51)$$

$$(\tilde{V}_n^{(s)'}(z,s))_q \equiv (\tilde{V}_n^{(s)'}(z,s)) + q(\tilde{Z}_{c_{n,m}}(s)) \cdot (\tilde{I}_n^{(s)'}(z,s))$$

One can also define the characteristic admittance matrix to be
the inverse of the characteristic impedance matrix, viz.,

$$(\tilde{Y}_{c_{n,m}}(s)) \equiv (\tilde{Z}_{c_{n,m}}(s))^{-1} \qquad (52)$$

Putting $q = +1$ and $q = -1$ in (51), one can obtain the following relations

$$(\tilde{V}_n(z,s)) = \frac{1}{2}[(\tilde{V}_n(z,s))_+ + (\tilde{V}_n(z,s))_-] \qquad (53)$$

$$(\tilde{Z}_{c_{n,m}}(s)) \cdot (\tilde{I}_n(z,s)) = \frac{1}{2}[(\tilde{V}_n(z,s))_+ - (\tilde{V}_n(z,s))_-] \qquad (54)$$

Thus, for forward traveling waves

$$(\tilde{V}_n(z,s))_+ = (\tilde{Z}_{c_{n,m}}(s)) \cdot (\tilde{I}_n(z,s))_+ \qquad (55)$$

and for backward traveling waves

$$(\tilde{V}_n(z,s))_- = -(\tilde{Z}_{c_{n,m}}(s)) \cdot (\tilde{I}_n(z,s))_- \qquad (56)$$

This is an important result which allows one to easily separate
the voltage and current vectors into forward and backward waves
and easily reconstruct the voltage and current vectors from the
waves.

3.2 Eigenmode Expansion

Expression (47) for $(\tilde{\gamma}_{c_{n,m}}(s))$ clearly indicates the

necessity of eigenmode expansion of the matrix product
$(Z'_{n,m}(s)) \cdot (\tilde{Y}'_{n,m}(s))$. The eigenvalues would yield the values of
the propagation constants of the eigenmodes. Corresponding to
each eigenmode, there is a complete set of eigenvectors for the
combined voltage. These properties are examined in this section.

3.2.1. Positive real properties.

a. Definitions. A rational function f(s) which is real for
real values of s, and whose real part is positive for all values
of s with a positive real part, is called a positive real function (p. r. function) [20]. A positive real matrix is one whose
eigenvalues are all p.r. functions. Let $(\tilde{P}_{n,m}(s))$ be a p.r.
matrix of size NxN, then the eigenvalue problem becomes

$$(\tilde{P}_{n,m}(s)) \cdot (\tilde{P}_n^{(r)}(s))_\delta = \tilde{P}_\delta(s)(\tilde{P}_n^{(r)}(s))_\delta$$

$$(\tilde{P}_n^{(\ell)}(s))_\delta \cdot (\tilde{P}_{n,m}(s)) = \tilde{P}_\delta(s)(\tilde{P}_n^{(\ell)}(s))_\delta$$

(57)

where $\delta = 1,2,\ldots,N$ is the eigenindex and $\tilde{p}_\delta(s)$ are the eigen-
values, and $(\tilde{P}_n^{(r)}(s))_\delta$ and $(\tilde{P}_n^{(\ell)}(s))_\delta$ are the eigenvectors.
Since $(\tilde{P}_{n,m}(s))$ is a p.r. matrix, $\tilde{p}_\delta(s)$ is a p.r. function of s.
If $\tilde{p}_\delta$ is independent of s, then $\tilde{p}_\delta$ is real and $\tilde{p}_\delta \geq 0$ for all δ.
For the oft encountered case of symmetric p.r. matrices we can
set

$$(\tilde{P}_n^{(r)}(s))_\delta \equiv (\tilde{P}_n^{(\ell)}(s))_\delta$$

$$(\tilde{P}_{n,m}(s)) = (\tilde{P}_{n,m}(s))^T$$

(58)

b. Eigenmode expansion. A p.r. matrix can be expanded as
follows [21]:

$$(\tilde{P}_{n,m}(s)) = \sum_\delta \tilde{p}_\delta(s)(P_n^{(r)}(s))_\delta (\tilde{P}_n^{(\ell)}(s))_\delta [(\tilde{P}_n^{(r)}(s))_\delta$$

$$\cdot (\tilde{P}_n^{(\ell)}(s))_\delta]^{-1}$$

(59)

If the matrix is the argument of a scalar function F, then

$$F[(\widetilde{P}_{n,m}(s))] = \sum_{\delta} F[\widetilde{p}_n(s)] (\widetilde{P}_n^{(r)}(s))_{\delta} (\widetilde{P}_n^{(\ell)}(s))_{\delta} [(\widetilde{P}_n^{(r)}(s))_{\delta}$$

$$\cdot (\widetilde{P}_n^{(\ell)}(s))_{\delta}]^{-1} \tag{60}$$

One assumes that the scalar function F is single-valued for all complex $\widetilde{p}_{\delta}(s)$. Otherwise the principal value of $F[p_{\delta}(s)]$ is defined to define the principal value (or matrix) of $F[(\widetilde{P}_{n,m}(s))]$.

For example, powers of the matrix can be expressed as

$$(\widetilde{P}_{n,m}(s))^{\varepsilon} = \sum_{\delta} [\widetilde{p}_{\delta}(s)]^{\varepsilon} (\widetilde{P}_n^{(r)}(s))_{\delta} (\widetilde{P}_n^{(\ell)}(s))_{\delta} [(\widetilde{P}_n^{(r)}(s))_{\delta}$$

$$\cdot (\widetilde{P}_n^{(\ell)}(s))_{\delta}]^{-1} \tag{61}$$

where for $\mathrm{Re}(s) > 0$ and $\mathrm{Im}(\varepsilon) = 0$, $\mathrm{Re}\{[\widetilde{p}(s)]^{\varepsilon}\} > 0$ defines the principal value of the ε-th power of a p.r. matrix.

c. <u>Normalized eigenvectors</u>. The normalized eigenvector is defined by

$$(\widetilde{p}_n^{\binom{r}{\ell}}(s))_{\delta} = \frac{(\widetilde{P}_n^{\binom{r}{\ell}}(s))_{\delta}}{\sqrt{(\widetilde{P}_n^{(r)}(s))_{\delta} \cdot (\widetilde{P}_n^{(\ell)}(s))_{\delta}}} \tag{62}$$

In terms of the normalized quantities, (59) and (60) become

$$(\widetilde{P}_{n,m}(s)) = \sum_{\delta} \widetilde{p}_{\delta}(s) (\widetilde{p}_n^{(r)}(s))_{\delta} (\widetilde{p}_n^{(\ell)}(s))_{\delta}$$

$$F[(\widetilde{P}_{n,m}(s))] = \sum_{\delta} F[\widetilde{p}_{\delta}(s)] (\widetilde{p}_n^{(r)}(s))_{\delta} (\widetilde{p}_n^{(\ell)}(s))_{\delta} \tag{63}$$

Note for symmetric p.r. matrices the normalized eigenvectors can be reduced to

$$(\widetilde{p}_n^{(r)}(s))_{\delta} = (\widetilde{p}_n^{(\ell)}(s))_{\delta} \equiv (\widetilde{p}_n(s))_{\delta} \tag{64}$$

d. <u>Transmission-line p.r. matrices</u>. Assume that the per-unit-length impedance and admittance matrices $(Z'_{n,m}(s))$, $(Y'_{n,m}(s))$ are passive. Then they are p.r. matrices.

In the special case of a lossless transmission line.

$$(Z'_{n,m}(s)) = s(L'_{n,m})$$

$$(Y'_{n,m}(s)) = s(C'_{n,m}) \tag{65}$$

where

$$(L'_{n,m}) \equiv \text{per-unit-length inductance matrix}$$

$$(C'_{n,m}) \equiv \text{per-unit-length capacitance matrix.} \tag{66}$$

which we also assume to be frequency independent (dispersionless) and, hence, constant matrices. The elements $L'_{n,m}$ and $C'_{n,m}$ are real, being derivable from quasi-static boundary-value problems (Laplace equation). The p.r. property of the per-unit-length impedance and admittance matrices then implies that the per-unit-length inductance and capacitance matrices are both p.r. and positive semidefinite. Thus, $(L'_{n,m})$ and $(C'_{n,m})$ have real non-negative eigenvalues.

If we further assume that $(L'_{n,m})$ is symmetric, then

$$(L'_{n,m}) \cdot (L'_n)_\delta = \ell'_\delta \ (L'_n)_\delta$$

$$(L'_n)_\delta (L'_{n,m}) = \ell'_\delta \ (L'_n)_\delta \tag{67}$$

In terms of the normalized eigenvectors $(\ell'_n)_\delta$, $(\tilde{Z}'_{n,m}(s))$ is expressible as

$$(\tilde{Z}'_{n,m}(s)) = s \sum_\delta \ell'_\delta \ (\ell'_n)_\delta (\ell'_n)_\delta \tag{68}$$

Similar expressions exist for symmetric $(\tilde{Y}'_{n,m}(s))$ and $(C'_{n,m})$.

3.2.2 Propagation matrix $(\tilde{\gamma}_{c_{n,m}}(s))$. The squared quantity of the propagation matrix is equal to the product of two matrices, each of which is typically symmetric (reciprocity), i.e.,

$$(\tilde{\gamma}_{c_{n,m}}(s))^2_V = (\tilde{Z}'_{n,m}(s)) \cdot (\tilde{Y}'_{n,m}(s)) \tag{69}$$

Let $(\tilde{V}_{c_n}(s))_\delta$ be the right eigenvector of $(\tilde{\gamma}_{c_{n,m}}(s))^2_V$

$$(\tilde{Z}'_{n,m}(s)) \cdot (\tilde{Y}'_{n,m}(s)) \cdot (\tilde{V}_{c_n}(s))_\delta = \tilde{\gamma}^2_{V_\delta}(s)(\tilde{V}_{c_n}(s))_\delta \qquad (70)$$

The corresponding quantity for the combined current mode is given by

$$(\tilde{\gamma}_{c_{n,m}}(s))^2_I = (\tilde{Y}'_{n,m}(s)) \cdot (\tilde{Z}'_{n,m}(s)) \qquad (71)$$

Let $(\tilde{I}_{c_n}(s))_\delta$ be the left eigenvector of $(\tilde{\gamma}_{c_{n,m}}(s))^2_V$

$$(\tilde{I}_{c_n}(s))_\delta \cdot (\tilde{Z}'_{n,m}(s)) \cdot (\tilde{Y}'_{n,m}(s)) = \tilde{\gamma}^2_{I_\delta}(s)(\tilde{I}_{c_n}(s))_\delta \qquad (72)$$

In this paper, the following simplified notation is used:

$$(\tilde{\gamma}_{c_{n,m}}(s)) \equiv (\tilde{\gamma}_{c_{n,m}}(s))_V \qquad (73)$$

Defining the normalized eigenvectors by

$$(\tilde{v}_{c_n}(s))_\delta = \frac{(\tilde{V}_{c_n}(s))_\delta}{\sqrt{(\tilde{V}_{c_n}(s))_\delta \cdot (\tilde{I}_{c_n}(s))_\delta}} \qquad (74)$$

and

$$(\tilde{i}_{c_n}(s))_\delta = \frac{(I_{c_n}(s))_\delta}{\sqrt{(\tilde{V}_{c_n}(s))_\delta \cdot (I_{c_n}(s))_\delta}} \qquad (75)$$

it is possible to expand the squared propagation matrix into

$$(\tilde{\gamma}_{c_{n,m}}(s))^2 = \sum_\delta \gamma^2_\delta(s)(\tilde{v}_{c_n}(s))(\tilde{i}_{c_n}(s)) \qquad (76)$$

and using (64) the propagation matrix is

$$(\tilde{\gamma}_{c_{n,m}}(s)) = \sum_\delta \tilde{\gamma}_\delta(s)(\tilde{v}_{c_n}(s))(\tilde{i}_{c_n}(s)) \qquad (77)$$

where $\tilde{\gamma}_\delta(s)$ is the principal value of $[\gamma^2_\delta(s)]^{1/2}$

Here principal value means:

$$\tilde{f}(s) \geq 0 \quad \text{for } s \geq 0 \tag{78}$$

and then

$$\tilde{f}^{1/2}(s) \geq 0 \quad \text{for } s \geq 0 \tag{79}$$

with analytic continuation away from the $s \geq 0$ axis. We then assume

$$\tilde{\gamma}_\delta^2(s) \geq 0 \quad \text{for } s \geq 0 \tag{80}$$

so that we may choose

$$\tilde{\gamma}_\delta(s) \geq 0 \quad \text{for } s \geq 0 \;, \; \delta = 1,2,\ldots,N \tag{81}$$

We further assume that the $\tilde{\gamma}_\delta(s)$ are p.r. functions so that

$$\left. \begin{array}{l} \text{Re}[\tilde{\gamma}_\delta(s)] \geq 0 \quad \text{for } s \geq 0 \\[2mm] \tilde{\gamma}_\delta(s) \text{ analytic for } s > 0 \end{array} \right\} \qquad \delta = 1,2,\ldots,N \tag{82}$$

Note that $\tilde{\gamma}_\delta(s)$ having this property corresponds to a + or right-going wave, since a positive real part corresponds to an attenuation in the + direction. A p.r. propagation constant is then a causal propagation constant. However, other more general forms are perhaps possible. For present purposes, p.r. propagation constants are assumed.

The necessity of choosing which square root to use for the propagation matrix is potentially troublesome. The matrix transmission-line equations may have buried in them certain mathematical problems, such as related to existence and uniqueness of solutions, representation of solutions, etc. The diagonalization of the square of the propagation matrix may depend on certain properties of the per-unit-length impedance and admittance matrices; this in turn influences the nature of the square root of the square of the propagation matrix. The problems in choice of the propagation constants may lead to some restrictions to situations that such choices are applicable or even possible. There are then some open questions requiring further research.

With the above definitions we obtain two sets of waves propagating in opposite directions along z . For all modes we have

$$\exp\left[-(\tilde{\gamma}_{c_{n,m}}(s))z\right] \quad \text{is + propagating}$$

$$\exp\left[(\tilde{\gamma}_{c_{n,m}}(s))z\right] \quad \text{is - propagating} \tag{83}$$

For a function F of $(\tilde{\gamma}_{c_{n,m}}(s))$, assuming nondegenerate modes, one can express

$$F[(\tilde{\gamma}_{c_{n,m}}(s))] = \sum_{\delta} F[\tilde{\gamma}_\delta(s)]\, (\tilde{v}_{c_n}(s))(\tilde{i}_{c_n}(s)) \qquad (84)$$

Specifically, we have

$$(\tilde{\gamma}_{c_{n,m}}(s)) = \sum_{\delta} \tilde{\gamma}_\delta(s)(\tilde{v}_{c_n}(s))_\delta(\tilde{i}_{c_n}(s))_\delta$$

$$(\tilde{\gamma}_{c_{n,m}}(s))^{-1} = \sum_{\delta} \tilde{\gamma}_\delta^{-1}(s)(\tilde{v}_{c_n}(s))_\delta(\tilde{i}_{c_n}(s))_\delta$$

$$e^{-q(\gamma_{c_{n,m}}(s))z} = \sum_{\delta} e^{-q\tilde{\gamma}_\delta(s)z}\, (\tilde{v}_{c_n}(s))_\delta(\tilde{i}_{c_n}(s))_\delta \qquad (85)$$

$$(\tilde{\gamma}_{c_{n,m}}(s))^0 = (1_{n,m}) = \sum_{\delta} (\tilde{v}_{c_n}(s))_\delta(\tilde{i}_{c_n}(s))_\delta \quad \text{(identity)}$$

3.2.3. Properties of $(\tilde{v}_{c_n}(s))_\delta$, $(\tilde{i}_{c_n}(s))_\delta$. The two normalized eigenvectors as defined in (69), (70), (74), and (75) possess unique properties, which are exploited in this section.

Rewriting (69) and (70)

$$(\tilde{Z}'_{n,m}(s))\cdot(\tilde{Y}'_{n,m}(s))\cdot(\tilde{v}_{c_n}(s))_\delta = \tilde{\gamma}_\delta^2(s)(\tilde{v}_{c_n}(s))_\delta$$

$$(\tilde{i}_{c_n}(s))_{\delta'}\cdot(\tilde{Z}'_{n,m}(s))\cdot(\tilde{Y}'_{n,m}(s)) = \tilde{\gamma}_{\delta'}^2(s)(i_{c_n}(s))_{\delta'} \qquad (86)$$

First, premultiply the first by $(\tilde{i}_{c_n}(s))_{\delta'}$, then postmultiply the second by $(\tilde{v}_{c_n}(s))_\delta$ (both in dot product sense). The difference of these two new equations becomes:

$$(\tilde{i}_{c_n}(s))_{\delta'} \cdot [(\tilde{Z}'_{n,m}(s)) \cdot (\tilde{Y}'_{n,m}(s))$$

$$- (\tilde{Z}'_{n,m}(s)) \cdot (\tilde{Y}'_{n,m}(s))] \cdot (\tilde{v}_{c_n}(s))_{\delta}$$

$$= [\tilde{\gamma}_{\delta}^2(s) - \tilde{\gamma}_{\delta'}^2(s)] \, (\tilde{i}_{c_n}(s))_{\delta'} \cdot (\tilde{v}_{c_n}(s))_{\delta}$$

$$= 0 \tag{87}$$

There are two possible cases. First, if $\gamma_\delta^2 \neq \gamma_{\delta'}^2$, then

$$(\tilde{i}_{c_n}(s))_{\delta'} \cdot (\tilde{v}_{c_n}(s))_{\delta} = 1_{\delta,\delta'} \tag{88}$$

where

$$1_{\delta,\delta'} = \begin{cases} 0 & \text{for } \delta \neq \delta' \\ 1 & \text{for } \delta = \delta' \end{cases} \tag{89}$$

are elements of the NxN identity matrix $(1_{\delta,\delta'})$ (or Kronecker delta). Equation (88) is called the biorthonormal relation; $(\tilde{i}_{c_n}(s))_{\delta}$ and $(\tilde{v}_{c_n}(s))_{\delta}$ are the biorthonormal eigenvectors. Second, if $\gamma_\delta^2 = \gamma_{\delta'}^2$, i.e., the degenerate case, the orthonormal vectors are constructed by other means such as the Gram-Schmidt procedure.

3.3 Solution of Combined Voltage Equations

3.3.1. Integration of combined voltage equation. The combined differential equation (48), i.e.,

$$\frac{d}{dz} (\tilde{V}_n(s))_q + q(\tilde{\gamma}_{c_{n,m}}(s)) \cdot (\tilde{V}_n(z,s))_q = (\tilde{V}_n^{(s)'}(z,s))_q \tag{90}$$

can be readily solved [22] to give

$$(\tilde{V}_n(z,s))_q = \exp\{-q(\tilde{\gamma}_{c_{n,m}}(s))[z-z_0]\} \cdot (\tilde{V}_n(z_0,s))_q$$

$$+ \int_{z_0}^{z} \exp\{-q(\tilde{\gamma}_{c_{n,m}}(s))[z-z']\} \cdot (\tilde{V}_n^{(s)'}(z',s))_q \, dz' \tag{91}$$

For a + wave (i.e., a wave propagating in the + z direction), let us assume that $(\tilde{V}_n(0,s))_+$ is specified, giving

$$(\tilde{V}_n(z,s))_+ = \exp\{-(\tilde{\gamma}_{c_{n,m}}(s))z\}\cdot(\tilde{V}_n(0,s))_+$$

$$+ \int_0^z \exp\{-(\tilde{\gamma}_{c_{n,m}}(s))[z-z']\}\cdot(\tilde{V}_n^{(s)'}(z',s))_+ dz' \quad (92)$$

Similarly for a - wave with $(\tilde{V}_n(L,s))_-$ assumed specified, we have

$$(\tilde{V}_n(z,s))_- = \exp\{(\tilde{\gamma}_{c_{n,m}}(s))[z-L]\}\cdot(\tilde{V}_n(L,s))_-$$

$$+ \int_L^z \exp\{(\tilde{\gamma}_{c_{n,m}}(s))[z-z']\}\cdot(\tilde{V}_n^{(s)'}(z',s))_- dz' \quad (93)$$

These results illustrate one aspect of the simplification introduced by the combined voltage in that the + wave depends only on the left boundary condition and the - wave depends only on the right boundary condition in a very compact way. Note that for the minus wave if we replace z by L-z as the coordinate variable, then the - wave has precisely the same form as the + wave, which one would expect by symmetry.

Using (85), equation (91) can be written in terms of eigenmodes, i.e.,

$$(\tilde{V}_n(z,s))_q = \sum_\delta \left\{ e^{-q\tilde{\gamma}_\delta(s)[z-z_0]}[(\tilde{i}_{c_n}(s))_\delta \right.$$

$$\cdot(\tilde{V}_n(z_0,s))_q](\tilde{v}_{c_n}(s))_\delta$$

$$+ \int_{z_0}^z e^{-q\tilde{\gamma}_\delta(s)(z-z_0)}[(\tilde{i}_{c_n}(s))_\delta$$

$$\left. \cdot(\tilde{V}_n^{(s)'}(z',s))]dz' \right\} (\tilde{v}_{c_n}(s))_\delta$$

$$= \sum_\delta [(\tilde{i}_{c_n}(s))_\delta\cdot(\tilde{V}_n(z,s))_q] (\tilde{v}_{c_n}(s))_\delta \quad (94)$$

Define coefficients of expansion as

$$\tilde{C}_{V_{\delta,q}}(z,s) = (\tilde{i}_{c_n}(s))_\delta \cdot (\tilde{V}_n(z,s))_q \tag{95}$$

For a + wave, i.e., one that travels from z=0 to z=L along the transmission line, let us assume that $(\tilde{V}_n(0,s))_+$ is specified. Then

$$(\tilde{V}_n(z,s))_+ = \sum_\delta C_{V_{\delta,+}}(z,s)(\tilde{v}_{c_n}(s))_\delta \tag{96}$$

and

$$C_{V_{\delta,+}}(z,s) = (\tilde{i}_{c_n}(s))_\delta \cdot (\tilde{V}_n(z,s))_+$$

$$= e^{-\tilde{\gamma}_\delta(s)z} [(\tilde{i}_{c_n}(s))_\delta \cdot (\tilde{V}_n(0,s))_+]$$

$$+ \int_0^z e^{-\tilde{\gamma}_\delta(s)(z-z')} [(\tilde{i}_{c_n}(s))_\delta \cdot (\tilde{V}_n^{(s)'}(z',s))_+] \, dz' \tag{97}$$

Similarly for a - wave with $(\tilde{V}_n(L,s))_-$ specified

$$(\tilde{V}_n(z,s))_- = \sum_\delta C_{V_{\delta-}}(z,s)(\tilde{v}_{c_n}(s))_\delta \tag{98}$$

and

$$C_{V_{\delta,-}}(z,s) = (\tilde{i}_{c_n}(s))_\delta \cdot (\tilde{V}_n(z,s))_-$$

$$= e^{\tilde{\gamma}_\delta(s)(z-L)} [(\tilde{i}_{c_n}(s))_\delta \cdot (\tilde{V}_n(L,s))_-]$$

$$+ \int_L^z e^{\tilde{\gamma}_\delta(s)(z-z')} [(\tilde{i}_{c_n}(s))_\delta$$

$$\cdot (\tilde{V}_n^{(s)'}(z',s))_-] \, dz' \tag{99}$$

Equations (96) and (98) show that there are 2N eigenwaves for a N-wire transmission line (plus a reference). These waves are characterized by $\tilde{C}_{V_{\delta,+}}(\tilde{v}_{c_n}(s))_\delta$ and $\tilde{C}_{V_{\delta,-}}(\tilde{v}_{c_n}(s))_\delta$, $\delta = 1$,

$2,\ldots,N$. One could define an eigenmatrix as follows

$$(\tilde{E}_{n,m}(s))_V = ((\tilde{v}_{c_n}(s))_1, (\tilde{v}_{c_n}(s))_2, \ldots, (\tilde{v}_{c_n}(s))_N) \qquad (100)$$

where the columns are the voltage eigenvectors. The eigenmode coefficient vector is defined by

$$(\tilde{C}_{V_n}(z,s))_q = (\tilde{C}_{V_{1,q}}(z,s), \tilde{C}_{V_{2,q}}(z,s), \ldots, \tilde{C}_{V_{N,q}}(z,s)) \qquad (101)$$

Equation (94) can be rewritten as

$$(V_n(z,s))_q = (E_{n,m}(s))_V (C_{V_n}(z,s))_q \qquad (102)$$

3.3.2 Semi-infinite transmission line. Many transmission-line properties can be learned by studying the semi-infinite line where the complications due to reflections do not exist. As discussed earlier, the per-unit-length electrical model of a transmission line is given in Figure 5.

Assuming that $(\tilde{V}_n(0,s))_+$ is given and there are no other sources along the line so that only + waves propagate, (92) gives

$$(\tilde{V}_n(z,s))_+ = \exp[-(\tilde{\gamma}_{c_{n,m}}(s))z] \cdot (\tilde{V}_n(0,s))_+ \qquad (103)$$

At $z = 0$,

$(\tilde{V}_n(0,s))_+$ is specified

$$(\tilde{V}_n(0,s))_- = (0_n) \qquad (104)$$

from (52)

$$(\tilde{V}_n(0,s))_- = (\tilde{V}_n(0,s)) - (\tilde{Z}_{c_{n,m}}(s)) \cdot (\tilde{I}_n(0,s)) = (0_n) \qquad (105)$$

Thus

$$(\tilde{V}_n(0,s)) = (\tilde{Z}_{c_{n,m}}(s)) \cdot (\tilde{I}_n(0,s)) \qquad (106)$$

Or, from (53)

$$(\tilde{I}_n(0,s)) = (\tilde{Y}_{c_{n,m}}(s)) \cdot (\tilde{V}_n(0,s)) \tag{107}$$

As is well-known, waves propagate in only one direction on a semi-infinite line driven at the one end, with voltage and current related by the characteristic impedance (Eq. 106). Thus, the effect of a semi-infinite multiconductor transmission line can be represented by an equivalent impedance network that is equivalent to the characteristic impedance matrix $(\tilde{Z}_{c_{n,m}}(s))$.

3.3.3 Normalization relation of $(\tilde{v}_{c_n}(s))_\delta$ and $(\tilde{i}_{c_n}(s))_\delta$ in terms of $(\tilde{Z}_{c_{n,m}}(s))$ and $(\tilde{Y}_{c_{n,m}}(s))$ and associated modal expansions.

For forward traveling waves only, (55) gives

$$(\tilde{V}_n(z,s)) = (\tilde{Z}_{c_{n,m}}(s)) \cdot (\tilde{I}_n(z,s)) \tag{108}$$

For $(\tilde{V}_n(z,s))$ chosen as a single mode, (96) gives

$$(\tilde{V}_n(z,s)) = C_{v_{\delta,+}} (\tilde{v}_{c_n}(s))_\delta \tag{109}$$

Hence

$$(\tilde{I}_n(z,s)) = C_{v_{\delta,+}} (\tilde{Y}_{c_{n,m}}(s)) \cdot (\tilde{v}_{c_n}(s))_\delta \tag{110}$$

Therefore, the δ-th mode for the current can be normalized as

$$(\tilde{i}_{c_n}(s))_\delta \equiv (\tilde{Y}_{c_{n,m}}(s)) \cdot (\tilde{v}_{c_n}(s))_\delta$$

$$(\tilde{v}_{c_n}(s))_\delta \equiv (\tilde{Z}_{c_{n,m}}(s)) \cdot (\tilde{i}_{c_n}(s))_\delta \tag{111}$$

Together with (88) this specifies $(\tilde{v}_{c_n}(s))$ and $(\tilde{i}_{c_n}(s))$ including their units

For nondegenerate modes (86), (88), and (111), give

$$(\tilde{v}_{c_n}(s))_\delta \cdot (\tilde{i}_{c_n}(s))_{\delta'} = (\tilde{v}_{c_n}(s))_\delta \cdot (\tilde{Y}_{c_{n,m}}(s))$$

$$\cdot (\tilde{v}_{c_n}(s))_{\delta'} = 1_{\delta,\delta'}$$

$$(\tilde{i}_{c_n}(s))_\delta \cdot (\tilde{v}_{c_n}(s))_{\delta'} = (\tilde{i}_{c_n}(s))_\delta \cdot (\tilde{Z}_{c_{n,m}}(s))$$

$$\cdot (\tilde{i}_{c_n}(s))_{\delta'} = 1_{\delta,\delta'} \tag{112}$$

For the first of these equations, left dyadic multiply by $(\tilde{v}_{c_n}(s))_\delta$ and right dyadic multiply by $(\tilde{i}_{c_n}(s))_\delta$, and sum over δ,δ':

$$\sum_{\delta,\delta'} (\tilde{v}_{c_n}(s))_\delta (\tilde{i}_{c_n}(s))_{\delta'} \, 1_{\delta,\delta'} = \sum_\delta (\tilde{v}_{c_n}(s))_\delta (\tilde{i}_{c_n}(s))_\delta$$

$$= (1_{n,m})$$

$$= \sum_{\delta,\delta'} (\tilde{v}_{c_n}(s))_\delta (\tilde{v}_{c_n}(s))_\delta \cdot (\tilde{Y}_{c_{n,m}}(s))$$

$$\cdot (\tilde{v}_{c_n}(s))_\delta, (\tilde{i}_{c_n}(s))_{\delta'} = \left[\sum_\delta (\tilde{v}_{c_n}(s))_\delta (\tilde{v}_{c_n}(s))_\delta \right]$$

$$\cdot (\tilde{Y}_{c_{n,m}}(s)) \cdot \left[\sum_{\delta'} (\tilde{v}_{c_n}(s))_{\delta'}, (\tilde{i}_{c_n}(s))_{\delta'} \right]$$

$$= \left[\sum_\delta (\tilde{v}_{c_n}(s))_\delta (\tilde{v}_{c_n}(s))_\delta \right] \cdot (\tilde{Y}_{c_{n,m}}(s)) \cdot (1_{n,m})$$

$$= \left[\sum_\delta (\tilde{v}_{c_n}(s))_\delta (\tilde{v}_{c_n}(s))_\delta \right] \cdot (\tilde{Y}_{c_{n,m}}(s)) \tag{113}$$

From which we conclude

$$(\tilde{Z}_{c_{n,m}}(s)) = (\tilde{Y}_{c_{n,m}}(s))^{-1} = \sum_\delta (\tilde{v}_{c_n}(s))_\delta (\tilde{v}_{c_n}(s))_\delta \tag{114}$$

In a similar manner from the second of (111) dyadic multiplication by $(\tilde{v}_{c_n}(s))_\delta$ on the left and $(\tilde{i}_{c_n}(s))_{\delta'}$ on the right and summing over δ,δ' gives

$$\sum_{\delta,\delta'} (\tilde{v}_{c_n}(s))_\delta (\tilde{i}_{c_n}(s))_{\delta'} 1_{\delta,\delta'} = \sum_\delta (\tilde{v}_{c_n}(s))_\delta (\tilde{i}_{c_n}(s))_{\delta'}$$

$$= (1_{,n,m})$$

$$= \sum_{\delta,\delta'} (\tilde{v}_{c_n}(s))_\delta (\tilde{i}_{c_n}(s))_\delta \cdot (\tilde{Z}_{c_{n,m}}(s))$$

$$\cdot (\tilde{i}_{c_n}(s))_{\delta'} (\tilde{i}_{c_n}(s))_{\delta'} = \left[\sum_\delta (\tilde{v}_{c_n}(s))_\delta (\tilde{i}_{c_n}(s))_\delta \right]$$

$$\cdot (\tilde{Z}_{c_{n,m}}(s)) \cdot \left[\sum_{\delta'} (\tilde{i}_{c_n}(s))_{\delta'} (\tilde{i}_{c_n}(s))_{\delta'} \right]$$

$$= (1_{n,m}) \cdot (\tilde{Z}_{c_{n,m}}(s)) \cdot \left[\sum_{\delta'} (\tilde{i}_{c_n}(s))_{\delta'} (\tilde{i}_{c_n}(s))_{\delta'} \right]$$

$$= (\tilde{Z}_{c_{n,m}}(s)) \cdot \left[\sum_\delta (\tilde{i}_{c_n}(s))_\delta (\tilde{i}_{c_n}(s))_\delta \right] \tag{115}$$

from which we conclude

$$(\tilde{Y}_{c_{n,m}}(s)) = (\tilde{Z}_{c_{n,m}}(s))^{-1} = \sum_\delta (\tilde{i}_{c_n}(s))_\delta (\tilde{i}_{c_n}(s))_\delta \tag{116}$$

These results are quite illuminating. Specifically, they show that the characteristic impedance matrices are symmetric, i.e.,

$$(\tilde{Z}_{c_{n,m}}(s))^T = (\tilde{Z}_{c_{n,m}}(s))$$

$$(\tilde{Y}_{c_{n,m}}(s))^T = (\tilde{Y}_{c_{n,m}}(s)) \tag{117}$$

This is evident from (114) and (116) which expand these as sums of symmetric dyads. The form in (117) is usually referred to as reciprocity, but this property was not assumed in the beginning, but is required by our results. Again, this is possibly associated with the assumed characteristics of the propagation matrix.

3.3.4 Expansion of $(\tilde{Z}'_{n,m}(s))$ and $(\tilde{Y}'_{n,m}(s))$ in terms of $(\tilde{v}_{c_n}(s))_\delta$ and $(\tilde{i}_{c_n}(s))_\delta$.

From (91) for the case of no sources along the semi-infinite line, $z \geq 0$, with only + waves, we have:

$$(\tilde{V}_n(z,s))_+ = \exp\{-(\tilde{\gamma}_{c_{n,m}}(s))z\} \cdot (\tilde{V}_n(0,s))_+$$

$$(\tilde{V}_n(z,s))_- = (0_n)$$

$$(\tilde{V}_n(z,s)) = (\tilde{Z}_{c_{n,m}}(s)) \cdot (\tilde{I}_n(z,s)) \tag{118}$$

Expanded in modal form we have

$$(\tilde{V}_n(z,s))_+ = \left\{ \sum_\delta e^{-\tilde{\gamma}_\delta(s)z} (\tilde{v}_{c_n}(s))_\delta (\tilde{i}_{c_n}(s))_\delta \right\} \cdot (\tilde{V}_n(0,s))_+$$

$$\frac{d}{dz}(\tilde{V}_n(z,s))_+ = -\left\{ \sum_\delta \tilde{\gamma}_\delta(s)\, e^{-\tilde{\gamma}_\delta(s)z} (\tilde{v}_{c_n}(s))_\delta (\tilde{i}_{c_n}(s))_\delta \right\}$$

$$\cdot (\tilde{V}_n(0,s))_+ \tag{119}$$

Noting the special relation between the voltage and current vectors we can write equations for the voltage vector the same as for the combined voltage vector, i.e.,

$$(\tilde{V}_n(z,s)) = \left\{ \sum_\delta e^{-\tilde{\gamma}_\delta(s)z} (\tilde{v}_{c_n}(s))_\delta (\tilde{i}_{c_n}(s))_\delta \right\} \cdot (\tilde{V}_n(0,s))$$

$$\frac{d}{dz}(\tilde{V}_n(z,s)) = -\left\{ \sum_\delta \tilde{\gamma}_\delta(s)\, e^{-\tilde{\gamma}_\delta(s)z} (\tilde{v}_{c_n}(s))_\delta (\tilde{i}_{c_n}(s))_\delta \right\} \tag{120}$$

$$\cdot (\tilde{V}_n(0,s))$$

Similarly, for the current vector we have by multiplication (dot product) by the characteristic admittance matrix

$$(\tilde{I}_n(z,s)) = (\tilde{Y}_{c_{n,m}}(s)) \cdot \left\{ \sum_\delta e^{-\tilde{\gamma}_\delta(s)z} (\tilde{v}_{c_n}(s))_\delta (\tilde{i}_{c_n}(s))_\delta \right\}$$

$$\cdot (\tilde{Z}_{c_{n,m}}(s)) \cdot (\tilde{I}_n(0,s)) \tag{121}$$

Then using the modal expansions for the characteristic admittance and impedance matrices in (116) and (114) respectively, together with the biorthonormal modal relation in (87) we have:

$$(\tilde{I}_n(z,s)) = \left\{ \sum_\delta e^{-\tilde{\gamma}_\delta(s)z} (\tilde{i}_{c_n}(s))_\delta (\tilde{v}_{c_n}(s))_\delta \right\}$$

$$\cdot (\tilde{I}_n(0,s))$$

$$\frac{d}{dz}(\tilde{I}_n(z,s)) = -\left\{ \sum_\delta \tilde{\gamma}_\delta(s)) e^{-\tilde{\gamma}_\delta(s)z} (\tilde{i}_{c_n}(s))_\delta (\tilde{v}_{c_n}(s))_\delta \right\}$$

$$\cdot (\tilde{I}_n(0,s)) \qquad (122)$$

Recall (34) and (35) without sources:

$$\frac{d}{dz}(\tilde{I}_n(z,s)) = -(\tilde{Y}'_{n,m}(s))\cdot(\tilde{V}_n(z,s))$$

$$\frac{d}{dz}(\tilde{V}_n(z,s)) = -(\tilde{Z}'_{n,m}(s))\cdot(\tilde{I}_n(z,s)) \qquad (123)$$

Comparing these to the above modal expansions of the derivatives we have first, considering the current derivative:

$$-\left\{ \sum_\delta \tilde{\gamma}_\delta(s) e^{-\tilde{\gamma}_\delta(s)z} (\tilde{i}_{c_n}(s))_\delta (\tilde{v}_{c_n}(s))_\delta \right\} \cdot (\tilde{Y}_{c_{n,m}}(s))$$

$$\cdot (\tilde{V}_n(0,s))$$

$$= -\left\{ \sum_\delta \tilde{\gamma}_\delta(s) e^{-\tilde{\gamma}_\delta(s)z} (\tilde{i}_{c_n}(s))_\delta (\tilde{i}_{c_n}(s))_\delta \right\} \cdot (\tilde{V}_n(0,s))$$

$$= -(\tilde{Y}'_{n,m}(s))\cdot(\tilde{V}_n(z,s)) \qquad (124)$$

Evaluating this at z=0 and noting that $(\tilde{V}_n(0,s))$ is an arbitrary N vector, we have

$$(\tilde{Y}_{n,m}(s)) = \sum_\delta \tilde{\gamma}_\delta(s) (\tilde{i}_{c_n}(s))_\delta (\tilde{i}_{c_n}(s))_\delta \qquad (125)$$

Similarly, using the voltage derivative equations

$$-\left\{ \sum_\delta \tilde{\gamma}_\delta(s) e^{-\tilde{\gamma}_\delta(s)z} (\tilde{v}_{c_n}(s))_\delta (\tilde{i}_{c_n}(s))_\delta \right\} \cdot (\tilde{Z}_{c_{n,m}}(s))$$

$$\cdot (\tilde{I}_n(0,s))$$

$$= - \left\{ \sum_\delta \tilde{\gamma}_\delta(s) \ e^{-\tilde{\gamma}_\delta(s)z} (\tilde{v}_{c_n}(s))_\delta (\tilde{v}_{c_n}(s))_\delta \right\} \cdot (\tilde{I}_n(0,s))$$

$$= (\tilde{Z}'_{n,m}(s)) \cdot (\tilde{I}_n(z,s)) \qquad\qquad (126)$$

Evaluating this at z=0 and noting that $(\tilde{I}_n(0,s))$ is an arbitrary N vector, we have

$$(\tilde{Z}'_{n,m}(s)) = \sum_\delta \tilde{\gamma}_\delta(s) \ (\tilde{v}_{c_n}(s))_\delta (\tilde{v}_{c_n}(s))_\delta \qquad\qquad (127)$$

Note that (125) explicitly illustrates that $(\tilde{Y}'_{n,m}(s))$ is symmetric and (127) does the same for $(\tilde{Z}_{n,m}(s))$, i.e.,

$$(\tilde{Z}'_{n,m}(s))^T = (\tilde{Z}'_{n,m}(s))$$

$$(\tilde{Y}'_{n,m}(s))^T = (\tilde{Y}'_{n,m}(s)) \qquad\qquad (128)$$

As in (117) for the characteristic impedance and admittance matrices, this symmetry is a statement of reciprocity for the impedance and admittance per-unit-length matrices. While this was not explicitly assumed at the start, it is a consequence of the development. This may be associated with the assumed diagonalization characteristics of the propagation matrix. Then let us consider the reciprocity (symmetry) of the impedance and admittance per-unit-length matrices as one of our assumptions for the present development.

Taking (127) and left or right dot-multiplying by a current eigenmode, we have

$$\tilde{\gamma}_\delta(s)(\tilde{v}_{c_n}(s))_\delta = (\tilde{Z}'_{n,m}(s)) \cdot (\tilde{i}_{c_n}(s))_\delta$$

$$= (\tilde{i}_{c_n}(s))_\delta \cdot (\tilde{Z}'_{n,m}(s)) \qquad\qquad (129)$$

A dot product with the δ'-th current mode gives

$$\tilde{\gamma}_\delta(s) \ 1_{\delta,\delta'} = (\tilde{i}_{c_n}(s))_\delta \cdot (\tilde{Z}'_{n,m}(s)) \cdot (\tilde{i}_{c_n}(s))_{\delta'} \qquad\qquad (130)$$

Given $\tilde{\gamma}_\delta(s)$ this normalizes the $(\tilde{i}_{c_n}(s))_\delta$ in terms of $(\tilde{Z}'_{n,m}(s))$.

Note also the relationship of the voltage and current modes via $(\tilde{Z}'_{n,m}(s))$ and $\tilde{\gamma}_\delta(s)$.

Similarly, dot multiplying (125) on left or right by a voltage eigenmode gives

$$\tilde{\gamma}_\delta(s) \; (i_{c_n}(s))_\delta = (\tilde{Y}'_{n,m}(s)) \cdot (\tilde{v}_{c_n}(s))_\delta$$

$$= (\tilde{v}_{c_n}(s))_\delta \cdot (\tilde{Y}'_{n,m}(s)) \tag{131}$$

A dot product with the δ'-th voltage mode gives

$$\tilde{\gamma}_\delta(s) \; 1_{\delta,\delta'} = (\tilde{v}_{c_n}(s))_\delta \cdot (\tilde{Y}'_{n,m}(s)) \cdot (\tilde{v}_{c_n}(s))_{\delta'} \tag{132}$$

This normalizes the $(\tilde{v}_{c_n}(s))_\delta$ in terms of $(\tilde{Y}'_{n,m}(s))$ and the $\tilde{\gamma}_\delta(s)$. Note also the relationship of the voltage and current modes via $(\tilde{Y}'_{n,m}(s))$ and $\tilde{\gamma}_\delta(s)$.

3.3.5 Termination condition of a tube. A transmission line is usually terminated at the two ends z=0 and z=L. The termination could be a lumped impedance, a distributed network, open-circuit, or short-circuit. If sources are included, these conditions can be represented by a generalized Thevenin equivalent network or a generalized Norton equivalent network.

Passive terminations can be specified as an impedance matrix $(\tilde{Z}_{T_{n,m}}(z,s))$ or an admittance matrix $(\tilde{Y}_{T_{n,m}}(z,s))$ where z=0 or L. The condition $(\tilde{Z}_{T_{n,m}}(L,s)) = (\tilde{Z}_{c_{n,m}}(s))$, or equivalently $(\tilde{Y}_{T_{n,m}}(L,s)) = (\tilde{Y}_{c_{n,m}}(s))$ specifies a perfectly matched line, and the transmission line behaves like a semi-infinite line for $0 \leq z \leq L$ (with an equivalent single end at z=0).

Alternatively, the terminating conditions can be specified by scattering matrices $(\tilde{S}_{n,m}(z,s))$ where z=0 or L. Consider at z=L (see Fig. 6); let the incoming waves be designated by a super-script - and the outgoing waves +. The scattering matrix is defined by

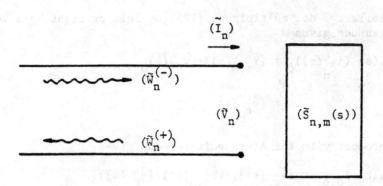

Fig. 6.   Incoming and outgoing wave at a junction.

$$(\widetilde{w}_n^{(+)}(s)) = (\widetilde{S}_{n,m}(z,s)) \cdot (\widetilde{w}_n^{(-)}(s)) \tag{133}$$

For the case illustrated in Figure 6, one observes that if this termination is taken as z=L, then

$$(\widetilde{w}_n^{(+)}(s)) = (\widetilde{V}_n(L,s))_-$$

$$(\widetilde{w}_n^{(-)}(s)) = (\widetilde{V}_n(L,s)_+ \tag{134}$$

One can then rewrite (127) as

$$(\widetilde{V}_n(L,s))_- = (\widetilde{S}_{n,m}(L,s)) \cdot (\widetilde{V}_n(L,s))_+ \tag{135}$$

which in this terminating case is the same as the definition of a reflection matrix given by

$$(\widetilde{S}_{n,m}(L,s)) = [(\widetilde{Z}_{T_{n,m}}(L,s)) + (\widetilde{Z}_{c_{n,m}}(s))]^{-1} \cdot [\widetilde{Z}_{T_{n,m}}(L,s))$$

$$- (\widetilde{Z}_{c_{n,m}}(s))] \tag{136}$$

Similarly, at z=0 the termination conditions are

$$(\widetilde{V}_n(0,s))_+ = (\widetilde{S}_{n,m}(0,s)) \cdot (\widetilde{V}_n(0,s))_- \tag{137}$$

and

$$(\tilde{S}_{n,m}(0,s)) = [(\tilde{Z}_{T_{n,m}}(0,s)) + (\tilde{Z}_{c_{n,m}}(s))]^{-1} \cdot [(\tilde{Z}_{T_{n,m}}(0,s))$$

$$- (\tilde{Z}_{c_{n,m}}(s))] \tag{138}$$

3.3.6 Solution of combined voltages. Combination of (91), (135) and (137) gives the solution of the combined voltage equation. Rewriting these equations

$$(\tilde{V}_n(z,s))_+ = \exp\{- (\tilde{\gamma}_{c_{n,m}}(s)z\} \cdot (\tilde{V}_n(0,s))_+$$

$$+ \int_0^z \exp\{-(\tilde{\gamma}_{c_{n,m}}(s))[z-z']\} \cdot (\tilde{V}_n^{(s)'}(z',s)_+ \, dz'$$

$$(\tilde{V}_n(z,s))_- = \exp\{(\tilde{\gamma}_{c_{n,m}}(s))[z-L]\} \cdot (\tilde{V}_n(L,s))_-$$

$$+ \int_L^z \exp\{(\tilde{\gamma}_{c_{n,m}}(s))[z-z']\} \cdot (\tilde{V}_L^{(s)'}(z',s))_- \, dz'$$

$$(\tilde{V}_n(0,s))_+ = (\tilde{S}_{n,m}(0,s)) \cdot (\tilde{V}_n(0,s))_-$$

$$(\tilde{V}_n(L,s))_- = (\tilde{S}_{n,m}(L,s)) \cdot (\tilde{V}_n(L,s))_+ \tag{139}$$

These equations can be solved by substitutions, or can be arranged in a matrix form, as described later in the BLT equation. As written here, these correspond to the special case of a transmission-line network consisting of two junctions (or terminations) connected by a single tube.

3.3.7 Reconstruction of total voltages and total currents. Once the combined voltages are evaluated, the total voltages and total currents are readily obtained.

From (51), one obtains

$$(\tilde{V}_n(z,s)) = \frac{1}{2}[(\tilde{V}_n(z,s))_+ + (\tilde{V}_n(z,s))_-]$$

$$(\tilde{I}_n(z,s)) = \frac{1}{2}(\tilde{Y}_{c_{n,m}}(s)) \cdot [(\tilde{V}_n(z,s))_+ - (\tilde{V}_n(z,s))_-] \tag{140}$$

Hence, if one knows $(\tilde{V}_n(z,s))_+$ and $(\tilde{V}_n(z,s))_-$ for a given tube, as well as $(\tilde{Y}_{c_{n,m}}(s))$ or $(\tilde{Z}_{c_{n,m}}(s))$ (being measurable or conceivably even calculable), then the measurable voltage and current vectors are directly reconstructable.

3.4 Sign Convention of q

It is noted that in the definitions of the combined voltage, the convention $q = +1$ is chosen to represent the wave propagating from $z=0$ to $z=L$. Correspondingly, $q = -1$ represents the wave propagating from $z=L$ to $z=0$.

Let us further denote the above quantities with a subscript u, i.e., $q_u = +1$ corresponds to wave propagating from left to right, i.e., from $z_u = 0$ to $z_u = L$. This is shown in Figure 7.

It is also permissible to choose, on the same tube, a different convention. Let $z_v = L - z_u$ be a new coordinate, as shown in Figure 7. The new q convention, q_v, is now opposite to q_u. Here, $q_v = +1$ corresponds to a wave traveling from right to left (i.e., $z_v = 0$ to $z_v = L$).

This convention will prove useful and is recalled in deriving the BLT equation.

The subsequent sections deal with multitube multiconductor transmission-line networks. More general notations, primarily in the form of additional subscripts denoting either the tube or the junction, are used.

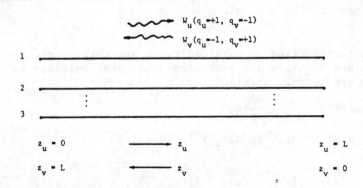

Fig. 7. Left and right traveling waves.

3.5 Summary

Since there are so many expressions, relations, etc., introduced in this section, it is useful to summarize them in tabular form, as presented in Tables 3.1-4.

4.   SUPERMATRICES AND SUPERVECTORS

Define a supermatrix, or more specifically, a dimatrix or tensor of rank four, as a partitioned matrix or matrix of matrices in the form

$$((D_{n,m})_{u,v}) \tag{141}$$

with elementary matrices or blocks

$$(D_{n,m})_{u,v} \tag{142}$$

and elements

$$D_{n,m;u,v} \tag{143}$$

such that the blocks or elementary matrices are $N_u \times M_v$, i.e.,

$$n = 1,2,\ldots,N_u$$

$$m = 1,2,\ldots,M_v \tag{144}$$

and the dimatrix is NxN, i.e.,

$$u = 1,2,\ldots,N$$

$$v = 1,2,\ldots,M \tag{145}$$

Note that this corresponds to a matrix with

$$\sum_{n=1}^{N} N_u \text{ rows}$$

$$\sum_{v=1}^{M} M_v \text{ columns} \tag{146}$$

which has been partitioned into blocks or elementary matrices by a partitioning of the row and column indices. Pictorially this corresponds to drawing horizontal and vertical lines completely through the matrix between selected adjacent rows and selected adjacent columns.

Table 3.1   Transmission-Line Equations

Name	Symbol	Relation
Trans. line eqns. (telegrapher eqns.)		$\frac{d}{dz}(\tilde{V}_n(z,s)) = -(\tilde{Z}'_{n,m}(s))\cdot(\tilde{I}_n(z,s)) + (\tilde{V}_n^{(s)'}(z,s))$
		$\frac{d}{dz}(\tilde{I}_n(z,s)) = -(\tilde{Y}'_{n,m}(s))\cdot(\tilde{V}_n(z,s)) + (\tilde{I}_n^{(s)'}(z,s))$
Combined Voltage Vector	$(\tilde{V}_n(z,s))_q$	$= (\tilde{V}_n(z,s)) + q(\tilde{Z}_{c_{n,m}}(s))\cdot(\tilde{I}_n(z,s))$
Combined Per-Unit-Length Source Vector	$(\tilde{V}_n^{(s)'}(z,s))_q$	$= (\tilde{V}_n^{(s)'}(z,s)) + q(\tilde{Z}_{c_{n,m}}(s))\cdot(\tilde{I}_n(z,s))$
Voltage Vector Reconstruction	$(\tilde{V}_n(z,s))$	$= \frac{1}{2}\left[(\tilde{V}_n(z,s))_+ + (\tilde{V}_n(z,s))_-\right]$
Current Vector Reconstruction	$(\tilde{I}_n(z,s))$	$= \frac{1}{2}(\tilde{Y}_{c_{n,m}}(s))\cdot\left[(\tilde{V}_n(z,s))_+ - (\tilde{V}_n(z,s))_-\right]$
Separation Index	q	$= \pm 1$
Combined Voltage Equation		$\left[(1_{n,m})\frac{d}{dz} + q(\tilde{\gamma}_{c_{n,m}}(s))\right]\cdot(\tilde{V}_n(z,s))_q = (\tilde{V}_n^{(s)'}(z,s))_q$
Propagation Matrix	$(\tilde{\gamma}_{c_{n,m}}(s))$	$= \left[(\tilde{Z}'_{n,m}(s))\cdot(\tilde{Y}'_{n,m}(s))\right]^{\frac{1}{2}}$ (principal or p.r. value)
Characteristic Impedance Matrix	$(\tilde{Z}_{c_{n,m}}(s))$	$= (\tilde{\gamma}_{c_{n,m}}(s))\cdot(\tilde{Y}'_{n,m}(s))^{-1} = (\tilde{Y}_{c_{n,m}}(s))^{-1}\cdot(\tilde{Z}'_{n,m}(s))$
Characteristic Admittance Matrix	$(\tilde{Y}_{c_{n,m}}(s))$	$= (\tilde{Z}_{c_{n,m}}(s))^{-1} = (\tilde{Y}'_{n,m}(s))\cdot(\tilde{\gamma}_{c_{n,m}}(s))^{-1}$
		$= (\tilde{Z}'_{n,m}(s))^{-1}\cdot(\tilde{\gamma}_{c_{n,m}}(s))$
General Solution (referenced to arbitrary position z_0)	$(\tilde{V}_n(z,s))_q$	$= \exp\left\{-q(\tilde{\gamma}_{c_{n,m}}(s))\left[z-z_0\right]\right\}\cdot(\tilde{V}_n(z_0,s))_q$
		$+ \int_{z_0}^{z}\exp\left\{-q((\tilde{\gamma}_{c_{n,m}}(s))\left[z-z'\right]\right\}\cdot(\tilde{V}_n^{(s)'}(z',s))_q\, dz'$
Solution for tube $0 \le z \le L$ in terms of boundary values	+ or right wave $(\tilde{V}_n(z,s))_+$	$= \exp\left\{-(\tilde{\gamma}_{c_{n,m}}(s))z\right\}\cdot(\tilde{V}_n(0,s))_+ +$
		$+ \int_0^z \exp\left\{-(\tilde{\gamma}_{c_{n,m}}(s))\left[z-z'\right]\right\}\cdot(\tilde{V}_n^{(s)'}(z',s))_+\, dz'$
	− or left wave $(\tilde{V}_n(z,s))_-$	$= \exp\left\{(\tilde{\gamma}_{c_{n,m}}(s))\left[z-L\right]\right\}\cdot(\tilde{V}_n(L,s))_-$
		$+ \int_L^z \exp\left\{(\tilde{\gamma}_{c_{n,m}}(s))\left[z-z'\right]\right\}\cdot(\tilde{V}_n^{(s)'}(z',s))_-\, dz'$

Table 3.2 Diagonalization of Propagation Matrix

Name	Symbol	Relation
Square of Propagation Matrix	$(\tilde{\gamma}_{c_{n,m}}(s))^2$	$= (Z'_{n,m}(s)) \cdot (Y'_{n,m}(s))$
Normalized Voltage Eigenvector	$(\tilde{v}_{c_n}(s))_\delta$	$(\tilde{\gamma}_{c_{n,m}}(s))^2 \cdot (\tilde{v}_{c_n}(s))_\delta = \tilde{\gamma}_\delta^2(s)\ (\tilde{v}_{c_n}(s))_\delta$
Normalized Current Eigenvector	$(\tilde{i}_{c_n}(s))_\delta$	$(\tilde{i}_{c_n}(s))_\delta \cdot (\tilde{\gamma}_{c_{n,m}}(s))^2 = \tilde{\gamma}_\delta^2(s)\ (\tilde{i}_{c_n}(s))_\delta$
Eigenvalue of Propagation Matrix	$\tilde{\gamma}_\delta(s)$	$= \left[\tilde{\gamma}_\delta^2(s)\right]^{\frac{1}{2}}$ (principal or p.r. value assumed)
Eigenindex	δ	$= 1, 2, \ldots, N$ (N×N matrices)
Biorthonormal Property (used for normalization)		$(\tilde{v}_{c_n}(s))_\xi \cdot (\tilde{i}_{c_n}(s))_{\xi'} = 1_{\delta,\delta'}$ (N independent eigenvectors assumed of both voltage and current types)
Function of Propagation Matrix	$F((\tilde{\gamma}_{c_{n,m}}(s)))$	$= \sum_\delta F(\tilde{\gamma}_\delta(s))\ (\tilde{v}_{c_n}(s))_\delta\ (\tilde{i}_{c_n}(s))_\delta$
Special Cases { Propagation Matrix	$(\tilde{\gamma}_{c_{n,m}}(s))$	$= \sum_\delta \tilde{\gamma}_\delta(s)\ (\tilde{v}_{c_n}(s))_\delta\ (\tilde{i}_{c_n}(s))_\delta$
Transpose	$(\tilde{\gamma}_{c_{n,m}}(s))^T$	$= \sum_\delta \tilde{\gamma}_\delta(s)\ (\tilde{i}_{c_n}(s))_\delta\ (\tilde{v}_{c_n}(s))_\delta$
Special Cases { Inverse	$(\tilde{\gamma}_{c_{n,m}}(s))^{-1}$	$= \sum_\delta \tilde{\gamma}_\delta^{-1}(s)\ (\tilde{v}_{c_n}(s))_\delta\ (\tilde{i}_{c_n}(s))_\delta$
Identity	$(1_{n,m})$	$= (\tilde{\gamma}_{n,m}(s))^0 = \sum_\xi (\tilde{v}_{c_n}(s))_\delta\ (\tilde{i}_{c_n}(s))_\delta$ $= \sum_\delta (\tilde{i}_{c_n}(s))_\delta\ (\tilde{v}_{c_n}(s))_\xi$

Table 3.3 Normalization of Voltage and Current Eigenmodes

Name	Symbol	Relation
Normalization via Characteristic Impedance and Admittance Matrices	Interrelation of voltage and current eigenmodes	$(\tilde{v}_{c_n}(s))_\delta = (\tilde{Z}_{c_{n,m}}(s)) \cdot (\tilde{I}_{c_n}(s))_\delta = (\tilde{I}_{c_n}(s))_\delta \cdot (\tilde{Z}_{c_{n,m}}$
		$(\tilde{I}_{c_n}(s))_\delta = (\tilde{Y}_{c_{n,m}}(s)) \cdot (\tilde{v}_{c_n}(s))_\delta = (\tilde{v}_{c_n}(s))_\delta \cdot (\tilde{Y}_{c_{n,m}}$
	Separate voltage and current eigenmode normalization	$1_{\delta,\delta'} = (\tilde{I}_{c_n}(s))_\delta \cdot (\tilde{Z}_{c_{n,m}}(s)) \cdot (\tilde{I}_{c_n}(s))_{\delta'}$
		$1_{\delta,\delta'} = (\tilde{v}_{c_n}(s))_\delta \cdot (\tilde{Y}_{c_{n,m}}(s)) \cdot (\tilde{v}_{c_n}(s))_{\delta'}$
Normalization via Per-Unit-Length Impedance and Admittance Matrices	Interrelation of voltage and current eigenmodes	$\tilde{\gamma}_\delta(s)(\tilde{v}_{c_n}(s))_\delta = (Z'_{n,m}(s)) \cdot (\tilde{I}_{c_n}(s))_\delta = (\tilde{I}_{c_n}(s))_\delta \cdot (Z'_n$
		$\tilde{\gamma}_\delta(s)(\tilde{I}_{c_n}(s))_\delta = (\tilde{Y}'_{n,m}(s)) \cdot (\tilde{v}_{c_n}(s))_\delta = (\tilde{v}_{c_n}(s))_\delta \cdot (\tilde{Y}'$
	Separate voltage and current eigenmode normalization	$1_{\delta,\delta'} = \tilde{\gamma}_\delta^{-1}(s)(\tilde{I}_{c_n}(s))_\delta \cdot (Z'_{n,m}(s)) \cdot (\tilde{I}_{c_n}(s))_{\delta'}$
		$1_{\delta,\delta'} = \tilde{\gamma}_\delta^{-1}(s)(\tilde{v}_{c_n}(s))_\delta \cdot (\tilde{Y}'_{n,m}(s)) \cdot (\tilde{v}_{c_n}(s))_{\delta'}$

Table 3.4 Representation of Other Matrices in Terms of Voltage and Current Normalized Eigenmodes (indicating assumed reciprocity)

Name	Symbol	Relation
Characteristic Impedance Matrix	$(\tilde{Z}_{c_{n,m}}(s))$	$= \sum_\delta (\tilde{v}_{c_n}(s))_\delta (\tilde{v}_{c_n}(s))_\delta$
Characteristic Admittance Matrix	$(\tilde{Y}_{c_{n,m}}(s))$	$= \sum_\delta (\tilde{I}_{c_n}(s))_\delta (\tilde{I}_{c_n}(s))_\delta$
Per-Unit-Length Impedance Matrix	$(Z'_{n,m}(s))$	$= \sum_\delta \tilde{\gamma}_\delta(s) (\tilde{v}_{c_n}(s))_\delta (\tilde{v}_{c_n}(s))_\delta$
Per-Unit-Length Admittance Matrix	$(\tilde{Y}'_{n,m}(s))$	$= \sum_\delta \tilde{\gamma}_\delta(s) (\tilde{I}_{c_n}(s))_\delta (\tilde{I}_{c_n}(s))_\delta$

For our purposes, the dimatrices will be square, i.e.,

$$N = M \tag{147}$$

Furthermore, the partitioning will be symmetric, i.e.,

$$N_u = M_v \quad \text{for } u = v \tag{148}$$

Hence the diagonal blocks

$$(D_{n,m})_{u,u} \text{ , size } N_u \times N_u \tag{149}$$

are square and off-diagonal blocks are symmetrically rectangular, i.e.,

$$(D_{n,m})_{u,v} \text{ , size } N_u \times N_v$$

$$(D_{n,m})_{v,u} \text{ , size } N_v \times N_u \tag{150}$$

Supervectors or divectors are similarly defined in the form

$$((V_n)_u) \tag{151}$$

with elementary vectors as

$$(V_n)_u$$

$$n = 1, 2, \ldots, N_u$$

$$u = 1, 2, \ldots, N \tag{152}$$

remembering that n,m and u,v are merely dummy indices. Note that the elements are designated as

$$V_{n;u} \text{ (not } V_{n,u}) \tag{153}$$

Define supermatrix multiplication in the dot product or contraction sense as

$$((A_{n,m})_{u,v}) : ((B_{n,m})_{u,v})$$

$$= \left(\sum_{u'=1}^{N} (A_{n,m})_{u,u'} \cdot (B_{n,m})_{u',v} \right)$$

$$= \left(\left(\sum_{u'=1}^{N} \sum_{n'=1}^{N_{u'}} A_{n,n';u,u'} \, B_{n',m;u',v} \right) \right)$$

$$= ((C_{n,m})_{u,v}) \tag{154}$$

Here we note contraction is done twice involving the second indices of the two pairs of indices for the first matrix, and the first indices of the two pairs of indices for the second matrix; this is denoted by two levels of dot product : , noting the two dots one above the other.

In (154) the two dimatrices are not necessarily square. It is merely required that the second indices m,v of the first dimatrix have the same range (hence same partitioning) as the first indices of the second dimatrix. Two dimatrices with this property are said to be of <u>compatible order</u>, for multiplication in the double dot product sense in this case, with order of multiplication specified.

In the present paper all dimatrices are taken as being of <u>symmetric compatible order</u>, i.e., (147, 148) apply and N and the N_u have the same values for all dimatrices in the particular discussion (i.e., describing a given physical situation). Furthermore, the directors are also taken as having the same compatible order. Thus we can form any such operations as

$$((A_{n,m})_{u,v}) + ((B_{n,m})_{u,v}) \qquad \text{dimatrix}$$

$$((A_{n,m})_{u,v}) : ((B_{n,m})_{u,v}) \qquad \text{diamatrix}$$

$$((B_{n,m})_{u,v}) : ((A_{n,m})_{u,v}) \qquad \text{dimatrix}$$

$$((A_{n,m})_{u,v}) : ((V_n)_u) \qquad \text{director}$$

$$((V_n)_u) : ((A_{n,m})_{u,v}) \qquad \text{director}$$

$$((V_n)_u) : ((W_n)_u) \qquad \text{scalar} \tag{155}$$

where dimatrix-director and director-director multiplication in the double dot product sense are obvious specializations of (154).

5.   IDENTITY SUPERMATRIX

Before continuing the supermatrices of the previous sections to yield an equation, it is necessary to define an identity supermatrix $((1_{n,m})_{u,v})$.

The identity supermatrix is such that its diagonal element matrices are all identity matrices, and all off-diagonal element matrices are zero matrices, i.e.,

$$(1_{n,m})_{u,v} = 1_{u,v}(1_{n,m})$$

$$1_{n,m} = \begin{cases} 1 \text{ for } n-m \\ 0 \text{ for } n \neq m \end{cases} \tag{156}$$

involving Kronecker deltas. The individual elements can be written as

$$1_{n,m;u,v} = \begin{cases} 1 \text{ for both } n=m \text{ and } u=v \\ 0 \text{ for either } n \neq m \text{ or } u \neq v \end{cases} \tag{157}$$

as a sort of super Kronecker delta or superidentity element. Note the identity supermatrix is then a symmetric dimatrix as in (147, 148).

For a supermatrix $((M_{n,m})_{u,v})$ of symmetric compatible order then

$$((1_{n,m})_{u,v}) : ((M_{n,m})_{u,v}) = ((M_{n,m})_{u,v})) : ((1_{n,m})_{u,v})$$

$$= ((M_{n,m})_{u,v}) \tag{158}$$

Also, an inverse $((M_{n,m})_{u,v})^{-1}$ of $((M_{n,m})_{u,v})$ exists such that

$$((M_{n,m})_{u,v})^{-1} : ((M_{n,m})_{u,v}) = ((M_{n,m})_{u,v}) : ((M_{n,m})_{u,v})^{-1}$$

$$= ((1_{n,m})_{u,v}) \tag{159}$$

provided

$$\det[((M_{n,n})_{u,v})] \neq 0 \tag{160}$$

Note that $((M_{n,m})_{u,v})^{-1}$ is of symmetric compatible order with $((M_{n,m})_{u,v})$.

6.   SCATTERING SUPERMATRIX

The concept of scattering matrices introduced in Section 3 for a terminated tube is extended here for junctions where more than one tube is connected. Collections and suitable ordering of scattering matrices at all junctions of the transmission-line network form a scattering supermatrix.

6.1 Junction Scattering Supermatrix

Consider the vth junction J_v with tube ends denoted by $J_{v;r}$ with index r as discussed in subsection 2.4. Let this junction be characterized by an impedance matrix

$$(\tilde{Z}_{n,m}(s))_v = (\tilde{Y}_{n,m}(s))_v^{-1} \tag{161}$$

The junction scattering matrix is defined so that

$$(\tilde{V}_n(s))_{v,+} = (\tilde{S}_{n,m}(s))_v \cdot (\tilde{V}_n(s))_{v,-} \tag{162}$$

where the subscripts + and - refer to the aggregate of respectively outgoing and incoming waves (N-waves) on the various tubes in the form of combined voltage vectors; remember that the current convention for outgoing waves is positive direction outward, and for incoming waves is positive direction inward.

In the supermatrix form, partition according to waves on the r_v tube ends connected to J_v as:

$$((\tilde{V}_n^{(0)}(s))_r)_v = ((\tilde{Z}_{n,m}(s))_{r,r'})_v : ((\tilde{I}_n^{(0)}(s))_r)_v$$

$$((\tilde{Y}_{n,m}(s))_{r,r'})_v \equiv ((\tilde{Z}_{n,m}(s))_{r,r'})_v^{-1} \tag{163}$$

where

$$(\tilde{V}_n^{(0)}(s))_{r;v} \quad , \quad (\tilde{I}_n^{(0)}(s))_{r;v}$$

$$r = 1,2,\ldots,r_v \tag{164}$$

are the voltage and current vectors on the rth tube ends at J_v with current convention into J_v.

The tube associated with the rth tube end at J_v has characteristic impedance and admittance matrices which can be put in supermatrix form for J_v as

$$((\tilde{Z}_{c_{n,m}}(s))_{r,r'})_v \equiv \text{tube-end characteristic-impedance super-}$$
$$\text{matrix for } J_v$$

$$((\tilde{Y}_{c_{n,m}}(s))_{r,r'})_v \equiv \text{tube-end characteristic-admittance}$$
$$\text{supermatrix for } J_v$$

$$= ((\tilde{Z}_{c_{n,m}}(s))_{r,r'})_v^{-1} \tag{165}$$

where

$$(\tilde{Z}_{c_{n,m}}(s))_{r,r';v} \equiv \begin{cases} \text{characteristic-impedance matrix for rth} \\ \text{tube end at } J_v \text{ for } r=r' \text{ (square} \\ (0_{n,m}) \text{ for } r \neq r' \text{ (rectangular)} \end{cases}$$

$$(\tilde{Y}_{c_{n,m}}(s))_{r,r';v} \equiv \begin{cases} \text{characteristic-admittance matrix for rth} \\ \text{tube end at } J_v \text{ for } r=r' \text{ (square)} \\ (0_{n,m}) \text{ for } r \neq r' \text{ (rectangular)} \end{cases}$$

$$(\tilde{Y}_{c_{n,m}}(s))_{r,r;v} = (\tilde{Z}_{c_{n,m}}(s))^{-1}_{r,r;v} \tag{166}$$

Thus, these impedance and admittance supermatrices for the tube ends at a given junction are block diagonal and may be represented in terms of the direct sum $\oplus$ as

$$((\tilde{Z}_{c_{n,m}}(s))_{r,r'})_v \equiv (\tilde{Z}_{c_{n,m}}(s))_{1,1;v} \oplus (\tilde{Z}_{c_{n,m}}(s))_{2,2;v}$$

$$\oplus \ldots \oplus (\tilde{Z}_{c_{n,m}}(s))_{r_v,r_v;v}$$

$$\equiv \overset{r_v}{\underset{r=1}{\oplus}} (\tilde{Z}_{c_{n,m}}(s))_{r,r;v}$$

$$((\tilde{Y}_{c_{n,m}}(s))_{r,r'})_v \equiv (\tilde{Y}_{c_{n,m}}(s))_{1,1;v} \oplus (\tilde{Y}_{c_{n,m}}(s))_{2,2;v}$$

$$\oplus \ldots \oplus (\tilde{Y}_{c_{n,m}}(s))_{r_v,r_v;v}$$

$$\equiv \overset{r_v}{\underset{r=1}{\oplus}} (\tilde{Y}_{c_{n,m}}(s))_{r,r;v} \tag{167}$$

where the convention used here is to maintain the partitioning according to the two pairs of indices (n,m and r,r') instead of combining them in one pair as in a regular matrix (or monomatrix). Note the subscript v on the supermatrices; the elementary matrices are also identified with v and the r,r' indices range over the tube ends at J_v, not over the wave indices u,v.

The scattering supermatrix for J_V is defined by

$$((\tilde{V}_n(s))_r)_{V,+} \equiv ((\tilde{S}_{n,m}(s))_{r,r'})_V : ((\tilde{V}_n(s))_r)_{V,-}$$

$$((\tilde{V}_n(s))_r)_{V,+} \equiv ((\tilde{V}_n^{(0)}(s))_r)_V - ((Z_{c_{n,m}}(s))_{r,r'})_V : ((\tilde{I}_n^{(0)}(s))_r)_V$$

$$\equiv \text{ outgoing wave supervector at } J_V$$

$$((\tilde{V}_n(s))_r)_{V,-} \equiv ((\tilde{V}_n^{(0)}(s))_r)_V + ((\tilde{Z}_{c_{n,m}}(s))_{r,r'})_V : ((\tilde{I}_n^{(0)}(s))_r)$$

$$\equiv \text{ incoming wave supervector at } J_V \qquad (168)$$

Again note that the J_V current convention is positive current into J_V so that the usual Ohm's law convention in (163) holds for J_V. Solving (168) for the voltage and current supervectors at J_V:

$$((\tilde{V}_n^{(0)}(s))_r)_V = \frac{1}{2}[((\tilde{V}_n(s))_r)_{V,+} + ((\tilde{V}_n(s))_r)_{V,-}]$$

$$((\tilde{I}_n^{(0)}(s))_r) = \frac{1}{2}((\tilde{Y}_{c_{n,m}}(s))_{r,r'})_V : [((\tilde{V}_n^{(0)}(s))_r)_{V,-}$$

$$- ((\tilde{V}_n^{(0)}(s))_r)_{V,+}] \qquad (169)$$

Now we can compute the junction scattering supermatrix for J_V by combining (168) and (169) with (163) to give

$$((\tilde{S}_{n,m}(s))_{r,r'})_V = [((\tilde{Z}_{n,m}(s))_{r,r'})_V : ((\tilde{Y}_{c_{n,m}}(s)_{r,r'})_V$$

$$+ ((1_{n,m})_{r,r'})_V]^{-1} : [((\tilde{Z}_{n,m}(s))_{r,r'})_V$$

$$: ((\tilde{Y}_{c_{n,m}}(s))_{r,r'})_V - ((1_{n,m})_{r,r'})_V]$$

$$= [((1_{n,m})_{r,r'})_V + ((\tilde{Z}_{c_{n,m}}(s))_{r,r'})_V$$

$$: ((\tilde{Y}_{n,m}(s))_{r,r'})_V]^{-1} : [((1_{n,m})_{r,r'})_V$$

$$- ((\tilde{Z}_{c_{n,m}}(s))_{r,r'})_V : ((\tilde{Y}_{n,m}(s))_{r,r'})_V] \qquad (170)$$

Note the identity supermatrix corresponding to J_v; it is, of course, partitioned in the same symmetric compatible order as are the various impedance and admittance supermatrices and the scattering supermatrix for J_v.

For the junction J_v let the rth tube end have $N_{v;r}$ conductors (plus the reference) so that the wave on this tube end is a vector of dimension $N_{v;r}$. The supervectors then have dimension for J_v as

$$N_v = \sum_{r=1}^{r_v} N_{v;r} \qquad (171)$$

The associated supermatrices are $r_v x r_v$ in terms of the blocks or elementary matrices; the corresponding matrices (unpartitioned) are $N_v x N_v$. The reader may consult Fig. 3 for an example of tube ends connecting to a junction.

6.2 Reindexing of Elementary Matrices in the Collection of Junction Scattering Supermatrices

Having considered the junction scattering supermatrix for J_v and noting that $v = 1,2,\ldots,N_J$ gives all the junctions, we then have the elementary scattering matrices from one tube to another wherever there is such a connection at any junction. The problem is one of rearranging the equations so as to combine the results for junctions and tubes to obtain a description of the overall transmission-line network.

To convert the junction scattering supermatrix to a network scattering supermatrix, consider the tube-end-wave matrix $(t_{r,u})_{v;E-W}$ which relates the tube ends (r) at junction J_v to the waves W_u on those tubes. Recall the definition from (26) of the elements of the tube-end-wave matrix as

$$t_{r,u;v;E-W} \equiv \begin{cases} 0 \text{ if } W_u \text{ does not connect to } J_v \text{ via the rth} \\ \text{tube end (i.e., } J_{v;r}) \\ -1 \text{ if } W_u \text{ enters } J_v \text{ via the rth tube end} \quad (172) \\ (i.e., J_{v;r,-}) \\ +1 \text{ if } W_u \text{ leaves } J_v \text{ via the rth tube end} \\ (i.e., J_{v;r,+}) \end{cases}$$

One can then construct this matrix for each J_v for $v = 1, 2, \ldots, N_J$ from the topological diagram (graph) for the transmission-line network giving the junction J_v numbering and wave W_u numbering (as in the example in Figure 2b), and from the corresponding diagram for each junction J_v including tube-end labeling (r to $J_{v;r}$) (as in the example for J_3 in Fig. 3).

Now to associate an elementary scattering matrix $(\tilde{S}_{n,m}(s))_{r,r';v}$ for two tube ends, $J_{v;r}$ and $J_{v;r'}$ with $(\tilde{S}_{n,m}(s))_{u,v}$ corresponding to two waves, W_u and W_v, in the over-all network is straightforward; one must associate

$$r' \text{ (or } J_{v;r'}) \rightarrow v \text{ (or } W_v) \text{ for the incoming wave}$$

$$r \text{ (or } J_{v;r}) \rightarrow u \text{ (or } W_u) \text{ for the outgoing wave} \qquad (173)$$

However, this is what the tube-end-wave matrix does.

Consider incoming waves corresponding to the second index, v, in $(\tilde{S}_{n,m}(s))_{u,v}$. For one and only one J_v there is a negative entry in $(t_{r',v})_{v;E-W}$ under the vth column; the corresponding row is the value of r'. Hence, for each v

$$r' \text{ (in } J_{v;r'}) \text{ is that } r' \ni t_{r',v;v;E-W} = -1 \qquad (174)$$

which is readily found and even automated on a computer. Said another way, v is a function (an integer function) of v and r'. To aid in the search for $J_{v;r'}$ the value of v (or junction J_v) is found from the junction-wave matrix $(t_{v,v})_{J-W}$ (as in 21)) by finding those values of v for which $t_{v,v;J-W}$ is nonzero; there are at most two such values of v corresponding to the W_v leaving one junction and entering another junction, except in the case of a self tube where W_v both leaves and enters the same junction. Considering the one or two possible J_v, the value of v and r' are readily found as in (174) or via a diagram. After going through $v = 1, 2, \ldots, N_W$ one can construct a table in the form,

v	v	r'
1		
2		
.		
.		
.		
N_W		

$$(175)$$

which is a table of correspondence of incoming waves W_v to junctions J_v and tube ends $J_{v;r'}$, with the values of v and r' filled in for every v.

Similarly, for outgoing waves corresponding to the first index u in $(\widetilde{S}_{n,m}(s))_{u,v}$, we have a value of r given by

$$r \text{ (in } J_{v;r}) \text{ is that } r \ni t_{r,u;v;E-W} = +1 \tag{176}$$

Hence, u is a function of v and r. Again utilizing the junction-wave matrix $(t_{v,j})_{J-W}$ and finding the values of $t_{v,u;J-W}$ which are nonzero, one reduces the consideration to at most two values of the junction index v. The values of v and r are then readily found from the tube-end-wave matrices as in (176). After going through $u = 1,2,\ldots,N_W$ one can construct a table in the form,

u	v	r
1		
2		
.		
.		
.		
N_W		

which is a table of correspondence of outgoing waves W_u to junctions J_v and tube ends $J_{v;r}$, with the values of v and r filled in for every u.

Hence, with each pair (u,v) we associate the pair $(J_{v_1;r}, J_{v_2;r'})$. Now we have:

$$v_1 = v_2 \equiv v \quad \text{for } W_v \text{ scattering into } W_u \text{ at junction } J_v$$

$$v_1 \neq v_2 \quad \text{for } W_v \text{ not scattering into } W_u \text{ (no interconnection) at any } J_v \tag{177}$$

Then we form the network elementary scattering matrices as

$$
(\tilde{S}_{n,m}(s))_{u,v} \equiv \begin{cases} (\tilde{S}_{n,m}(s))_{r,r';v} & \text{for } v_1 = v_2 = v \text{ or } W_v \\ & \text{scattering into } W_u \text{ at } J_v \\ (0_{n,m}) = (0_{n,m})_{u,v} & \text{for } v_1 \neq v_2 \text{ or } W_v \\ & \text{not scattering into } W_u \end{cases}
$$

$$(178)$$

This gives an explicit algorithm for constructing the $(\tilde{S}_{n,m}(s))_{u,v}$ from the collection of junction scattering super-matrices $((\tilde{S}_{n,m}(s))_{r,r'})_v$.

The reader will also note the correspondence of these results with the wave-wave matrix $(W_{u,v})$ as defined in (24) as

$$
W_{u,v} = \begin{cases} 1 & \text{for } v_1 = v_2 \equiv v \text{ and } W_v \text{ scattering into} \\ & W_u \text{ at } J_v \\ 0 & \text{for } v_1 \neq v_2 \text{ or } W_v \text{ not scattering} \\ & \text{into } W_u \end{cases}
$$

$$(179)$$

The wave-wave matrix then indicates which (u,v) pairs must be considered for finding nonidentically zero scattering matrices, thereby simplifying the search among the elementary matrices comprising the junction scattering supermatrices.

6.3 Scattering Supermatrix

The proper ordering of all the junction scattering matrices into one large matrix forms the system (or network) scattering supermatrix $((\tilde{S}_{n,m}(s))_{u,v})$. This supermatrix is a collection of the junction scattering matrices, which themselves are collections of individual tube scattering matrices. The latter are matrices containing reflection and transmission coefficients of individual wires within the tubes. Thus, $((\tilde{S}_{n,m}(s))_{u,v})$ is a dimatrix (or tensor of rank four).

The wave-wave matrix $(W_{u,v})$ gives the structure of the scattering supermatrix since the scattering supermatrix is in general block sparse as:

$$(\widetilde{S}_{n,m}(s))_{u,v} = (0_{n,m})_{u,v} \quad \text{for } W_{u,v} = 0 \tag{180}$$

Hence, also, the scattering supermatrix is $N_W \times N_W$ in terms of the u,v indices, i.e.,

$$u,v = 1,2...,N_W \tag{181}$$

The elementary scattering matrices $(\widetilde{S}_{n,m}(s))_{u,v}$ are $N_u \times N_v$; i.e.,

$$n = 1,2,...,N_u$$

$$m = 1,2,...,N_v \tag{182}$$

where

N_u = number of conductors (not including reference) on the
    tube with uth wave   (183)

and likewise for N_v.

As a special case, it is interesting to note that if there are no self tubes (with both ends connected to the same junction), then

$$W_{u,u} = 0 \quad \text{for } u = 1,2,...N_W \quad \text{for no self tubes} \tag{184}$$

$$(\widetilde{S}_{n,m}(s))_{u,u} = (0_{n,m})_{u,u} \quad \text{for } n,m = 1,2,...,N_u \text{ (square)}$$

In this case the scattering supermatrix has zero matrices for its diagonal blocks; this will complement the identity supermatrix which has, as its only nonzero elementary matrices, the diagonal blocks which are identity matrices (as discussed in Section 5). This case is anticipated to be quite common in practice.

7.   DEFINITIONS OF SOME IMPORTANT SUPERMATRIX AND SUPERVECTOR
      QUANTITIES BASED ON RESULTS FOR WAVES ON A TUBE

This section takes the results for the combined voltages on a tube and separates them into the wave variables for the network. The resulting equation for a general combined voltage wave W_u is used to relate the combined voltages at both ends of the tube with the sources along the tube. Each term is generalized to a form appropriate to the transmission-line network, i.e., supermatrices and supervectors, by aggregating the results for all W_u for $u = 1,2,...,N_W$.

7.1 Common Equation for the Two Waves on a Tube

Let us take the results for the propagation on a single tube developed in subsection 3.3 from (92) and (93) as

$$(\tilde{V}_n(z,s))_+ = \exp\{-(\tilde{\gamma}_{c_{n,m}}(s))z\}\cdot(\tilde{V}_n(0,s))_+$$

$$+ \int_0^z \exp\{-(\tilde{\gamma}_{c_{n,m}}(s))[z-z']\}\cdot(\tilde{V}_n^{(s)'}(z',s))_+ \, dz'$$

$$(\tilde{V}_n(z,s))_- = \exp\{(\tilde{\gamma}_{c_{n,m}}(s))[z-L]\}\cdot(\tilde{V}_n(L,s))_-$$

$$\qquad\qquad\qquad\qquad\qquad\qquad\qquad\qquad\qquad\qquad\qquad (185)$$

$$+ \int_L^z \exp(\tilde{\gamma}_{c_{n,m}}(s))[z-z']\}\cdot(\tilde{V}_n^{(s)'}(z',s))_- \, dz'$$

Then, as discussed in subsection 3.4, let us identify the two waves on the tube with two waves of the transmission-line network, say W_u and W_v.

Consider the + wave; call this W_u and set the coordinate and dimension variables as

$L_u \equiv L \equiv$ length of path for W_u

$z_u \equiv z \equiv$ wave coordinate for W_u

$0 \le z_u \le L_u$

$N_u \equiv N \equiv$ number of conductors (less reference) on tube and dimension of vectors for W_u $\qquad$ (186)

The wave and source conventions are then

$$(\tilde{V}_n(z_u,s))_u \equiv (\tilde{V}_n(z,s))_+ = (\tilde{V}_n(z_u,s))$$

$$+ (\tilde{Z}_{c_{n,m}}(s))_u \cdot (\tilde{I}_n(z_u,s))$$

$$\equiv \text{combined voltage for } W_u$$

$$(\tilde{V}_n^{(s)'}(z_u,s))_u \equiv (\tilde{V}_n^{(s)'}(z,s))_+ = (\tilde{V}_n^{(s)'}(z_u,s))$$

$$+ (\tilde{Z}_{c_{n,m}}(s))_u \cdot (\tilde{I}_n^{(s)'}(z_u,s))$$

$\equiv$ combined voltage source per unit length for W_u

$(\tilde{Z}_{c_{n,m}}(s))_u \equiv (\tilde{Y}_{c_{n,m}}(s))_u^{-1} \equiv$ characteristic impedance matrix for W_u

$(\tilde{\gamma}_{c_{n,m}}(s))_u \equiv (\tilde{\gamma}_{c_{n,m}}(s)) \equiv$ propagation matrix for W_u (187)

Note that for W_u the current $(\tilde{I}_n(z_u,s))$ convention is taken as positive in the direction of increasing z_u (as in Fig. 5). Likewise, the voltage source per unit length $(\tilde{V}_n^{(s)'}(z_u,s))$ (including any discrete voltage sources) is taken as positive increasing in the direction of increasing z_u. The first of (185) takes the form for W_u as

$$(\tilde{V}_n(z_u,s))_u = \exp\{-(\tilde{\gamma}_{c_{n,m}}(s))_u z_u\} \cdot (\tilde{V}_n(0,s))_u$$

$$+ \int_0^{z_u} \exp\{-(\tilde{\gamma}_{c_{n,m}}(s))_u [z_u - z_u']\}$$

$$\cdot (\tilde{V}_n^{(s)'}(z_u',s))_u \, dz_u' \tag{188}$$

Next consider the - wave; call this W_v and set the coordinate dimension variables as

$L_v \equiv L \equiv$ length of path for W_v

$z_v \equiv L - z \equiv$ wave coordinate for W_v

$0 \leq z_v \leq L_v$

$N_v \equiv N \equiv$ number of conductors (less reference) on tube and dimension of vectors for W_v (189)

The wave and source conventions are then

$$(\tilde{V}_n(s_v,s))_v \equiv (\tilde{V}_n(z,s))_- = (\tilde{V}_n(z_v,s))$$

$$+ (\tilde{Z}_{c_{n,m}}(s))_v \cdot (\tilde{I}_n(z_v,s))$$

$$\equiv \text{combined voltage for } W_v$$

$$(\tilde{V}_n^{(s)'}(z_v,s))_v \equiv -(\tilde{V}_n^{(s)'}(z,s))_- = (\tilde{V}_n^{(s)'}(s_v,s))$$

$$+ (\tilde{Z}_{c_{n,m}}(s))_v \cdot (\tilde{I}_n^{(s)'}(z_v,s))$$

$$\equiv \text{combined voltage source per unit length} \\ \text{for } W_v$$

$$(\tilde{Z}_{c_{n,m}}(s))_v \equiv (\tilde{Y}_{c_{n,m}}(s))_v^{-1} \equiv \text{characteristic impedance matrix} \\ \text{for } W_v$$

$$(\tilde{\gamma}_{c_{n,m}}(s))_v \equiv (\tilde{\gamma}_{c_{n,m}}(s)) \equiv \text{propagation matrix for } W_v \quad (190)$$

Now for W_v the current $(\tilde{I}_n^{(s)}(z_v,s))$ convention is taken as positive in the direction of increasing z_v and, hence, of decreasing z (opposite to that for W_u). Similarly, the voltage source per unit length is taken as positive in the direction of increasing z_v, which is the direction of decreasing z. This is so that for W_v the conventions are defined with respect to z_v, in the same manner as for W_u they have been defined with respect to z_u. The second of (185) then takes the form for W_v as

$$(\tilde{V}_n(z_v,s))_v = \exp\{-(\tilde{\gamma}_{c_{n,m}}(s))_v z_v\} \cdot (\tilde{V}_n(0,s))_v$$

$$+ \int_0^{z_v} \exp\{-(\tilde{\gamma}_{c_{n,m}}(s))_v[z_v-z_v']\}$$

$$\cdot (\tilde{V}_n^{(s)'}(z_v',s))_v \, dz_v' \quad (191)$$

which is exactly the same as for W_u in (188). Hence, only one such equation need be considered; it is applicable for all $u = 1,2,\ldots,N_W$, thereby applying to all waves in the transmission-line network.

The chosen conventions for W_u and W_v to have the same form with respect to z_u and z_v are then important to simplifying the formulation of the network equations. With these choices we have relations between the two waves W_u and W_v on the same tube for coordinates and dimensions as:

$$L_u \equiv L_v \equiv L$$

$$z_u + z_v = L$$

$$N_u = N_v = N \qquad (192)$$

The wave and source relations are (for uncombined quantities)

$$(\tilde{V}_n(z_u,s)) = (\tilde{V}_n(z_v,s))$$

$$(\tilde{I}_n(z_u,s)) = -(\tilde{I}_n(z_v,s))$$

$$(\tilde{V}_n^{(s)'}(z_u,s)) = -(\tilde{V}_n^{(s)'}(z_v,s))$$

$$(\tilde{I}_n^{(s)'}(z_u,s)) = (\tilde{I}_n^{(s)'}(z_v,s))$$

$$(\tilde{Z}_{c_{n,m}}(s))_u = (\tilde{Z}_{c_{n,m}}(s))_v = (\tilde{Z}_{c_{n,m}}(s))$$

$$(\tilde{V}_{c_{n,m}}(s))_u = (\tilde{\gamma}_{c_{n,m}}(s))_v = (\tilde{\gamma}_{c_{n,m}}(s)) \qquad (193)$$

7.2 Relation of Combined-Voltage Waves on Both Ends of a Tube

Now in (188) (or equivalently, (191)) we have the combined voltage at any z_u in terms of the value (boundary condition) at $z_u = 0$. Setting $z_u = L_u$ we introduce the boundary value there as giving

$$(\tilde{V}_n(L_u,s))_u = \exp\{-(\tilde{\gamma}_{c_{n,m}}(s))_u L_u\} \cdot (\tilde{V}_n(0,s))_u$$

$$+ \int_0^{L_u} \exp\{-(\tilde{\gamma}_{c_{n,m}}(s))_u [L_u - z_u']\}$$

$$\cdot (\tilde{V}_n^{(s)'}(z_u',s))_u \, dz_u' \qquad (194)$$

This evidently relates $(\tilde{V}_n(0,s))_u$, which is an outgoing wave from the junction at $z_u = 0$, to $(\tilde{V}_n(L_u,s))$, which is an incoming wave

to the junction at $z_u = L_u$. This is used later with the scattering supermatrix to form the BLT equation for the transmission-
line network.

As a matter of convention, let all sources be considered as
being present in the tubes instead of the junctions. If one has
a junction with an equivalent circuit containing sources, as for
example in Fig. 4, then the sources can be moved just across the
terminals into the tube, a movement of zero distance. Note then
that the boundary values $(\tilde{V}_n(0,s))_u$ and $(\tilde{V}_n(L_u,s))_u$ are combined
voltages on the junction "side" of the connections to the junc-
tion. Given this convention again, note the different conventions for sources for the two different waves on a tube, as discussed above.

7.3 Propagation Characteristic Supermatrix

Considering the various terms in (194), let us first aggregate all the propagation terms not associated with the sources
into a block diagonal propagation supermatrix as

$$
((\tilde{\Gamma}_{n,m}(s))_{u,v})
$$

$$
\equiv \exp\{-(\tilde{\gamma}_{c_{n,m}}(s))_1 L_1\} \oplus \exp\{-(\tilde{\gamma}_{c_{n,m}}(s))_2 L_2\}
$$

$$
\oplus \ldots \oplus \exp\{-(\tilde{\gamma}_{c_{n,m}}(s))_{N_W} L_{N_W}\}
$$

$$
\equiv \overset{N_W}{\underset{u=1}{\oplus}} \exp\{-(\tilde{\gamma}_{c_{n,m}}(s))_u L_u\}
$$

$$
\equiv \text{propagation supermatrix} \tag{195}
$$

where the elementary matrices (blocks) are given by

$$
(\tilde{\Gamma}_{n,m}(s))_{u,v} =
\begin{cases}
\exp\{-(\tilde{\gamma}_{c_{n,m}}(s))_u L_u\} & \text{for } u = v \\
(0_{n,m}) & \text{for } u \neq v
\end{cases}
$$

$$
= 1_{u,v} \exp\{-(\tilde{\gamma}_{c_{n,m}}(s))_u L_u\} \tag{196}
$$

7.4 Source Supervector and Supermatrix Integral Operator

Again from (194) let us define a source vector for W_u in traveling from $z_u = 0$ to $z_u = L_u$ as

$$(\tilde{v}_n^{(s)}(s))_u \equiv \int_0^{L_u} \exp\{-(\tilde{\gamma}_{c_{n,m}}(s))_u[L_u-z_u']\}$$

$$\cdot (\tilde{v}_n^{(s)'}(z_u',s))_u \; dz_u' \tag{197}$$

The source supervector is then merely

$$((\tilde{v}_n^{(s)}(s))_u) \equiv \left(\int_0^{L_u} \exp\{-(\tilde{\gamma}_{c_{n,m}}(s))_u[L_u-z_u']\} \right.$$

$$\left. \cdot \; (\tilde{v}_n^{(s)'}(z_u',s))_u \; dz_u' \right) \tag{198}$$

One can factor the above result by the use of a supermatrix integral operator. Define the elementary matrix blocks of this operator as

$$(\tilde{\Lambda}_{n,m}(z_u',s;(\cdot)))_{u,v}$$

$$\equiv \begin{cases} \displaystyle\int_0^{L_u} \exp\{-(\tilde{\gamma}_{c_{n,m}}(s))_u[L_u-z_u']\}(\cdot) \; dz_u' & \text{for } u = v \\[2em] (0_{n,m}) & \text{for } u \neq v \end{cases}$$

$$= 1_{u,v} \int_0^{L_u} \exp\{-(\tilde{\gamma}_{c_{n,m}}(s))_u[L_u-z_u'](\cdot) \; dz_u' \tag{199}$$

where the argument $(\cdot)$ indicates the place to put the expression following the operator in order to perform the operation. This is defined so that:

$$(\tilde{\Lambda}_{n,m}(z_u',s;(\cdot)))_{u,v} \cdot (\tilde{V}_n^{(s)'}(z_u',s))_v$$

$$= 1_{u,v} \; (\tilde{V}_n^{(s)}(s))_v$$

$$= 1_{u,v} \int_0^{L_v} \exp\{-(\tilde{\gamma}_{c_{n,m}}(s))_v[L_v-z_v']\}$$

$$\cdot(\tilde{V}_n^{(s)'}(z_v',s)) \; dz_v' \tag{200}$$

with the multiplication, which is part of the operation, taken in the dot product sense. We can then readily form

$$((\tilde{V}_n^{(s)}(s))_u) = ((\tilde{\Lambda}_{n,m}(z_u',s;(\cdot)))_{u,v}):((\tilde{V}_n^{(s)'}(z_u',s))_u)$$

$$((\tilde{\Lambda}_{n,m}(z_u',s;(\cdot)))_{u,v})$$

$$\equiv \bigoplus_{u=1}^{N_W} (\tilde{\Lambda}_{n,m}(z_u',s;(\cdot)))_{u;v}$$

$$\equiv \bigoplus_{u=1}^{N_W} \int_0^{L_u} \exp\{-(\tilde{\gamma}_{c_{n,m}}(s))_u[L_u-z_u']\}(\cdot) \; dz_u'$$

$$\equiv \text{propagation supermatrix integral operator}$$

$$((\tilde{V}_n^{(s)'}(z_u',s))_u) \equiv \text{distributed source supervector} \tag{201}$$

Note that the propagation supermatrix integral operator in (201) is a generalization of the propagation supermatrix in (195) to allow for continuous combined voltage sources along the wave coordinates instead of just the boundary conditions (equivalent sources) at the set of $z_u = 0$.

7.5 Combined Voltage Supervector

For completeness we have the aggregate of combined voltage vectors in (194) as

$((\tilde{V}_n(0,s))_u) \equiv$ combined voltage supervector of outgoing
waves at the junctions

$((\tilde{V}_n(L_u,s))_u) \equiv$ combined voltage supervector of incoming
waves at the junctions   (202)

In this paper we will formulate the BLT equation in terms of the outgoing waves at the junctions, but other forms are also possible.

8.   BLT EQUATION

Combining the results of the previous derivations we can write the BLT equation for the description of the transmission-line network. We begin with the scattering supermatrix in Section 6 which relates the incoming waves to the outgoing waves as

$$((\tilde{V}_n(0,s))_u) = ((\tilde{S}_{n,m}(s))_{u,v}):((\tilde{V}_n(L_u,s))_u) \qquad (203)$$

using the combined voltage supervectors from subsection 7.5. Note the distinction between incoming waves $(z_u = L_u)$ and outgoing waves $(z_u = 0)$ at the set of junctions or tube ends.

Next, relate the incoming waves at the output ends of the tubes $(z_u = L_u)$ to the same waves at the input end of the same tubes $(z_u = L_u)$, albeit at different junctions in general. Taking (194) in supermatrix form, we have

$((\tilde{V}_n(L_u,s))_u)$

$\quad = ((\tilde{\Gamma}_{n,m}(s))_{u,v}):((\tilde{V}_n(0,s))_u) + ((\tilde{V}_n^{(s)}(s))_u)$

$\quad = ((\tilde{\Gamma}_{n,m}(s))_u):((\tilde{V}_n(0,s))_u)$

$\quad + ((\tilde{\Lambda}_{n,m}(z_u',s;(\cdot)))_{u,v}):((\tilde{V}_n^{(s)\,'}(z_u',s))_u) \qquad (204)$

Combining (203) and (204) we have

$((\tilde{V}_n(0,s))_u)$

$\quad = ((\tilde{S}_{n,m}(s))_{u,v}):((\tilde{\Gamma}_{n,m}(s))_{u,v}):((\tilde{V}_n(0,s))_u)$

$\quad + ((\tilde{S}_{n,m}(s))_{u,v}):((\tilde{V}_n^{(s)}(s))_u) \qquad (205)$

That is rearranged by use of the supermatrix identity as

$$
\begin{aligned}
[((1_{n,m})_{u,v}) &- ((\tilde{S}_{n,m}(s))_{u,v}) : ((\tilde{\Gamma}_{n,m}(s))_{u,v})] : ((\tilde{V}_n(0,s))_u) \\
&= ((\tilde{S}_{n,m}(s))_{u,v}) : ((\tilde{V}_n^{(s)}(s))_u) \\
&= ((\tilde{S}_{n,m}(s))_{u,v}) : ((\tilde{\Lambda}_{n,m}(z_u',s;(\cdot)))_{u,v}) \\
&\quad : ((\tilde{V}_n^{(s)'}(z_u',s))_u)
\end{aligned}
$$

(206)

This is one form of the BLT equation with the unknowns taken as the combined voltage waves leaving the junctions. Note again that all sources are given a convention as being on the tubes in the wave coordinates $0 \le z_u < L_u$ so they are picked up in the integration along the wave coordinates and are not included in the combined voltages at the junctions $((\tilde{V}_n(0,s))_u)$ which are being computed.

For computational purposes the BLT equation is one large matrix equation with square matrices of size NxN and vectors of dimension N where

$$
N = \sum_{u=1}^{N_W} N_u
$$

(207)

Noting, however, the sparse nature of these matrices with blocks of zeros, one may be able to take advantage of the partitioning used to construct the supermatrices to simplify computations.

9.   RECONSTRUCTION OF VOLTAGES AND CURRENTS

Having solved the BLT equation as in (206) in some form or other, we have a set of combined voltages such as the outgoing combined voltages $((\tilde{V}_n(0,s))_u)$ at the junctions. From these one can find voltages and currents essentially everywhere, including at the junction terminals (tube ends) and at arbitrary positions on the tubes.

Consider the important case of voltages and currents at the tube ends (junctions). Let the two waves on a particular tube be W_u and W_v as in subsection 7.1. Using the conventions established there we have

$$N_u = N_v = N \equiv \text{dimension of vectors}$$

$$L_u = L_v = L \equiv \text{length}$$

$$z_u + z_v = L \equiv \text{relation between two wave coordinates}$$

$$z = z_u = L - z_v \equiv \text{tube coordinate} \tag{208}$$

Then we have at z=0

$$(\tilde{V}_n(0,s)) = \frac{1}{2} [(\tilde{V}_n(0,s))_u + (\tilde{V}_n(L_v,s))_v]$$

$$(\tilde{I}_n(0,s)) = \frac{1}{2} (\tilde{Y}_{c_{n,m}}(s)) \cdot [(\tilde{V}_n(0,s))_u - (\tilde{V}_n(L_v,s))_v] \tag{209}$$

with the current positive in the +z direction or out of the junction at z=0. At the other end with z=L we have

$$(\tilde{V}_n(L,s)) = \frac{1}{2} [(\tilde{V}_n(L_u,s))_u + (\tilde{V}_n(0,s))_v]$$

$$(\tilde{I}_n(L,s)) = \frac{1}{2} (\tilde{Y}_{c_{n,m}}(s)) \cdot [(\tilde{V}_n(L_u,s))_u - (\tilde{V}_n(0,s))_v] \tag{210}$$

with current positive in the +z direction or into the junction at z=L. For substitution into the above equations, one uses

$$(\tilde{V}_n(L_u,s))_u = (\tilde{\Gamma}_{n,m}(s))_{u,u} \cdot (\tilde{V}_n(0,s))_u$$

$$+ (\tilde{\Lambda}_{n,m}(z_u',s;(\cdot)))_{u,u} \cdot (\tilde{V}_n^{(s)'}(z_u',s))_u$$

$$(\tilde{V}_n(L_v,s))_v = (\tilde{\Gamma}_{n,m}(s))_{v,v} \cdot (\tilde{V}_n(0,s))_v$$

$$+ (\tilde{\Lambda}_{n,m}(z_v',s;(\cdot)))_{v,v} \cdot (\tilde{V}_n^{(s)'}(z_v',s))_v \tag{211}$$

so that W_u has $z_u = L_u$ related to $z_u = 0$ and W_v has $z_v = L_v$ related to $z_v = 0$. In this form the current and voltage vectors

equation (206), and the combined sources along the tube via (211)
for the two waves on the tube.

For more general positions along the tube of interest

$$(\tilde{V}_n(z,s)) = \frac{1}{2} [(\tilde{V}_n(z_u,s))_u + (\tilde{V}_n(z_v,s))_v]$$

$$(\tilde{I}_n(z,s)) = \frac{1}{2} (\tilde{Y}_{c_{n,m}}(s)) \cdot [(\tilde{V}_n(z_u,s))_u - (\tilde{V}_n(z_v,s))_v] \qquad (212)$$

with current positive in the +z direction which is equivalent to
the $+z_u$ direction and to the $-z_v$ direction. For substituting
into (212), one uses (188) and (191) repeated here as

$$(\tilde{V}_n(z_u,s))_u = \exp\{-(\tilde{\gamma}_{c_{n,m}}(s))_u z_u\} \cdot (\tilde{V}_n(0,s))_u$$

$$+ \int_0^{z_u} \exp\{-(\tilde{\gamma}_{c_{n,m}}(s))_u [z_u - z_u']\}$$

$$\cdot (\tilde{V}_n^{(s)'}(z_u',s))_u \, dz_u'$$

$$(\tilde{V}_n(z_v,s))_v = \exp\{-(\tilde{\gamma}_{c_{n,m}}(s))_v z_v\} \cdot (\tilde{V}_n(0,s))_v$$

$$+ \int_0^{z_v} \exp\{-(\tilde{\gamma}_{c_{n,m}}(s))_v [z_v-z_v']\}$$

$$\cdot (\tilde{V}_n^{(s)'}(z_v',s))_v \, dz_v' \qquad (213)$$

In this form the combined voltage supervectors $((\tilde{V}_n(0,s))_u)$
leaving the junctions as computed from the BLT equation (206) and
the combined sources $((\tilde{V}_n^{(s)'}(z_u',s))_u)$ along the tubes (or waves)
can be used to compute the combined voltages, and thereby the
voltages and currents at any position $z = z_u = L_v - z_v$ along any
tube of interest.

10.   SOME FORMS OF SOLUTIONS OF BLT EQUATIONS

Having formulated the BLT equation (206), one can represent
its solution in various ways. The reader should note that the

10. SOME FORMS OF SOLUTIONS OF BLT EQUATIONS

Having formulated the BLT equation (206), one can represent its solution in various ways. The reader should note that the particular form in (206) is only one of many forms the BLT equation can take; in this case the unknowns are the combined voltages scattered from (outward propagation from) the junctions.

Since the BLT equation has been cast in the form of a supermatrix equation, the solution can be written directly as

$$((\widetilde{V}_n(0,s))_u)$$

$$= [((1_{n,m})_{u,v}) - ((\widetilde{S}_{n,m}(s))_{u,v}) : ((\widetilde{\Gamma}_{n,m}(s))_{u,v}]^{-1}$$

$$: ((\widetilde{S}_{n,m}(s))_{u,v}) : ((\widetilde{\Lambda}_{n,m}(z'_u,s;(\cdot)))_{u,v}) : ((\widetilde{V}_n^{(s)'}(z'_u,s))_u) \quad (214)$$

For each complex frequency, s, this solution can be directly computed via integration (for the distributed sources), supermatrix multiplication, and supermatrix inversion, typically by computer. However, this approach may have limited utility for some kinds of problems due to a desire for the transient behavior and/or the characterization of the solution (such as bounding it) for a large class of excitations $((\widetilde{V}_n^{(s)'}(z'_u,s))_u)$.

Considerable work has been done in representing the solution of electromagnetic scattering problems, as formulated in integral equations, in terms of the eigenmode expansion method (EEM) and the singularity expansion method (SEM). The literature on SEM and EEM is quite extensive and the reader can consult two review book chapters [23, 24] concerning this subject and obtain a bibliography. While the SEM and EEM concepts have been cast in terms of electromagnetic integral equations, there is a direct connection to matrix equations because of the moment method (MoM) which is used to matricize the integral equations, i.e., put the integral equations in a form for numerical evaluation as on a computer [25]. In fact, some of the original developments in SEM and EEM theory and application used matrix concepts to arrive at the needed ideas and techniques [26-28]. Hence, SEM and EEM are directly applicable to the BLT equation, as in (206) or in other related forms of it. A few of the results are presented here to indicate the forms of some of the basic results as applicable to the supermatrix BLT equation.

In EEM form one defines eigenmode supervectors and eigenvalues via

$$[((1_{n,m})_{u,v}) - ((\tilde{S}_{n,m}(s))_{u,v}):((\tilde{\Gamma}_{n,m}(s))_{u,v})]:((\tilde{V}_n(s))_u)_\beta$$

$$= \tilde{\lambda}_\beta(s)((\tilde{V}_n(s))_u)_\beta$$

$$((\tilde{L}_n(s))_u)_\beta:[((1_{n,m})_{u,v}) - ((\tilde{S}_{n,m}(s))_{u,v}):((\tilde{\Gamma}_{n,m}(s))_{u,v})]$$

$$= \tilde{\lambda}_\beta(s)((\tilde{L}_n(s))_u)_\beta$$

$$\beta = 1,2,\ldots,N \tag{215}$$

where for distinct eigenvalues we have the biorthogonal property

$$((\tilde{L}_n(s))_u)_\beta:((\tilde{V}_n(s))_u)_{\beta'} = 0 \quad \text{for} \quad \beta \neq \beta' \tag{216}$$

This result also applies in the weaker case of independent eigen-supervectors. From (207) we have N eigenvalues and assume the existence of N independent eigensupervectors (of both left and right kinds separately). The right eigenmodes are used to ex-pand $(\tilde{V}_n(0,s))$ which gives the outgoing waves at the junctions. The left eigenmodes appear to be related to the incoming waves at the junctions, and this aspect will hopefully be considered in a future paper.

Defining normalized eigensupervectors as

$$((\tilde{v}_n(s))_u)_\beta \equiv [((\tilde{L}_n(s))_u)_\beta:((\tilde{V}_n(s))_u)_\beta]^{-\frac{1}{2}} ((\tilde{V}_n(s))_u)_\beta$$

$$((\tilde{\ell}_n(s))_u)_\beta \equiv [((\tilde{L}_n(s))_u)_\beta:((\tilde{V}_n(s))_u)_\beta]^{-\frac{1}{2}} ((\tilde{L}_n(s))_u)_\beta \tag{217}$$

we have the biorthonormal property

$$((\tilde{\ell}_n(s))_u)_\beta:((\tilde{v}_n(s))_u)_{\beta'} = 1_{\beta,\beta'} \tag{218}$$

This allows us to write the expansion

$$((1_{n,m})_{u,v}) - ((\tilde{S}_{n,m}(s))_{u,v}):((\tilde{\Gamma}_{n,m}(s))_{u,v})$$

$$= \sum_\beta \tilde{\lambda}_\beta(s) [((\tilde{L}_n(s))_u)_\beta:((\tilde{V}_n(s))_u)_\beta]^{-1}((\tilde{V}_n(s))_u)_\beta((\tilde{L}_n(s))_u)_\beta$$

$$= \sum_\beta \tilde{\lambda}_\beta(s) ((\tilde{v}_n(s))_u)_\beta ((\tilde{\ell}_n(s))_u)_\beta \tag{219}$$

which is an example of a dyadic expansion using a dyadic or outer product of supervectors. The inverse is

$$[((1_{n,m})_{u,v}) - ((\widetilde{S}_{n,m}(s))_{u,v}):((\widetilde{\Gamma}_{n,m}(s))_{u,v})]^{-1}$$

$$= \sum_{\beta} \widetilde{\lambda}_{\beta}^{-1}(s)\ ((\widetilde{v}_n(s))_u)_{\beta}\ ((\widetilde{\ell}_n(s))_u)_{\beta} \tag{220}$$

and the identity is

$$((1_{n,m})_{u,v}) = \sum_{\beta} ((\widetilde{v}_n(s))_u)_{\beta}\ ((\widetilde{\ell}_n(s))_u)_{\beta}$$

$$= \sum_{\beta} ((\widetilde{\ell}_n(s))_u)_{\beta}\ ((\widetilde{v}_n(s))_u)_{\beta} \tag{221}$$

Combining (219) with (221) also gives

$$((\widetilde{S}_{n,m}(s))_{u,v}):((\widetilde{\Gamma}_{n,m}(s))_{u,v})$$

$$= \sum_{\beta} [1 - \widetilde{\lambda}_{\beta}(s)]\ ((\widetilde{v}_n(s))_u)_{\beta}\ ((\widetilde{\ell}_n(s))_u)_{\beta} \tag{222}$$

The solution of the BLT equation (206) can then be written as a sum of eigensupervector contributions as

$$((\widetilde{V}_n(s))_u)$$

$$= \sum_{\beta} \widetilde{\lambda}_{\beta}^{-1}(s)[((\widetilde{\ell}_n(s))_u)_{\beta}:((\widetilde{S}_{n,m}(s))_{u,v})$$

$$:((\widetilde{\Lambda}_{n,m}(z'_u,s;(\cdot)))_{u,v}):((\widetilde{V}_n^{(s)}(z'_u,s))_u)]$$

$$((\widetilde{v}_n(s))_u)_{\beta} \tag{223}$$

Note that this solution expresses the outgoing combined voltages at the junctions in terms of eigenmodes at the junctions. These eigenmodes can be extended throughout the tubes of the transmission-line network by the techniques discussed in Section 9; these extended eigenmodes can then be used to construct the combined voltages and voltages and currents throughout the tubes. However, these eigenmodes are not anticipated to be simply related to the tube eigenmodes (Section 3), which may be more appropriate for extending the combined voltages at the junctions to the combined voltages, voltages, and currents throughout the network tubes.

Concerning the SEM representation of the solution, there is much that can be adapted from the work on electromagnetic scattering and antenna problems. The general form of the solution of the BLT equation in the form expressed in (206) is

$$((\tilde{V}_n(0,s))_u) = \sum_\alpha \tilde{f}(s_\alpha) \, \tilde{\eta}_\alpha(s) \, ((v_n)_u)_\alpha \, (s-s_\alpha)^{-n_\alpha}$$

$$+ \text{ other singularity terms} \tag{224}$$

where $\tilde{f}(s)$ (or $f(t)$) is some excitation waveform which appears in the combined sources $((\tilde{V}_n^{(s)\,'}(z_u,s))_u)$ and which is taken out so as to give equivalent delta-function response in defining the coupling coefficients $\tilde{\eta}_\alpha(s)$. For present purposes, we can set $\tilde{f}(s) = 1$, assuming that the excitation has been appropriately normalized. The order of the pole is $n_\alpha = 1,2,\ldots$, but here only the first order is considered; second order can be adapted from [24].

The natural mode supervectors are found

$$[((1_{n,m})_{u,v}) - ((\tilde{S}_{n,m}(S_\alpha))_{u,v}) : ((\tilde{\Gamma}_{n,m}(s_\alpha))_{u,v}]$$
$$: ((v_n)_u)_\alpha = ((0_n)_u)$$
$$((\ell_n)_u)_\alpha) : [((1_{n,m})_{u,v}) - ((\tilde{S}_{n,m}(s_\alpha))_{u,v})$$
$$: ((\tilde{\Gamma}_{n,m}(s_\alpha))_{u,v})] = ((0_n)_u) \tag{225}$$

where the left mode supervectors are also referred to as the coupling supervectors. The natural frequencies are found as the solutions of

$$\det[((1_{n,m})_{u,v}) - ((\tilde{S}_{n,m}(s_\alpha))_{u,v}) : ((\tilde{\Gamma}_{n,m}(s_\alpha))_{u,v})] \tag{226}$$
$$= 0$$

These can be related to the eigenvalues via

$$D(s,\lambda) = \det[(1-\lambda)((1_{n,m})_{u,v}) - ((\tilde{S}_{n,m}(s))_{u,v})$$
$$: ((\tilde{\Gamma}_{n,m}(s))_{u,v})] \tag{227}$$

for which we have

$$D(s,\tilde{\lambda}_\beta(s)) = 0$$

$$D(s_\alpha,0) = 0 \tag{228}$$

from which we set

$$\alpha = (\beta,\beta'), \tag{229}$$

such that

$$\tilde{\lambda}_\beta(s_{\beta,\beta'}) = 0 \tag{230}$$

associates the natural frequencies with the zeros of eigenvalues. The natural modes are similarly related to the eigenmodes as

$$((v_n)_u)_{\beta,\beta'} = N_{\beta,\beta'}((\tilde{v}_n(s_{\beta,\beta'}))_u)_\beta$$

$$((\ell_n)_u)_{\beta,\beta'} = M_{\beta,\beta'}((\tilde{\ell}_n(s_{\beta,\beta'}))_u)_\beta \tag{231}$$

where $N_{\beta,\beta'}$ and $M_{\beta,\beta'}$ are complex constants related to the normalization chosen for the natural modes.

The class 1 coupling coefficients are given by

$$\tilde{\eta}_\alpha(s)$$

$$= \frac{((\ell_n)_u)_\alpha : ((\tilde{\Lambda}_{n,m}(z'_u,s_\alpha;(\cdot)))_{u,v}) : ((\tilde{v}_n^{(s)'}(z'_u,s_\alpha))_u) \; e^{-(s-s_\alpha)t_o}}{((\ell_n)_u)_\alpha : \frac{\partial}{\partial s}[((1_{n,m})_{u,v}) - ((\tilde{S}_{n,m}(s))_{u,v}):((\tilde{\Gamma}_{n,m}(s))_{u,v})]|_{s-s_\alpha} : ((v_n)_u)_\alpha} \tag{232}$$

$$= N_{\beta,\beta'}^{-1}\left[\frac{\partial}{\partial s}\lambda_\beta(s)\Big|_{s=s_{\beta,\beta'}}\right]^{-1}((\tilde{\ell}_n(s_{\beta,\beta'}))_u):((\tilde{\Lambda}_{n,m}(z'_u,s_{\beta,\beta'};(\cdot)))_{u,v})$$

$$:((\tilde{v}_n^{(s)'}(z'_u,s_{\beta,\beta'}))_u) \; e^{-(s-s_{\beta,\beta'})t_o}$$

where the turn-on time t_o can be taken as a function of position (n and u indices) in the network. With this class 1 coupling coefficient, the time-domain form of (224) is

$$((V_n(0,t))_u) = \sum_\alpha \tilde{f}(s_\alpha) \; \tilde{\eta}_\alpha(s_\alpha)((v_n)_u)_\alpha \; e^{s_\alpha t} \; u(t-t_o)$$

$$+ \text{ other singularity terms} \tag{233}$$

The class 2 coupling coefficients (corresponding to the SEM representation of the inverse matrix in (214)) are given by

$$
\tilde{\eta}_\alpha(s)
$$

$$
= \frac{((e^{-(s-s_\alpha)t_o}\ell_n)_u)_\alpha : ((\tilde{\Lambda}_{n,m}(z'_u,s;(\cdot)))_{u,v}) : ((\tilde{v}^{(s)\,\prime}_n(z'_u,s))_u)}{((\ell_n)_u)_\alpha : \frac{\partial}{\partial_s}[((1_{n,m})_{u,v}) - ((S_{n,m}(s))_{u,v}) : ((\tilde{\Gamma}_{n,m}(s))_{u,v})]\Big|_{s=s_\alpha} : ((V_n)_u)_\alpha} \tag{234}
$$

$$
= N^{-1}_{\beta,\beta'}\left[\frac{\partial}{\partial_s}\tilde{\lambda}_\beta(s)\Big|_{s=s_{\beta,\beta'}}\right]^{-1}((e^{-(s-s_{\beta,\beta'})t_o}\ell_n(s_{\beta,\beta'}))_u) : ((\tilde{\Lambda}(z'_u,s;(\cdot)))_{u,v})
$$

$$
: ((\tilde{v}^{(s)\,\prime}_n(z'_u,s)))
$$

where the turn-on time t_o can be taken as a function of two sets of position variables (n and u) in the network corresponding to both the summation with the left mode supervectors and the position of observation. In time domain the class 2 coupling coefficients give more complicated results than (233) for class 1 due to the appearance of a time convolution.

Like the eigenmodes, the natural modes can be extended throughout the transmission-line network and made a function of the z_u coordinates. These can then be used for representing voltage and current supervectors throughout the tubes in the network.

This section has merely indicated some of the properties of BLT equations, particularly due to their formal similarity to electromagnetic integral equations. This analogy should provide much insight and future results.

11. CONCLUSION

This has been a long quest. While we have found a few things of apparent significance, the quest is not finished. As with many results the answers raise as many, if not more, questions. There are several general areas for future development that come to mind.

The BLT equation (including its alternate forms) expresses the characteristics of a multiconductor transmission-line network in a single supermatrix equation. In this form various properties of the network can be explored. Various properties related to energy and reciprocity can be formulated. In this regard, the symmetry properties of the various impedance and admittance matrices in the network need to be explored. This appears to have some relation to the diagonalizability properties of the propagation matrices.

A development parallel to transmission-line network topology is scatterer topology. In scatterer topology a hierarchical topology related to shielding concepts has been introduced. Perhaps this hierarchical topology can be introduced into some kinds of transmission-line networks to simplify their analysis and/or synthesis. Turning the question around, perhaps the transmission-line network topology and the BLT equation can aid in developing new insights into scatterer topology and the associated equations describing the electromagnetic scattering.

REFERENCES

1. O. Heaviside, Electromagnetic Theory, New York: Dover, 1950 (from 3 vols., 1893, 1899, and 1912).

2. J. A. Cooper, D. E. Merewether, and R. L. Parker (eds.), "Electro-magnetic Pulse Handbook for Missiles and Aircraft in Flight", EMP Interaction 1-1, Sept. 1972.

3. F. M. Tesche and T, K. Liu, "Selected Topics in Transmission-Line Theory for EMP Internal Interaction Problems", Interaction Note 318, March 1977.

4. S. Frankel, Multiconductor Transmission-Line Analysis, Artech House, 1977.

5. C. Paul, "On Uniform Multimode Transmission Lines", IEEE Trans. MTT, vol. MTT-21, Aug. 1973, pp. 556-558.

6. C. E. Baum, "Coupling into Coaxial Cables from Currents and Charges on the Exterior", USNC/URSI Meeting, Amherst, Massachusetts, Oct. 1976.

7. F. M. Tesche, "A General Multiconductor Transmission-Line Model", USNC/URSI Meeting, Amherst, Massachusetts, Oct. 1976.

8. C. E. Baum, "The Treatment of the Problem of Electromagnetic Interaction with Complex Systems", IEEE 1978 International Symposium on Electromagnetic Compatibility, Atlanta, Georgia, June 1978.

9. See, for example, N. Balabanian, Fundamentals of Circuit Theory (Boston, Allyn and Bacon, Inc., 1962).

10. C. E. Baum, "How to Think About EMP Interaction," Proc. 1974 Spring FULMEN Meet., pp. 12-23, April 1974.

11. C. E. Baum, "The Role of Scattering Theory in Electromagnetic Interference Problems," in P.L.E. Uslenghi (ed.) Electromagnetic Scattering, Academic Press, 1978.

12. F. M. Tesche, M. A. Morgan, B. Fishbine and E. R. Parkinson, "Internal Interaction Analysis: Topological Concepts and Needed Model Improvements," Interaction Note 248, July 1975.

13. E. F. Vance, "Shiedling and Grounding Topology for Interference Control," Interaction Note 306, April 1977.

14. F. M. Tesche, "Topological Concepts for EMP Internal Interaction," in Special Issue on the Nuclear Electromagnetic Pulse, IEEE Trans. on Antennas and Propagat., Vol. AP-26, January 1978, and IEEE Trans. on Electromagnetic Compatibility, col. EMC-20, February 1978.

15. C. E. Baum, T. K. Liu, F. M. Tesche and S. K. Chang, "Numerical Results for Multiconductor Transmission-Line Networks," Interaction Note 322, September 1977.

16. F. M. Tesche and T. K. Liu, "An Electric Model for a Cable Clamp," Interaction Note 307, December 1976.

17. S. Coen, T. K. Liu and F. M. Tesche, "Calculation of the Equivalent Capacitance of a Rib near a Single Wire Transmission Line," Interaction Note 310, February 1977.

18. K.S.H. Lee and F. C. Yang, "A Wire Passing by a Circular Aperture in an Infinite Ground Plane," Interaction Note 317, February 1977.

19. C. R. Paul, "Efficient Numerical Computation of the Frequency Response of Cables Illuminated by an Electromagnetic Field," IEEE Trans. on Microwave Theory and Technique, vol. MII-22, pp. 456-457, April 1974.

20. E. A. Guillemin, The Mathematics of Circuit Analysis. MIT Press, 1949.

21. C. E. Baum, "On the Eigenmode Expansion Method for Electro-
     magnetic Scattering and Antenna Problems, Part I: Some
     Basic Relations for Eigenmode Expansions and Their Relation
     to the Singularity Expansion," Interaction Note 229,
     January 1975.

22. E. A. Coddington and H. Levinson, Theory of Ordinary Dif-
     ferential Equations, McGraw-Hill, 1965.

23. C. E. Baum, "The Singularity Expansion Method", in L. Felsen
     (ed.), Transient Electromagnetic Fields, Springer Verlag,
     1976.

24. C. E. Baum, "Toward an Engineering Theory of Electromagnetic
     Scattering: The Singularity and Eigenmode Expansion Meth-
     ods", in P.L.E. Uslenghi (ed.), Electromagnetic Scattering,
     Academic Press, 1978.

25. R. F. Harrington, Field Computation by Moment Methods,
     MacMillian, 1968.

26. C. E. Baum, "On the Singularity Expansion Method for the
     Solution of Electromagnetic Interaction Problems", Inter-
     action Note 88, December 1971.

27. F. M. Tesche, "On the Singularity Expansion Method as
     Applied to Electromagnetic Scattering from Thin Wires",
     Interaction Note 102, April 1972.

28. C. E. Baum, "On the Eigenmode Expansion Method for Electro-
     magnetic Scattering and Antenna Problems, Part I: Some
     Basic Relations for Eigenmode Expansions and Their Relation
     to the Singularity Expansion", Interaction Note 229,
     January 1975.

BASIC PRINCIPLES OF GROUNDING AND SHIELDING WITH RESPECT TO EQUIVALENT CIRCUITS

Johannes Wiesinger

Hochspannungstechnik und Elektrische Anlagen
8014 Neubiberg
West Germany

1. PRESUPPOSITION

One can work with lumped parameters, if the wave length of the electromagnetic field is much greater than the dimensions of the components of the diagnostic devices considered. If a diagnostic device is relatively large it may be divided into sufficient small parts in which quasi-stationary fields can be assumed. The advantage of this kind of consideration is that calculations can be done immediately in the time domain.

2. INTRODUCTION

Figures 1 and 2 show two typical diagnostic devices, resulting from high-voltage research arrangements.

In Figure 1 the principle of a lightning trigger station is shown. The characteristics are:

- The EMP radiated from the largely extended lightning channel is not correlated to the measuring signal, which is proportional to the lightning current.

- Due to the ground impedance balance an essential part of the lightning current flows along the metal conduit, creating a longitudinal coupling voltage along the inside of the conduit.

- Partial lightning currents are led successively to ground along the conduit so that the conduit current varies.

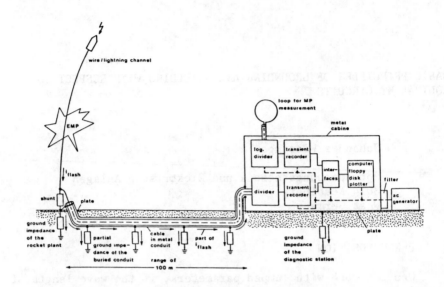

Fig. 1.   Diagnostic system in a lightning triggering station.

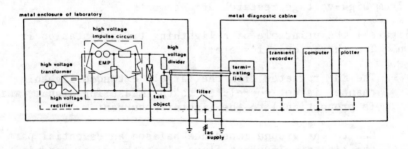

Fig. 2.   Diagnostic system in a high voltage laboratory.

Figure 2 shows the principle of impulse voltage testing. The characteristics are:

- The discharge current of the test circuit flows in a closed loop, and there are only small currents along the cable shield to the diagnostic cabine.

- The EMP of the local source is prohibited from the measuring system by the shields of the laboratory and the diagnostic cabine.

From these two examples we can deduce the principal arrangement of diagnostic devices, shown in Fig. 3.

Interference voltages and currents are induced into several parts of the diagnostic device: Fig. 4. The voltages u_H are caused by the changing magnetic field of the EMP due to shielding failures, and the currents i_E are caused by the changing electric field. The coupling voltage u_c along the cable shield results from the shield current due to the ground impedance imbalance (the impedance of the power supply also influences the shield current according to the values of the stray inductance and capacitance, respectively, of the filter transformer).

3.   COAX CABLE IN A CONDUIT

The largely extended coax cable is a very important part of the diagnostic device with regard to the interference. In order to calculate the coupling voltage, the <u>current</u> distribution along the shield must be known. This current results from the partial ground current as well as the electrically and magnetically induced current caused by the EMP.

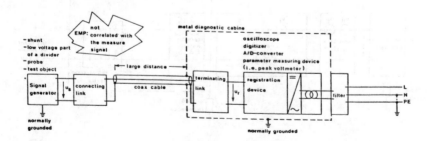

Fig. 3.   Principal arrangement of diagnostic devices.

3.1 <u>Equivalent Circuit with Lumped Parameters for Calculation of Cable Shield Current</u>

With respect to the coupling voltage, the cable shield must be divided into pieces of sufficient small length Δl, forming π-links. Δl should be smaller than about 1/10 the wavelength λ considered ($\frac{\Delta l}{v}$ should not be greater than about 1/10 of the rise time of the shield current). For a cable above ground, with height h and the shield radius r:

$$L' \cong 0.2 \cdot \ln\frac{2h}{r} \; (\frac{\mu H}{m}), \quad R' \cong 0 \quad \text{and} \quad v = c_o$$

$$C' \cong \frac{56}{\ln\frac{2h}{r}} \; (\frac{pF}{m}), \qquad G' \cong 0 \quad \text{(All dimension in meters.)}$$

For a cable buried in the ground with the length 1:

$$L' \cong 0.2 \; \ln\frac{1}{r} \; (\frac{\mu H}{m}), \qquad R' \cong 0 \quad \text{and} \quad v = \sqrt{\frac{4\pi f}{L' \cdot G'}}$$

$$C' \cong \frac{56}{\ln\frac{1}{r}} \; (\frac{pF}{m}), \qquad G' \cong \frac{\pi}{\rho_E} \cdot \frac{1}{\ln\frac{1}{r}} \; (S/m)$$

Fig. 5 shows the equivalent circuit of a cable sheet (conduit) with the above parameters. $\bar{Z}_b$ and $\bar{Z}_e$ are the ground impedances at the beginning and at the end.

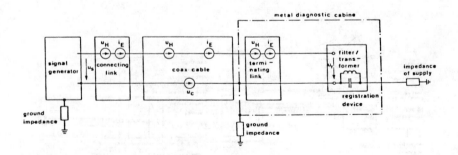

Fig. 4.   Interference sources in a diagnostic device.

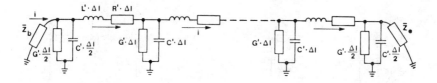

Fig. 5.   Equivalent circuit of a cable sheet.

In addition to the ground current, electrically and magnetically induced currents flow on the shield, corresponding to
Fig. 6.   There is a partial displacement current i_v, impressed by
the electrical component of the EMP, where

$$i_v = \varepsilon_o \int_{\Delta A} \dot{E} \cdot dA , \qquad \Delta A = 2\pi r \cdot \Delta l \quad \text{(Fig. 6)}$$

The displacement current related to a length of 1 meter and to a
field change of $1 \frac{MV/m}{\mu s}$ is shown in Fig. 7 as a function of geometrical parameters.

There is also a partial induced voltage u_v, impressed by the magnetic component of the EMP:

$$u_v = \varepsilon_o \int_{\Delta A} \dot{H} \cdot dA , \qquad \Delta A: \text{ see Fig. 6.}$$

3.2   Equivalent Circuit with Lumped Parameters for Calculation of Coupling Voltage

If the current along the cable shield (divided into sufficient small partial lengths Δl) has been calculated according to
Section 3.1, the longitudinal coupling voltage can be determined.
In order to do so, conduit with wall thickness s and radius r
(according to Fig. 8) is divided into sufficient small concentric
cylinders with the wall thickness Δr.   The longitudinal resistance of a partial cylinder is R_μ and R_v, respectively, and
the transversal interior inductance between two neighboring cylinders is $L_{\mu v}$.

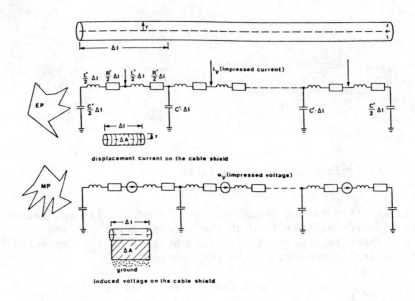

Fig. 6. Induced currents on the cable sheet.

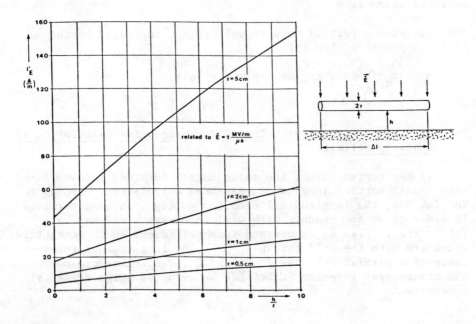

Fig. 7. Displacement current on a cable shield.

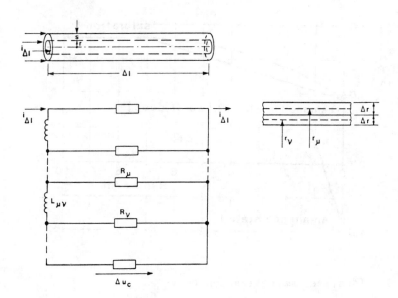

Fig. 8.   Equivalent circuit of a conduit.

$$R_\mu = \frac{\rho}{2\pi r_\mu \cdot \Delta r} \cdot \Delta l \qquad R_\nu = \frac{\rho}{2\pi r_\nu \Delta r} \cdot \Delta l$$

$$L_{\mu\nu} = \frac{\mu_o \cdot \mu_r}{2\pi} \cdot \ln \frac{r_\mu}{r_\nu} \cdot \Delta l$$

Note that for ferromagnetic materials $\mu_r = f(i_{\mu\nu})$!

It is sufficient to divide the conduit into about 10 partial cylinders in the case of non-ferromagnetic materials, and into about 100 cylinders in the case of ferromagnetics. The current on the shield, $i_{\Delta l}$, calculated by the methods given in Section 3.1, is inserted into the exterior cylinder; the coupling voltage Δu_c drops at the interior cylinder. This equivalent circuit is well known from shunt theory and was formally discussed in detail by Schwab [1]. As Steinbigler has shown, for common ferromagnetic materials an idealized magnetization curve can be assumed according to Fig. 9 [2]. The worst case is the remanance state 2.

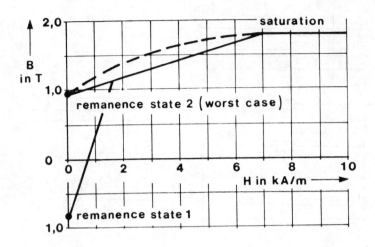

Fig. 9.   Idealized magnetization curve.

3.3 <u>Worst Case Calculation</u>

For a conduit with the length 1, radius r, and wall thickness s, according to Fig. 10 the worst case of the coupling voltage is given by the assumption of an unit step current i_o along
the conduit (infinite fast rise to the value $\hat{i}_o$ with infinite
duration).

The principal shape of the corresponding coupling voltage
$u_{c/o}$ is shown in Fig. 10. It can be defined by the delay time
T_o, the rise time T_r, the maximum steepness S_{max}, and the maximum
value $\hat{u}_{c/o}$. For practical purposes the definition of T_o, given
in Fig. 10, seems acceptable.

For non-ferromagnetic materials the following formulas apply [3]:

$$\hat{u}_{c/o} = R_{c/o} \cdot \hat{i}_o \quad (kV),$$

$$DC \text{ resistor } R_{c/o} = \frac{\rho \cdot 1}{\pi \cdot s(s + 2r)} \quad (\Omega),$$

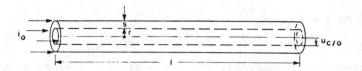

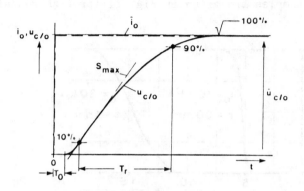

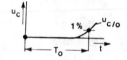

Fig. 10. Worst-case coupling voltage of a conduit.

$$T_r = 0.05 \cdot \frac{s^2}{\rho} \ (\mu s),$$

$$T_r = 0.3 \cdot \frac{s^2}{\rho} \ (\mu s),$$

$$S_{max} = 1.4 \ \frac{\rho^2 \cdot 1}{s^3 \cdot (s + 2r)} \cdot \hat{i}_o \ (\frac{kV}{\mu s}),$$

where $\hat{i}_o$ (kA), s(mm), $\rho(\frac{\Omega mm^2}{m})$, 1(m), r(mm).

For example, for a copper conduit with r = 10 mm, s = 1 mm, l = 100 m, and $\hat{\imath}_o$ = 1 kA:

$u_{c/o}$ = 27 V, T_o = 2.8 μs, T_r = 17 μs, S_{max} = 2.2 V/μs.

For ferromagnetic materials the shape of $u_{c/o}$ depends on the value of $\hat{\imath}_o$! Examples are given in Fig. 11 for iron conduits

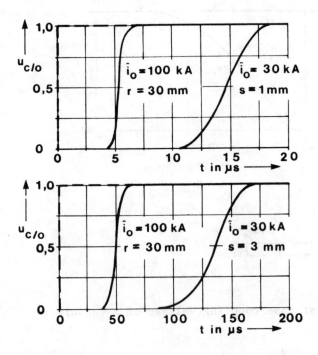

Fig. 11. Worst-case coupling voltage of ferromagnetic conduits.

with r = 30 mm, and s = 1 mm and 3 mm, respectively. $\hat{\imath}_o$ was assumed to be 30 kA and 100 kA, respectively [2].

Most important for submicrosecond measuring tasks is the delay time T_o, which should be greater than 1 μs. In this case the current along the conduit will not influence the measuring signal.

3.4 Meshed Cable Shields

According to Fig. 12 meshed cable shields could be considered as conduits with holes. Therefore, a magnetically induced voltage Δu arises between the shield and the interior conductor. As can be seen from Fig. 13 [4], the related coupling resistor R_c exceeds the d.c. value $R_{c/o}$ for frequencies greater than 10 MHz. R_c and $R_{c/o}$ are defined as the quotients of the maximum value of the coupling voltage and the maximum current value on the shield, respectively.

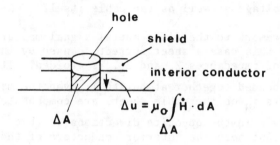

Fig. 12. Conduit with holes.

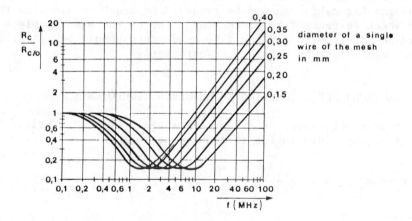

Fig. 13. Coupling resistors of meshed shields.

3.5 Some Practical Conclusions

First, a remark on the double shielding technique (Fig. 14). In this case, a shielded cable lays inside a conduit. T_o of the conduit should be greater than the signal time. Then one can bond the cable sheet additionally to a possible bonding at point ⓑ and to the conduit at the entrance to the cabine at point ⓔ. If the measured time is greater than T_o of the conduit, the cable sheet should only be bonded at point ⓑ. But in this case, the coupling voltage u_c arises between the cable sheet, and the cabine and the electronic system must be insulated against this voltage as well as the cable itself.

A second remark to the differential signal measuring method (Fig. 15). In this case a sheet current, caused by the imbalance of the grounding impedances $\bar{Z}_b$ and $\bar{Z}_e$, is avoided. If the cables are also bonded together at point ⓑ, even the magnetically induced currents i_H on the cable sheets are compensated, causing coupling voltages in the opposite directions so that their sum becomes zero. But here the interior conductors of the cables must be insulated against the cable sheets according to the voltage drops on $\bar{Z}_b$, which can be of quite high values in high-voltage test facilities.

In Fig. 16 it is shown at which points the test circuit and the diagnostic cabine should be grounded in order to minimize the cable sheet currents (the ground conductor of the main supply should be interrupted and the life conductors should be isolated against the cabine by a safety transformer).

4.   ELECTROMAGNETIC COUPLING TO MEASURING COMPONENTS

If an electromagnetic field penetrates a hole in a shield (Fig. 17), the field inside is reduced as follows:

$$E_s = \frac{2}{3\pi} \cos\theta \cdot E_o \cdot \left(\frac{r}{s}\right)^3$$

$$H_s = \frac{4}{3\pi} \sin\theta \cdot H_o \left(\frac{r}{s}\right)^3$$

The field inside the shield induces voltages and displacement currents into the circuit.

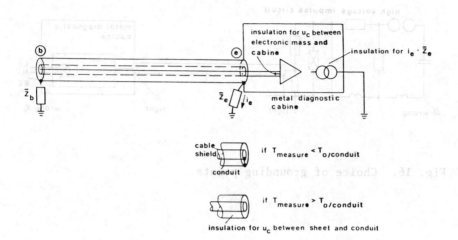

Fig. 14. Double shielding.

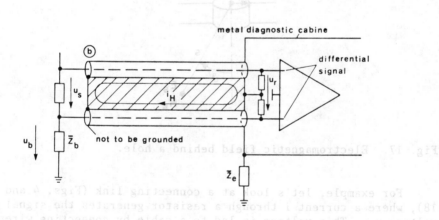

Fig. 15. Differential signal measuring.

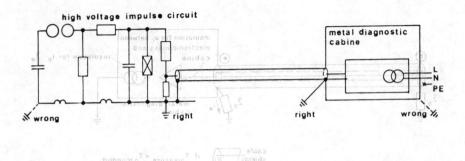

Fig. 16. Choice of grounding points.

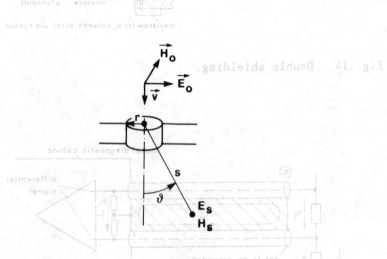

Fig. 17. Electromagnetic field behind a hole.

        For example, let's look at a connecting link (Figs. 4 and
18), where a current i through a resistor generates the signal
voltage u_s. This voltage is led to a cable by connecting wires,
which forms a loop with the mutual inductance M. In this loop
voltages are induced by the changes of the magnetic fields,
namely, the "internal" field, caused by the current i, and the
"external" field, penetrating the hole of the shield. Additionally, the "external" electric field causes a displacement current

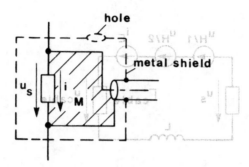

Fig. 18.   Connecting link.

on the wires. For this case the equivalent circuit is shown in
Fig. 19, where L is the self inductance of the loop and Γ_{cable} is
the transfer impedance of the cable. $u_{H/1}$ and $u_{H/2}$ are the induced voltages generated by the "internal" and "external" magnetic fields, respectively, and i_E is the displacement current of
the "external" electric field.

Fig. 20 shows an arrangement of a loop near an undefined
extended conductor, with the current i as shown. For this case
one can calculate the mutual inductance M and the self induc-
tanced L of the loop as follows, if r is the radius of the loop
wire [3]:

$$M = 0.2 \cdot b \cdot \ln \frac{a \cdot s}{s} \; (\mu H)$$

$$L = 0.4 \left[2 \sqrt{a^2 + b^2} - (a + b) + a \cdot \ln \frac{2b}{r(1+\sqrt{1+(b/a)^2})} + b \cdot \ln \frac{2a}{r(1+\sqrt{1+(a/b)^2})} \right] (\mu H)$$

The equivalent circuit of the loop of Fig. 20 is given in
Fig. 21, where $u_H = M \cdot di/dt$ is the induced voltage and R is the
loop resistance.

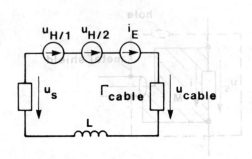

Fig. 19.   Interference to a connecting link.

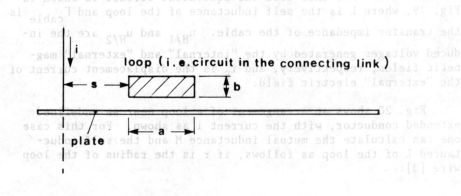

Fig. 20.   Magnetic induction into a loop.

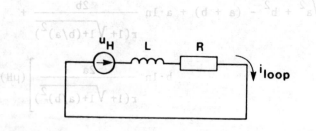

Fig. 21.   Equivalent circuit of a loop.

5.   TEST FOR DETECTION OF GROUNDING AND SHIELDING FAILURES
(Figure 22)

The voltage to be measured drops at the impedance $\bar{Z}$ of the signal generator. If the connection to the cable is interrupted or if $\bar{Z}$ is short circuited, no signal should be registered. This test procedure is well known as an "open-circuit test" (preferably an induced-voltage test and a test for currents on the cable shield). These two tests can be combined into one "hybrid-zero circuit test", giving the real values of all combined interferences. Therefore this test can also be used for software correction of interferences in the measuring signal. For this test the signal generator is short-circuited and its impedance is then in series with the connecting line immediately at the signal generator.

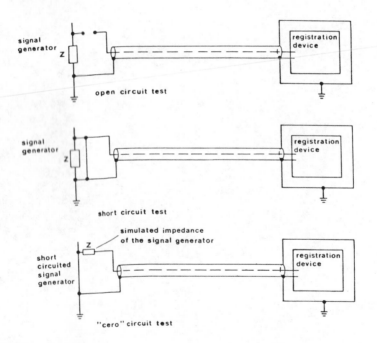

Fig. 22.   Test for grounding and shielding failures.

REFERENCES

1. A. Schwab, Hochspannungsmeβtechnik, Springer-Verla
    Heidelberg, New York, 1981.

2. H. Steinbigler, J. Wiesinger: "Voltage Response of Screen-
    ing Tubes to a Unit Lightning Current with Regard to Ferro-
    magnetic and Non-Ferromagnetic Materials", 5th Intern. Symp.
    on Electromagnetic Compatibility, 1980, Paper 42 K4.

3. P. Hasse, J. Wiesinger, Handbuch für Blitzschutz and Erdung,
    Richard Pflaum Verlag München, 1982.

4. Kabel and Draht, "Kabel für die Datenübertragung - Teil 2",
    Rhenocompkabel, 1971.

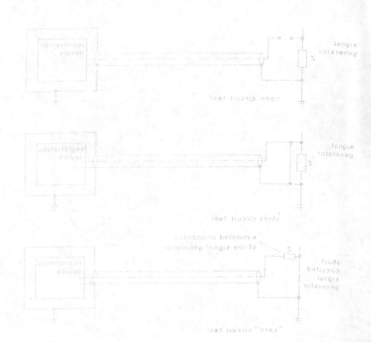

SYSTEM DESIGN: PRACTICAL SHIELDING AND GROUNDING TECHNIQUES BASED ON ELECTROMAGNETIC TOPOLOGY

Werner Graf

SRI International
Menlo Park, California
USA

1. BACKGROUND

The basic principles of electromagnetic topology are discussed in a previous paper by C. Baum. In this paper we will focus on the practical implications of those principles.

The fundamental problem of interference control is the separation of a source of interference from a potential victim (Fig. 1). The source could be the pulse power experiment and the victim could be the diagnostics. In order that the diagnostic equipment not pick up some spurious signals but rather the quantities to be measured, we must somehow separate the experiment and the diagnostics electromagnetically. Because the experiments of interest involve fast transients, the separation must be broadband. Furthermore, because most of the experiments also involve high power levels, the separation must be large.

Ad hoc solutions to interfacing problems abound in the literature, and almost all these approaches are narrowband remedies. Such remedies are usually insufficient for an effective electromagnetic separation in a transient environment. Interference control problems are basically electromagnetic problems, but they must be solved by circuit designers who are knowledgeable about circuit elements. Hence, these problems are often addressed as equivalent circuits with lumped parameters. If the equivalent circuit is a good model of reality, and if the lumped parameters can be well defined, such an approach is efficient and useful. However, a real system is often only approximated by the equivalent circuit, which does not lend itself to a good understanding of the underlying physics.

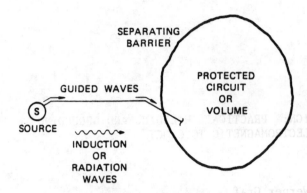

FIGURE 1   THE BASIC INTERFERENCE CONTROL
            PROBLEM

Fig. 1.   The basic interference control problem.

This problem was recognized when protection of military
systems against external transients became important. Power line
transients, the effects of a high-altitude electromagnetic pulse
(HEMP), and lightning, are examples of such transients. These
broad-spectrum, high-amplitude sources of interference require an
isolation between the source and a potential victim that is
equally broadband. The redesign required to accommodate broadband immunity measures, and the recognition that the procedures
used to control the effects of a HEMP are applicable to the
control of any broadband, high-amplitude source, led to an extensive review of existing interference control techniques and
practices [6].

If we postulate that interference control is a broadband
electromagnetic separation between an objectional source and a
potential victim, electromagnetic topology will provide practical
insights and the necessary tools to solve interference problems
in a unified way. That is, whatever means are used to isolate a
pulse power source from the associated diagnostic instrumentation
will not conflict with any other isolation techniques that might
be used for the experiment. Electromagnetic topology results in
an overall system approach to interference control, with high
confidence in meeting expected performance requirements.

The fundamental principle is simple: to achieve broadband
electromagnetic isolation of a source and a victim, the two must
be separated by a barrier that is effectively impervious to electromagnetic energy. Although such a statement may appear overly
simple, a thorough understanding of the implications of this
simple statement leads to a very effective and practical approach

to the stated problem. First we define an ideal barrier and then
examine the compromises found in practical barriers. Whatever
steps we take to ameliorate the compromises in the real barrier
will help to achieve the required isolation; whatever steps we
take that exacerbate the compromises will be detrimental.

In the next section, we describe the ideal barrier. Although simple in concept, a thorough understanding of the ideal
barrier is necessary to understand the compromises in a real barrier. This description will also help to clarify commonly used
and misused terms in interference control (e.g., grounding,
shielding, and configuration control).

We continue with a discussion of the practical barrier and
compromises of its ideal character in three areas: diffusion,
apertures, and penetrating conductors. An understanding of these
compromises is basic to the design of an effective barrier.

The final section in this lecture discusses the inconsistent
techniques found most frequently in grounding and cable shield
terminations. The topological principles will give clear guid-
ance in these areas. Some simple laboratory experiments are
described that illustrate these concepts. A summary of interference control principles and some corollaries conclude this section.

2.   THE IDEAL BARRIER

The ideal barrier is a physically and topologically closed
surface that completely encloses either the source or the victim.
The surface must have infinite conductivity and finite thickness,
with no apertures or wire penetrations. Such a barrier completely separates two volumes of space in the electromagnetic
sense: no electromagnetic information can be transmitted from
one side of the barrier to the other side. The latter statement
merely underscores one of the limitations of the ideal barrier;
no distinction is made between signal and noise.

The primary reason the ideal barrier is so effective is because it is closed, not because it is made of a superconducting
surface. We will illustrate this point below when we discuss
diffusion in a practical barrier. Also note that a barrier and a
shield are not the same, although a shield is usually a major
part of a barrier, as we will see below.

As stated in the introduction, the ideal barrier is just
that: an ideal that may never be achieved in reality, but that
can nevertheless serve as a guide to the design of a practical
barrier that is effectively impervious to electromagnetic energy.

The following discussion complements the mathematical approach described in the lecture by C. Baum; that is, all of the interference control techniques could be expressed in mathematical terms.

3.   COMPROMISES IN THE PRACTICAL BARRIER

A practical barrier consists of many elements. The major portion is usually a metal shield. Other elements include filters, surge arrestors, limiters, and common-mode rejection devices. All of the elements of the barrier are designed to get as close to the ideal barrier as possible, that is, to minimize the impact of the necessary compromises.

It is convenient and instructive to separate all compromises of the ideal barrier into three categories: These categories are

  a. Diffusion (because conductivity is not infinite)

  b. Apertures (because diagnostics and people need air)

  c. Penetrations (because the sensors need to communicate with the instrumentation).

3.1 Diffusion

Continuous, closed sheet-metal shields are by far the most effective electromagnetic shields because they severely limit the penetration of energy in the frequency range where the skin effect is important, and because they are good reflectors of propagating waves throughout the spectrum. For example, we can estimate the voltage induced on conductors inside a spherical shield from the peak voltage induced in the largest loop that fits inside the shield. This peak voltage has been calculated for a sphere made of copper, aluminum, and steel, with a radius of 10 m, for various wall thicknesses, and with the source of interference a HEMP [5]. If the wall thickness is only 0.2 mm, then the peak voltage induced in the largest loop inside the 10-m radius sphere is less than 1 V in all three cases. If the wall thickness is 1 mm, the peak voltage is only a few millivolts. Thus, the amount of electromagnetic energy that diffuses through a closed metal shield is negligible as long as the shield is at least of structural thickness and a reasonably good conductor. For this reason, shield thickness and material selection may be determined by other considerations, rather than by the amount of shielding effectiveness required.

The adequacy of the barrier is not limited by the shielding capability of finitely conducting metals of structural thicknesses; however, it is limited by the opening made to accommodate the diagnostic system.

3.2 Apertures

A practical barrier will contain apertures in many forms: doors, windows, and air-conditioning openings, as well as seams. Apertures may be important paths for electromagnetic energy to couple to the inside of the barrier. However, in the presence of untreated penetrating conductors (signal and power lines), apertures are often negligible. The ideal remedy for this compromise is to weld the aperture shut with a metal plate. This is possible only if the aperture is needed only during construction. If the aperture is necessary for ventilation, then a honeycomb panel consisting of an array of waveguides beyond cutoff is a simple and effective solution. Such a panel closes the barrier below the cutoff frequency. Other possible solutions include wire meshes and screens. In all cases, the goal is to subdivide a large aperture into many small ones.

3.3 Penetrations

The openings in the shield required for insulated conductors (such as power and signal wiring) almost completely defeat the barrier. Interference currents can propagate through the barrier virtually unattenuated along these conductors. To minimize this type of compromise, every wire penetration must be treated with an appropriate element, such as a filter, limiter, surge arrestor, or common-mode rejection device. To penetrate the barrier with even one untreated insulated conductor (including grounding conductors) defeats the most important goal of the barrier. To be effective as an electromagnetic separator of two volumes, the barrier must be closed.

4. DESIGNING AN EFFECTIVE BARRIER

An effective barrier includes the aperture and penetration treatments necessary to make the barrier a closed, substantially impervious surface. A metal shield with typical holes and penetrations does not form an adequate barrier. Because the openings required in a shield defeat the barrier, the concept of shield topology has been used to identify locations where the shield is compromised [1]. Since the ideal barrier is a topologically closed, continuous, impervious surface between the source and the

victim, any deviation from this ideal must be examined closely to
ascertain the effectiveness of the barrier.

The essence of interference control, then, is the definition
of the topology of the barrier surface and the identification of
weak spots in the barrier. When the barrier topology coincides
with a conducting shield, the fortification of weak spots is
analogous to closing the holes in the shield. That is, special
treatment is given to apertures and to insulated conductors penetrating the shield in order to limit the interference that can
pass through the shield at these openings.

The barrier does not need to be a metal shield surface, only
a closed surface impervious to electromagnetic interference.
However, the use of metal sheet or plate for the majority of the
barrier has obvious advantages, because metal shields are discrete and easily identified, controlled, and maintained. Furthermore, the region of greatest concern in a metal shield is
limited to the few easily identified openings in the shield; on
the other hand, a barrier topology that does not coincide with a
metal surface is apt to be less well defined, but it still must
be controlled and maintained to be impervious to all forms of
electromagnetic waves (those propagating through space, as well
as those guided along wires or other waveguides). The physical
shape of the barrier is not important, but the barrier must form
a topologically closed surface surrounding the protected zone.

Signal and power line penetrations present the most difficult problem in designing a practical barrier, since these penetrations are intentional violations of the barrier. Nevertheless, with proper treatments, it is possible to close the barrier
at least outside the frequency band of interest with filters, or
above some voltage or current threshold with surge arrestors or
limiters. In many cases, it might be possible to substitute
fiber optics for signal lines.

5.   INCONSISTENT TECHNIQUES

Attempts to control electromagnetic interference often involve techniques that are inconsistent with the physical principles outlined so far. The two techniques most often misunder-
stood are grounding and cable shield terminations. We will
discuss both of these and present some experimental data to
support our recommendations.

5.1 Grounding

Grounding is more a safety consideration than an interference control technique. Indeed, it is difficult to envision how grounding could be an element of an electromagnetic barrier. However, if done incorrectly, grounding can completely defeat a barrier and become an effective interference distribution system. Correcting an ill-conceived grounding scheme will improve performance, and this is why proper grounding is often credited with interference reduction. Examining grounding from the perspective of electromagnetic topology immediately gives a simple rule: no ground conductor should be permitted to penetrate a barrier surface. Instead, it should be terminated on one side of the barrier and, if necessary or desired, regenerated on the other side. We know of no application that requires the penetration of a ground wire through a barrier. (If it is desired to do so, the ground wire must be filtered just as a signal or power line would be.) Within any given zone between two barriers, any ground system that is appropriate for the particular application may be chosen. This could be a single-point ground, a multiple-point ground, or an equipotential plane (a single-point ground is difficult to maintain in any system where people have access). Ground loops appear to be inevitable in even the best designs as the system evolves. An exception might be inside a sealed equipment unit where the circuit designer has complete control over the physical layout and the cabling. But ground loops are usually a problem only where the barrier has been compromised; if the integrity of the electromagnetic barrier is preserved, the importance of such loops diminishes greatly or even vanishes.

It is important to emphasize that broadband interference control is not achievable if grounding conductors penetrate barrier surfaces. To determine how much attenuation is lost if such a penetration is permitted, we designed a simple experiment to demonstrate quantitatively the superior nature of the topological ground. Early measurements [6] showed that the topological ground is preferable in the high-frequency regime. However, most barrier compromises appear to arise from low-frequency considerations. The following experiments were therefore designed to reveal the behavior of ground systems at frequencies below 100 kHz. Continuous-wave signals were used instead of transients to obtain the necessary dynamic range [2].

A small die-case instrumentation box was used to simulate an equipment enclosure. Inside the box, a battery-powered operational amplifier measured the open-circuit voltage induced on the wall by an interference source on the outside. Two configurations were tested (Fig. 2). The first simulates a penetrating signal ground (Fig. 2a); the measured voltage was set to 0 dB by definition. The second configuration simulates the topologically

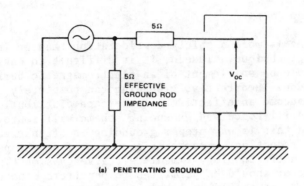

(a) PENETRATING GROUND

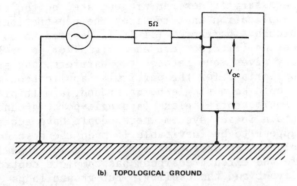

(b) TOPOLOGICAL GROUND

Fig. 2. Test setup (symbolic).

proper ground (Fig. 2b); the voltage measured in this case was
normalized to the one measured with the penetrating ground.

The results of the measurements are shown in Figure 3. With
the penetrating ground voltage set to 0 dB, the topologically
proper ground gives an open-circuit voltage of -115 dB, even at
the lowest frequency measured. Above about 20 kHz, the shielding
inherent in the metal walls of the box begins to be effective.
The topological ground provides better isolation than the penetrating ground, even with a large aperture (without penetrations). Curve 2 in Figure 3 shows the same measurements as curve
1, but with the lid removed. At the lowest frequencies shown,
there is no difference in the open-circuit voltage induced inside
the box. However, at the frequencies where the walls become effective as shields, the open aperture begins to show its effect.
At frequencies above 100 kHz, the effectiveness of closing the
box can be seen in the difference between curves 1 and 2.

The exact level of the open-circuit voltage depends primarily on two factors: the geometry of the measurement setup and
the ground rod impedance. In the first case, the voltage will be

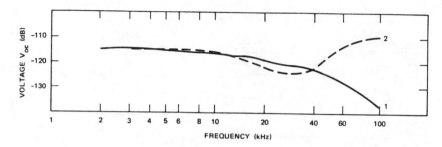

Fig. 3.   Open-circuit voltage, v_{oc}, as a function of frequency
         (Setup as in Fig. 2b):   (1) Box closed, (2) Lid removed.

different if the measurement point is at a different location.
We found a variation of up to 6 dB due to this factor. In the
second case, the ground rod impedance could be as low as 1Ω, or
higher than 20Ω; we chose 5Ω as a representative value of the impedance of a typical ground rod. The higher the impedance, the
more important it is to use a topological ground system. However, even with a 1-Ω ground rod, the difference between the two
ground systems is more than 100 dB.

5.7 Cable Shield Terminations

We performed several simple experiments to demonstrate the
effectiveness of topological cable shield terminations. The frequency range of the experiments was chosen to reflect the broadband nature of many transient interference sources, as discussed
in the previous sections. (For a complete description of the
experiments, see [1], [2]).

Two small instrumentation boxes, made of die-cast aluminum,
were separated by a distance of 2 m. Each contained a simple
"circuit" (Fig. 4a) consisting of a resistor equal to the characteristic impedance of the coaxial cable (RG62/U) used to connect
the two boxes. Box 1 was insulated from the ground plane, while
Box 2 was connected to it.

In practical cases, both boxes would be connected to the
ground plane. A varying magnetic field produced by an interfering source then would induce an emf and, therefore, a current in
the loop formed by the cable shield, the ground plane, and the
shield terminations. If the shield is not terminated at one end
to interrupt the loop current, the same emf would be induced in
the center conductor of the coaxial cable. The setup in Figure 4a is equivalent to the practical case in that the test

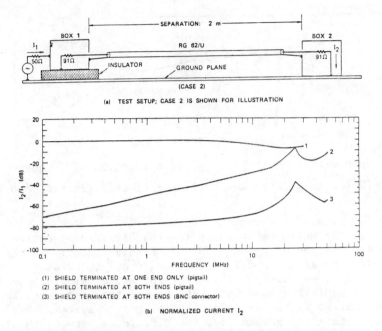

Fig. 4. Experiments with coaxial cable (RG62/U).

source drives a current directly on the cable shield. Therefore,
any conclusions drawn from these experiments regarding shield
terminations also apply to the case where magnetic fields inter-
act with the loop and induce a shield current.

The effects of three different cable shield terminations are
shown in Figure 4b. Resonance effects dominate above 10 MHz; the
data should be interpreted only below that frequency. Curve 1,
which shows the case for the cable shield terminated only at the
source with a 10-cm pigtail, indicates that the current in the
receiving circuit is equal to the source current. An analysis at
dc confirms that this is expected. However, Curve 1 implies only
that the shield provides no electrodynamic shielding; the cable
shield still would provide electrostatic protection. Such an
arrangement, however, provides no protection against interference.

The second case (Curve 2) involves pigtails at both ends of
the cable shield. Shielding is effective for the length of the
cable, and interference leaks into the receiving circuit only at
the two pigtails. The slope of the curve is 20 dB per decade,
which indicates that the dominating effect is inductive coupling
(which is proportional to frequency).

For the third case (Curve 3), BNC connectors were used.
Only a small dependence on frequency can be seen (below 10 MHz;
above that frequency the resonance effect dominates). This small
frequency dependence is probably caused by leakage through the
holes in the braided shield.

The type of shield termination used often reflects the operating frequency band of the circuit or systems involved; such
considerations do not account for the possibility of high-frequency or transient interference, because such interference is
considered to be outside the frequency band to which the circuit
or system would normally respond. Nevertheless, such a circuit
can be interfered with if the noise level is high enough, as is
the case for pulse power experiments.

Similar experiments were conducted with a shielded twisted
pair replacing the coaxial cable (Fig. 5). The results shown in
Figure 5 lead us to conclusions similar to those discussed above.
In addition, Case 5 shows the differential-mode current for a
balanced configuration. Figure 5 shows that the balanced configuration offers more attenuation than any of the other practices
shown, up to at least 5 MHz in this case. (In some cases, it may
be difficult to balance a circuit at high frequencies.) At low
frequencies, a balanced circuit, even without the shield termination, offers 20 dB more attenuation than any of the unbalanced
cases with shield termination. The poorest performance is obtained with an unbalanced load and no shield termination (Case
1). In this case, the use of shielded twisted pair provides no
benefit (beyond electrostatic protection) over the use of a
single wire.

Similar conclusions have been reported by other researchers
(e.g., [3], [4]).

6.   CONCLUSIONS

We have given a practical interpretation of electromagnetic
topology and how it can be applied to interference control in a
transient environment where broadband separation of source and
victim is essential. A separating barrier is used to prevent the
pulse power experiment from interacting with the diagnostics.
The goal is to stop both guided waves on conductors and waves
propagating through space. The most important principles derived
from the application of electromagnetic topology are summarized
below. (For theoretical background see the lecture by C. Baum.)

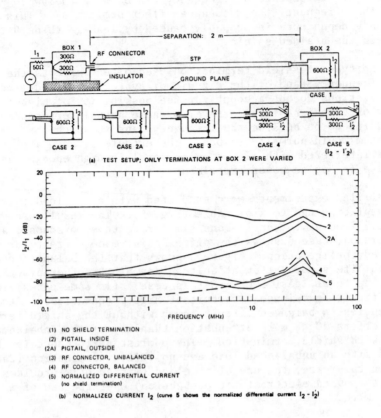

(a) TEST SETUP; ONLY TERMINATIONS AT BOX 2 WERE VARIED

(1) NO SHIELD TERMINATION
(2) PIGTAIL, INSIDE
(2A) PIGTAIL, OUTSIDE
(3) RF CONNECTOR, UNBALANCED
(4) RF CONNECTOR, BALANCED
(5) NORMALIZED DIFFERENTIAL CURRENT
    (no shield termination)

(b) NORMALIZED CURRENT I_2 (curve 5 shows the normalized differential current $I_2 - I'_2$)

Fig. 5.   Experiments with shielded twisted pair.

Signal and power conductors serving the protected circuit
should be treated with filters, common-mode rejection circuits,
or other means of preventing source currents (induced or conducted) from entering the protected volume (Fig. 6).

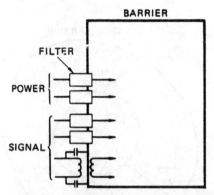

**(a) BARRIER CLOSURE AT POWER AND SIGNAL
      CONDUCTOR PENETRATION**

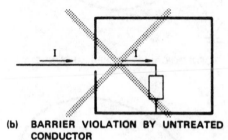

**(b) BARRIER VIOLATION BY UNTREATED
      CONDUCTOR**

Fig. 6.   Barrier closure about signal and power conductors.

Grounding conductors should not cross barrier surfaces, because any conductor that penetrates a barrier without treatment
provides a path for interference waves to propagate through the
barrier (Fig. 7).

Configuration control should be used to minimize the number
of penetrating signal and power conductors that must be treated
(Fig. 8) and to group these conductors so that external source
currents are concentrated in one local region of the barrier,
rather than distributed over the entire barrier surface (Fig. 9).

Barrier effectiveness must at least ensure that the source
does not damage the protected circuit; on the other hand, if the
barrier reduces the source-generated transients in the protected
volume to a level that is smaller than the internally produced
transients, no benefit accrues from further improvement in barrier effectiveness (Fig. 10).

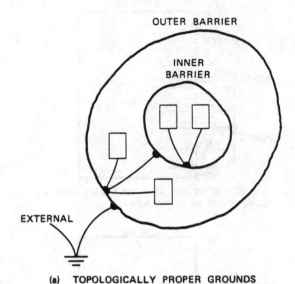

(a) TOPOLOGICALLY PROPER GROUNDS

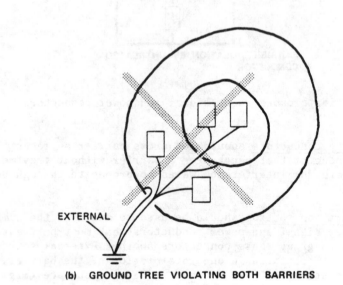

(b) GROUND TREE VIOLATING BOTH BARRIERS

Fig. 7. Grounding topology.

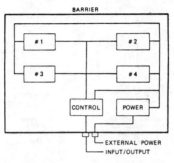

(a) MINIMUM PENETRATIONS (interconnections inside barrier)

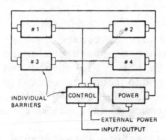

(b) LARGE NUMBER OF PENETRATING CONDUCTORS AND TREATMENTS

Fig. 8.   Limiting the number of penetrations.

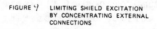

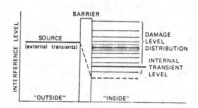

Fig. 9.   Limiting shield excitation by concentrating external connections.

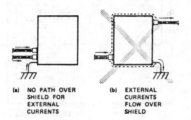

Fig. 10. Barrier effectiveness requirement.

    Cable shields should be closed, not grounded. The proper
configuration is closure of the electromagnetic barrier by eliminating any gap between the cable shield and the equipment housing
(Fig. 11). If the ground loop presents a problem, then the
shield current should be interrupted (for example, by isolating
one equipment unit from ground), but not the shield itself.

    A systematic approach to interference control based on elec-
tromagnetic topology as outlined in this chapter will result in a
diagnostic that is separated from the pulse power experiment by
an effectively impervious electromagnetic barrier. The integrity
of the barrier surface must be preserved in spite of requirements
for insulated power and signal conductors to pass through the
barrier, or of requirements for doors, access hatches, etc., to
be cut in the barrier wall. Broadband interference control means
accommodating these compromises of the shield without unneces-
sarily degrading the ability of the barrier to separate the internal environment from the external environment.

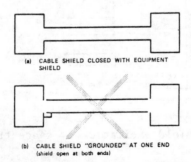

Fig. 11. Cable shield termination.

REFERENCES

1. Graf, W. and Vance, E. F., 1982, in: "Proceedings of the
    IEEE 1982 National Aerospace and Electronics Conference,"
    p. 4.

2. Graf, W., Hamm, J. M., and Vance, E. F., 1983, Defense Nu-
    clear Agency Final Report DNA 5433F-2.

3. Paul, C. R. and McKnight, J. W., 1979, IEEE Trans. Electro-
    magn. Compat., 21:92.

4. Paul, C. R., 1980, IEEE Trans. Electromagn. Compat., 22:161.

5. Vance, E. F., 1980, IEEE Trans. Electromagn. Compat.,
    22:319.

6. Vance, E. F., Graf, W., and Nanevicz, J. E., 1980, Defense
    Nuclear Agency Final Report DNA 5433F-1.

A SYSTEMATIC, PRACTICAL APPROACH TO THE DESIGN OF SHIELDED
ENCLOSURES FOR DATA ACQUISITION AND CONTROL

George Chandler

Los Alamos National Laboratory
Albuquerque, NM
USA

1.   INTRODUCTION

The design of shielded enclosures is usually an adjunct to
more pressing and interesting activities - the glamor is usually
associated with what goes into them, or next to them. The process is often obscure and smacks of the black arts, because the
ability to predict the shielding effectiveness of real enclosures, for example, is not high. This lecture will cover the
practical design of shielded enclosures for magnetic fusion experiments at Los Alamos. We will discuss models for the prediction of shielding effectiveness, data acquisition cabling
schemes, layout of the power and data cables, air conditioning,
and ground schemes inside the screen room.

2.   SYSTEM VIEWPOINT

The most important phase in screen room design occurs at
the beginning, when the attitude of the designer is set. The
screen room is a system which must accommodate diverse needs
without compromise, and the potential users and designers should
agree on what the needs are, and how they are to be served. Most
important, when there are several users, as on a controlled
fusion experiment where many diagnosticians use the same screen
room, there must be agreement on rules to maintain the integrity
of the product. The paramount importance of this agreement cannot be over-emphasized.

The screen room designer must provide for the following:

Control of the effects of externally generated ambient elec-
tromagnetic fields.
Control of the effects of internally generated fields.
Cable routing.
Power.
Penetrations for data lines.
Cooling and ventilation.
Lighting.
Fire protection
Expansion.

These must be supplied within the economic means available,
which makes good prediction methods desirable.

3.   CONTROL OF EXTERNAL EFFECTS

3.1 Grounding

This is somewhat outside the purview of this talk, but the
screen room should be part of a thoughtfully designed ground system, usually at a corner of the ground star, connected by a
ground highway, and constructed on an isolated base.

3.2 Shielding

This is the most obvious function of a shielded enclosure,
and the most difficult to predict satisfactorily. The first step
is to determine the shielding required. A model of the primary
field source must be developed, such as Figure 1, which is a
crude model of the FRX-C fusion experiment at Los Alamos. The
model should give an estimate of the significiant spectral peaks
of the source as shown in Figure 2, where the voltage of the
peaks is that which would be induced in a loop the size of the
screen room. The next exercise is to determine the acceptable
levels of voltage at the input terminals of the equipment in the
room. The designer then estimates the amount of reduction available by reducing the loop area of possible cable paths by proper
routing, twisting, location near ground planes, etc., and finally
calculates the shielding required from the enclosure itself.

Most shielding prediction texts teach the use of the so-
called transmission-line approach, which calculates the effect of
a sheet of shielding material of infinite extent on an impinging
electromagnetic wave. This method, well treated in White [1], is

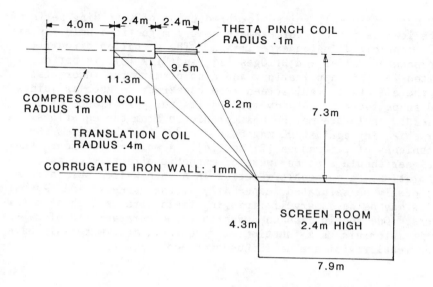

Fig. 1. Dimensions of the FRX-C source-screen room model.

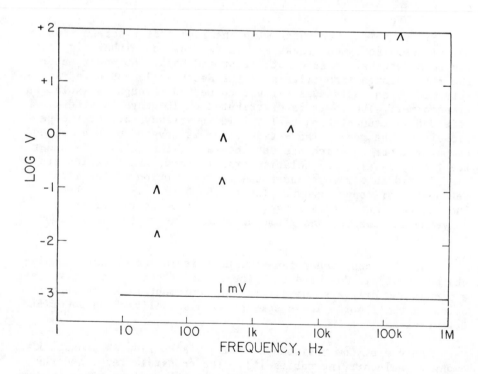

Fig. 2. Location of FRX-C power peaks.

not especially well suited for prediction of shielding performance in near fields, for example near a magnetic dipole, as are
most controlled fusion experiments. The so-called circuit
approach of Miller and Bridges [2], and Rizk [3], is better
suited for this application, and also permits the incorporation
of the effects of real screen room construction such as joints
and shape factors. At low frequencies the performance of a
shielded enclosure may be best predicted from the equations
for shielding against DC magnetic fields [4]. We prefer the
techniques of references [2] and [4]. A word of caution: the
designer should also be aware of magnetic effects such as
saturation - a permeable material in a thin-walled screen room
may easily be saturated, especially at the corners, and several
walls may be necessary if impinging fields are as high as a few
gauss. One should also be aware that the permeability of magnetic materials is a function of frequency; most materials have
a permeability of one at frequencies above 300 KHz.

4.   CONTROL OF INTERNAL FIELDS.

4.1 Layout

     Figure 3 is a sketch showing the layout of a screen room at
Los Alamos. Equipment racks are in two rows on either side of
the room, accessible from both front and back. AC power enters
the room through line filters on the South wall, and each rack is
serviced by an individual circuit connected through an isolation
transformer. The power is distributed on 10-gauge twisted
shielded instrumentation cable, laid in wireways around the pe-
riphery of the room. The entire floor is covered with a raised
computer floor on which are set the racks; the cabling is under
the floor, as are the isolation transformers. Data cables are
distributed in wireways under the center section of the floor,
set on top of copper ground planes. Each rack is connected to
the ground plane with a strap, isolated from nearby racks by
polyethylene sheets, and from the power line by the isolation
transformers.

     The air conditioner fan-coil unit is in one corner and dis-
charges cool air downward into the raised floor, which serves as
a plenum and delivers the air to the equipment racks through
holes in the floor. These same holes also deliver the power and
data cables.

     The use of low resistance ground planes reduces crosstalk
among signal-carrying cables [5]. The crosstalk level and the
maximum allowable noise level establish the maximum signal which
should be permitted to enter or be generated in the screen room.

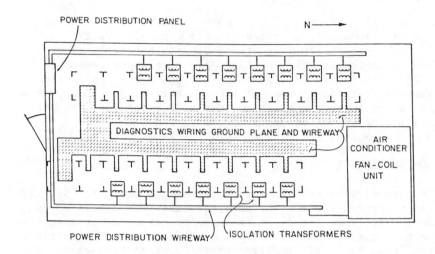

POWER DISTRIBUTION PANEL   N ———▶

DIAGNOSTICS WIRING GROUND PLANE AND WIREWAY

AIR
CONDITIONER

FAN - COIL
UNIT

POWER DISTRIBUTION WIREWAY   ISOLATION TRANSFORMERS

Fig. 3.   Sketch of screen room layout.

4.2   Cable Routing

The most important effect of using wireways and ground highways to carry cables is to reduce the loop area exposed to external fields.   This is a lot cheaper than shielding.   One rule that
should be enforced is that all cables within the screen room
should follow the highways; experimenters should not be permitted
to connect cables between adjacent racks, for example, except by
going through the below-floor wireway.   Discipline is important
here.

4.3   Power

Design of the power distribution system should give consideration to reducing the fields induced in nearby data cables.
This is accomplished by twisting the power leads (we use twisted
shielded pair, using the "drain wire" as the third, ground lead).
Isolation transformers are constructed deliberately to have leakage flux, so the designer should also consider putting them inside a magnetic shield.   We have found a box of 45-mil mu-metal
completely enclosing the transformer to be satisfactory for 2.5-
KVA units.   The second problem posed by the power distribution
system is the creation of ground loops.   The use of isolation
transformers at every equipment station eliminates the loops
formed by the power leads with data cables and other power leads.

4.4 Penetrations for Data Lines.

Figure 4 is a sketch of the scheme we have used successfully
to bring data acquisition cables (coaxial, usually RG223) into
the room. Consideration is given to crosstalk between cables;
the ferrite cores and low-resistance ground plane accommodate
that. The elimination of signals picked up by the cables from
the experiment itself is accomplished by careful grounding, passing the cables through a bulkhead at the junction box, and carrying them inside sealed conduits to the screen room.

5.   COOLING AND VENTILATION

The use of a raised floor as a plenum and duct for cool air
has been described above. The designer must choose whether to
include the air conditioner in the room or not. In one installation we have used a split system with the fan-coil unit inside
the room (Fig. 5). This requires carrying Freon through the
screen room walls. The only way to penetrate a screen room wall
is to pass a non-conducting tube through a section of pipe which
serves as a waveguide-beyond-cutoff; but no commercial Freon-
rated hoses are available which do not contain conductors. We
simply made a joint from PVC tubing, which is not recommended for
use with freon, but seems to work anyway. Figure 5 includes a
sketch of this arrangement.

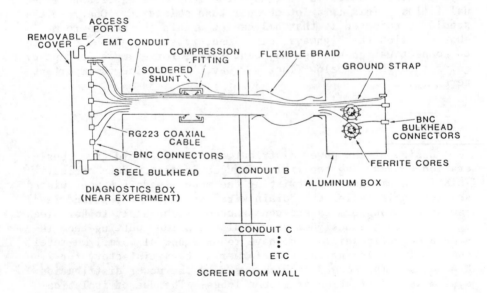

Fig. 4. Diagnostics cable wiring.

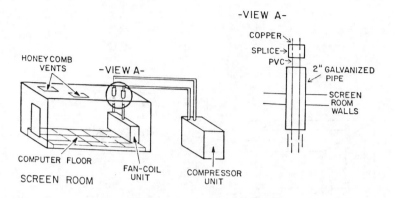

Fig. 5.   Air conditioner with fan-coil unit inside screen room.
         Sketch of Freon line penetration.

A second approach is to mount the fan-coil unit outside the
screen room and bring the cool air in through honeycomb vents,
as shown in Figure 6.   The vents can be very expensive.

6.   LIGHTING

Lights can generate a considerable amount of interference.
Power leads should be twisted, and for severely sensitive rooms
the designer may want to consider installing a DC incandescent
lighting system.

7.   SAFETY

If a fire protection system must be installed, we prefer the
scheme illustrated in Figure 7.   This is a dry-standpipe water
sprinkler system.   Halon systems were popular for awhile, but
they are a one-shot extinguisher which may drain out of the room
as soon as the door is opened; for this reason our safety person-
nel do not recommend it.

8.   EXPANSION

The best way to maintain the integrity of the screen room is
to anticipate the needs of the experimenters who are likely to
destroy it if their needs are not met.   This means allowing for
expansion.   It should be possible to add new diagnostics cable
conduits, for example, or to add new power circuits.   If the
designer doesn't allow for that, the experimenter will find a
way.

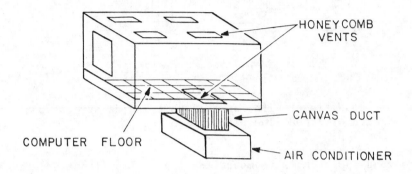

Fig. 6. Fan-coil unit outside screen room.

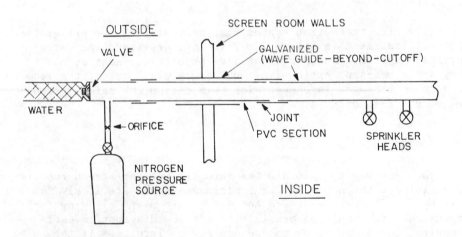

Fig. 7. Dry-standpipe fire protection system.

9.   CONCLUSION

The method above has been developed over a number of years by experimenters at Los Alamos, whose contributions I freely acknowledge. I include a number of useful references in the list below for the novice [6-9].

REFERENCES

1. Donald R. J. White, Electromagnetic Shielding Materials and Performance, Don White Consultants Inc., 14800 Springfield Road, Germantown, MD 20767 USA, 1975.

2. D. A. Miller and J. E. Bridges, "Geometrical Effects on Shielding Effectiveness at Low Frequencies," IEEE Trans. on EMC 8, 174-186, 1966.

3. F. A. M. Rizk, "Low Frequency Shielding Effectiveness of a Double Cylinder Enclosure," IEEE Trans. on EMC 19, 14-21, 1977.

4. A. K. Thomas, "Magnetic Shielded Enclosure Design in the DC and VLF region," IEEE Trans. on EMC 10, 142-152, 1968.

5. R. J. Mohr, "Coupling Between Open and Shielded Wires Over a Ground Plane," IEEE Trans, on EMC 9, 34-45, 1967.

6. Interference Technology Engineer's Master (ITEM, similar to EEM), P.O. Box 328, Plymouth Meeting, PA 19462 USA.

7. J. E. Bridges, D. A. Miller, and R. B. Schultz, "Comparison of Shielding Calculations," IEEE Trans. on EMC 10, 175-177, 1968.

8. Ralph Morrison, Grounding and Shielding Techniques in Instrumentation, John Wiley and Sons, 1967.

9. Henry W. Ott, Noise Reduction Techniques in Electronic Systems, J. Wiley and Sons, 1976.

INDEX